solar age catalog

A GUIDE TO SOLAR ENERGY KNOWLEDGE AND MATERIALS FROM THE EDITORS OF SOLAR AGE

Printed in the United States of America

First printing: September 1977

Second printing: November 1977

Published by SolarVision, Inc.
Distributed to the book trade by Cheshire Books, Church Hill,
Harrisville, N.H. 03450

ISBN 0-918984-00-9

Library of Congress Catalog Number: 79-79117
Main entry under title:
Solar age catalog: a guide to solar energy knowledge and materials
Includes Index.

A foreword from the sponsors

This is the year of the sun. This year the idea of using the sun's energy directly, matching it to the myriad needs for energy that we have developed, is no longer an eccentric notion of possible value, maybe, after the year 2000. It's here. Now. Congress knows it: as we write, there are bills in both House and Senate to encourage the use of solar energy through tax incentives and through low-interest loans for solar equipment. The President knows it: solar has a substantial place in his new energy policy. Agencies of the Executive branch have been ordered to open the windows on sunlight for others—and to use it themselves. Every federal building to be built from now on must either use solar energy or come up with good reasons why they don't use it. The states know it: more than half of them have passed state legislation that will help citizens to use solar energy.

But most important, we, the people, know it. Across the country solar meetings, fairs, exhibitions, and courses are being presented to crowds many times the size of those expected. People once concerned with energy only as they set thermostats on cold winter nights are suddenly struck by the idea, caught by the immediate practicality and, yes, the romance of using the sun. Clean, non-polluting, infinitely reliable (at least renewable as long as our stellar fusion furnace lasts, another six billion years or so), as much your own as a suntan is, as close as your window, as familiar as your roof, the sun is *here*. The idea that all you have to do is stretch out your hand, spread out your collector, to gather it in is a very warming one.

It is, of course, not quite as simple as that. Intelligent use of the sun involves knowledge and careful, sophisticated re-examination of physical laws, the way heat and light act and interact, the nature of materials, the cycles of weather, the wholeness of climate, the totality and construction of working ecological systems, and more. It also involves investigations of, and appreciation for, elegant uses of technology—man's unique contribution to his own still-shaky niche in the wholeness of the natural world. ⫸→

"Wholeness" may be the key to the appeal of solar energy. It combines elements of do-it-yourself —because the principles are simple, the materials are at hand, and the fuel is almost insistent in its democratic availability to everyone—with elements of some of the highest technology (consider photovoltaic cells, that nevertheless can still be made, after a fashion, in a basement workshop). The use of solar energy offers the opportunity for individuals to gain the kind of wholeness and control over elements of their individual lives that other animals, fitted into well-defined ecological niches, routinely enjoy.

How do you make the move from idea to action, though? We're pioneers, those of us involved in solar energy, and pioneers are usually on the road long before the roadmaps are printed. *Solar Age*, the magazine of the sun, has been around for a year and a half now. We have helped spread the word on the charting of some of the trails. We see the growing need for a guide, not only to the *things* of solar energy—the collectors, the controls, the materials— but also to some of the specific knowledge that goes with their use. Pioneers need to know a bit more than tourists about the territory into which they venture. The word to those who want to use solar energy still is: learn before you buy.

The more you learn, the simpler and less costly the use of the sun becomes.

This catalog is the product of an intensive process of collecting information, and arranging it in a form we think·most likely to be useful both to the general public interested in solar, and to people already actively in the field, whose needs for information are specific and detailed. It benefits enormously from knowledge generously shared by some of the most knowledgeable people we know. We wish to thank all of our authors, who when asked to write on the areas of their expertise, responded uniformly with great clarity and helpfulness.

They have assisted to make this catalog unique. But while it is the first of its kind, it won't be the last. We are planning new editions at yearly intervals, to be expanded as the field grows. For month-by-month news, the New Products section of *Solar Age* is being expanded to include a catalog update.

From all of us, then: sunshine.

Publisher   Executive Editor   Editor

solar age

A MAGAZINE OF THE SUN

Cover by Gesine Ehlers

PUBLISHER
Kurt J. Wasserman
EXECUTIVE EDITOR
Bruce Anderson
EDITOR
Sandra Oddo
ASSISTANT EDITOR
Martin McPhillips
EDITORIAL ASSISTANT
Janis Kobran
ADVERTISING MANAGER
Richard Livingstone
ART DIRECTOR
Linda M. Rath

EDITORIAL ADVISORY BOARD
Fred J. Dubin
Pres., Dubin Bloome Assoc.
N.Y., N.Y.
Erich A. Farber
Dir., Solar Energy and Energy Conversion Lab
University of Florida
Jan F. Kreider
Pres., Environmental Consulting Services, Inc.
Boulder, Colo.
Frank Kreith
Prof. of Engineering, University of Colorado
Hans Meyer
Windworks, Inc., Mukwonago, Wisc.
P. Richard Rittelmann
Burt, Hill Assoc., Architects
Pittsburgh, Pa.
Roland Winston
Enrico Fermi Laboratory
University of Chicago
David Wright
Environmental Architect, Sea Ranch, Cal.

Contents

CONTENTS

PHOTOCONVERSION

MEASUREMENT AND DATA COLLECTION

COMPARATIVE DATA TABLES

PERFORMANCE CURVES

DIRECTORY OF MANUFACTURERS

RESOURCES LISTINGS

DIRECTORIES

A GUIDE TO USING THE CATALOG

Certainly the most ambitious project we at *Solar Age* have attempted, this catalog has been since its conception something more than ordinary. We wanted a mixture of articles and information that would provide anyone wishing to work intelligently with solar energy a place of reference for products and services, technical know-how, and practical ideas and solutions. Therefore, knowing how to use this catalog is at least as important as owning a copy. Here is a concise outline of how it works:

1. Products are listed individually, by category. The product-by-product listings include product descriptions and essential information provided by manufacturers in response to fairly elaborate forms.

———————

2. Most of the products (including flat plate collectors, controls, complete systems: heating, cooling, hot water and pool heating, etc.) appear together on comparative data tables in a section starting on page 148. These tables and the information contained in them are not in any sense "official," but are intended to present manufacturers' responses on the technical specifications of their products in a form in which they may be easily compared and considered. In preparing these charts we have been most careful to transfer as accurately as possible the information that we received usually from manufacturers on our own forms, and occasionally from product brochures and the like.

———————

3. Introducing many product categories are articles there to provide intensive knowledge of the category from one or more perspectives. For instance, the how-to-build-it article in the flat plate collector section will teach you how to build one, *and how to understand one.* This is true of Steve Baer's piece on heat exchangers: without being highly technical it succeeds as an unusually clear and imaginative work of explanatory prose. The piece by Dan Ward on sizing heating systems is quite technical and definitely aimed at our technical audience, but we insist that it is not beyond the scope of the highly interested layperson. The same goes for Jan Krieder's more philosophical article on the second law of thermodynamics, which argues that we need a major shift in our whole approach to energy use and the accounting of it. These are but a sample, and we suggest that you roam to articles that can help you the most.

4. The resource and service directory includes architects, engineers, designers, builders, information people and other firms and individuals active in the solar field. We have listed them by state, but many of them also work on a national and regional range (this is generally stated in the listing).

———————

5. This whole catalog is loaded with names and addresses, telephone numbers and people to contact. The inclusion of products is generally based on whether or not manufacturers provided us with the information that we requested. This is the first attempt to present solar products in depth and, we hope, the beginning of a strong tradition of openness and comprehension within the solar community.

———————

6. From this first edition of the Solar Age Catalog to the next we will continue our coverage of products in a regular "New Products and Catalog Supplement" section of the monthly *Solar Age* magazine.

In producing this catalog we received invaluable assistance from Douglas Taff of Garden Way Laboratories and from Dan Scully and Charles Michal of Total Environmental Action. Their insights and suggestions led to considerable refinement of method of product presentation. We also extend our thanks to Kathy and Peter Burmeister of Mid-Hud Communications, and Stella Golden of Boro Hall Graphics, our very patient typesetters.

Martin McPhillips

Catalog Editor

GLOSSARY

*Short definitions of some of the names and terms
that crop up in solar energy* *

Absorbent—the less volatile of the two working fluids used in an absorption cooling device.

Absorber—the blackened surface in a solar collector that absorbs the solar radiation and converts it to heat energy.

Absorptance—the ratio of solar energy absorbed by a surface to the solar energy striking it.

Active system—a solar heating or cooling system that requires external mechanical power to move the collected heat.

Air-type collector—a solar collector that uses air as the heat transfer fluid.

Altitude—the angular distance from the horizon to the sun.

ASHRAE—abbreviation for the American Society of Heating, Air-conditioning and Refrigerating Engineers.

Auxiliary heat—the extra heat provided by a conventional heating system for periods of cloudiness or intense cold, when a solar heating system cannot provide enough.

Azimuth—the angular distance between true south and the point on the horizon directly below the sun.

Btu, or British thermal unit—the quantity of heat needed to raise the temperature of 1 pound of water 1°F.

Calorie—the quantity of heat needed to raise the temperature of 1 gram of water 1°C.

Coefficient of heat transmission, or U-value—the rate of heat loss in Btu per hour through a square foot of a wall or other building surface when the difference between the indoor and outdoor air temperatures is 1°F.

Collector—any of a wide variety of devices (flat-plate, concentrating, vacuum tube, greenhouse, etc.) used to collect solar energy and convert it to heat.

Collector efficiency—the ratio of heat energy extracted from a collector to the quantity of solar energy striking the cover, expressed in percent.

Concentrating collector—a device that uses reflective surfaces to concentrate the sun's rays onto a smaller area, where they are absorbed and converted to heat energy.

Conductance—a property of a slab of material equal to the quantity of heat in Btu per hour that flows through one square foot of the slab when a 1°F temperature difference is maintained between the two sides.

Conduction—the transfer of heat energy through a material by the motion of adjacent atoms and molecules.

Conductivity—a measure of the ability of a material to permit conduction heat flow through it.

Convection—the transfer of heat energy from one location to another by the motion of fluids that carry the heat.

Cover plate—a sheet of glass or transparent plastic that sits above the absorber in a flat-plate collector.

Degree-Day—a unit that represents a 1°F deviation from some fixed reference point (usually 65°F) in the mean daily outdoor temperature. If the outdoor temperature is 40°F for one day, then twenty-five (65° minus 40°) Degree Days result. Used to determine the demand of a heating season for different locales.

Design heat load—the total heat loss from a house under the most severe winter conditions likely to occur.

Design temperature—a temperature close to the lowest expected for a location, used to determine the design heat load.

Diffuse radiation—sunlight that is scattered from air molecules, dust, and water vapor and comes from the entire sky vault.

Direct methods—techniques of solar heating in which sunlight enters a house through the windows and is absorbed inside.

Direct radiation—solar radiation that comes straight from the sun, casting shadows on a clear day.

Double-glazed—covered by two panes of glass or other transparent material.

ERDA—U.S. Energy Research and Development Administration, responsible for commercial, institutional, and industrial solar demonstration program; and for general coordination of federal solar energy research and development.

Emittance—a measure of the propensity of a material to give off thermal radiation.

Continued

with thanks to Bruce Anderson, Michael Riordan, and The Solar Home Book.

GLOSSARY *continued*

Eutectic salts—a group of materials that melt at low temperatures, absorbing large quantities of heat and then, as they re-crystallize, release that heat. One method used for storing solar energy as heat.

FEA—Federal Energy Administration, responsible for federal energy policy.

Flat-plate collector—a solar collection device in which sunlight is converted to heat on a plane surface, usually made of metal or plastic. A heat transfer fluid is circulated through the collector to transport heat to be used directly or to be stored.

Forced convection—the transfer of heat by the flow of warm fluids, driven by fans, blowers, or pumps.

Glaubers salt—sodium sulfate ($Na_2SO_4 \cdot 10H_2O$). a eutectic salt that melts at 90°F and absorbs about 104 Btu per pound as it does so.

Gravity convection—the natural movement of heat through a body of fluid that occurs when a warm fluid rises and cool fluid sinks under the influence of gravity,

HUD—U.S. Department of Housing and Urban Development, responsible for residential solar demonstration program.

Header—the pipe that runs across the top (or bottom) of an absorber plate, gathering (or distributing) the heat transfer fluid from (or to) the grid of pipes that runs across the absorber surface.

Heat capacity—a property of a material, defined as the quantity of heat needed to raise one cubic foot of material 1°F.

Heat exchanger—a device, such as a coiled copper tube immersed in a tank of water, that is used to transfer heat from one fluid to another through an intervening metal surface.

Heating season—the period from about October 1 to about May 1, during which additional heat is needed to keep a house warm.

Heat pump—a mechanical device that transfers heat from one medium (called the heat source) to another (the heat sink), thereby cooling the first and warming the second.

Heat sink—a medium or container to which heat flows (see **heat pump**).

Heat source—a medium or container from which heat flows (see **heat pump**).

Heat storage—a device or medium that absorbs collected solar heat and stores it for periods of inclement or cold weather.

Heat storage capacity—the ability of a material to store heat as its temperature increases.

Indirect system—a solar heating or cooling system in which the solar heat is collected outside the building and transferred inside using ducts or piping and, usually, fans or pumps.

Infiltration—the movement of outdoor air into the interior of a building through cracks around windows and doors or in walls, roofs, and floors.

Infrared radiation—electromagnetic radiation, whether from the sun or a warm body, that has wavelengths longer than visible light.

Insolation—the total amount of solar radiation—direct, diffuse and reflected—striking a surface exposed to the sky.

Insulation—a material with high resistance or R-value that is used to retard heat flow.

Integrated system—a solar heating or cooling system in which the solar heat is absorbed in the walls or roof of a dwelling and flows to the rooms without the aid of complex piping or ducts.

Langley—a measure of solar radiation, equal to one calorie per square centimeter.

Life-cycle costing—an estimating method in which the long-term costs such as energy consumption, maintenance, and repair can be included in the comparison of several system alternatives.

Liquid-type collector—a collector with a liquid as the heat transfer fluid.

Natural convection—see **gravity convection**.

Nocturnal cooling—the cooling of a building or heat storage device by the radiation of excess heat into the night sky.

psi—pounds per square inch (pressure).

Passive system—a solar heating or cooling system that uses no external mechanical power to move the collected solar heat.

Percentage of possible sunshine—the percentage of daytime hours during which there is enough direct solar radiation to cast a shadow.

Photosynthesis—the conversion of solar energy to chemical energy by the action of chlorophyll in plants and algae.

Photovoltaic cells—semi-conductor devices that convert solar energy into electricity.

Radiant panels—panels with integral passages for the flow of warm fluids, either air or liquids. Heat from the fluid is conducted through the metal and transferred to the rooms by thermal radiation.

Radiation—the flow of energy across open space via electromagnetic waves, such as visible light.

Reflected radiation—sunlight that is reflected from surrounding trees, terrain or buildings onto a surface exposed to the sky.

Refrigerant—a liquid such as freon that is used in cooling devices to absorb heat from surrounding air or liquids as it evaporates.

Resistance, or R-value—the tendency of a material to retard the flow of heat.

Retrofitting—the application of a solar heating or cooling system to an existing building.

Risers—the flow channels or pipes that distribute the heat transfer liquid across the face of an absorber.

R-value—see **resistance**.

Seasonal efficiency—the ratio of solar energy collected and used to that striking the collector, over an entire heating season.

Selective surface—an absorber coating that absorbs most of the sunlight hitting it but emits very little thermal radiation.

Shading coefficient—the ratio of the solar heat gain through a specific glazing system to the total solar heat gain through a single layer of clear, double-strength glass.

Shading mask—a section of a circle that is characteristic of a particular shading device. This mask is superimposed on a circular sun path diagram to determine the time of day and the months of the year when a window will be shaded by the device.

Solar house, or solar tempered house—a dwelling that obtains a large part, though not necessarily all, of its heat from the sun.

Solar radiation—electromagnetic radiation emitted by the sun.

Specific heat—the quantity of heat, in Btu, needed to raise the temperature of 1 pound of a material 1°F.

Sun path diagram—a circular projection of the sky vault, similar to a map, that can be used to determine solar positions and to calculate shading.

Thermal capacity—the quantity of heat needed to warm a collector up to its operating temperature.

Thermal mass, or thermal inertia—the tendency of a building with large quantities of heavy materials to remain at the same temperature or to fluctuate only very slowly; also, the overall heat storage capacity of a building.

Thermal radiation—electromagnetic radiation emitted by a warm body.

Thermosiphoning—see **gravity convection**.

Tilt angle—the angle that a flat collector surface forms with the horizontal.

Trickle-type collector—a collector in which the heat transfer liquid flows down open channels in the front face of the absorber.

Tube-in-plate absorber—an aluminum or copper sheet metal absorber plate in which the heat transfer fluid flows through passages formed in the plate itself.

Tube-type collector—a collector in which the heat transfer liquid flows through metal tubes that are wired, soldered, or clamped to the absorber plate.

Ultraviolet radiation—electromagnetic radiation, usually from the sun, with wavelengths shorter than visible light.

Unglazed collector—a collector with no transparent cover plate.

U.S. Department of Energy—proposed cabinet department that will take over U.S. energy programs, including the functions of ERDA and FEA.

U-value—see **coefficient of heat transmission**.

☀

Solar heating in SMALL BUILDINGS

By Steve Baer

In New Mexico it is easy to heat small buildings with the sun. The most interesting questions don't concern solutions, but rather how problems manage to survive. Some puzzling attitudes are in the way. It is almost as if house-heating were a small but important outpost of some more difficult problem—a major problem that periodically sends reinforcements to the minor problems so that it can hang on unsolved to bother and confuse us.

If you are looking for answers to conservation problems it seems to me a very bad beginning to label the interested public as "consumers." This encourages an attitude incapable of solving most problems. What can a consumer do? He can complain if it doesn't taste good; he can yell if he wants more; he can burp when he gets enough. None of this has much to do with direct solutions.

The direct gain building

One layer of glass allows about 87 percent of the incident sunlight to pass through it. Two layers allow about 75 percent. Suppose I replace a south-facing insulated wall with a window? During the night I'm sorry I did this because the window allows heat to escape. During the day sun passes through the window and I am glad, for the house receives free warmth from the sun. How do they balance? What if the sun doesn't shine one day? Consider the possibility of being uncomfortable at night because it is too cold, and being uncomfortable during the day because the window admits so much sun that the house gets too hot.

Your south window salesman has a hard time telling you how the window will work. After listening to his/her hemming and hawing do you want to hear more? Perhaps better to listen to

Steve Baer is the president of Zomeworks Corporation.

the electric salesman who points to the knob and tells you to turn it this way when you are hot and that way when you are cold. But every time you turn that knob you move something else, perhaps hundreds of miles away. The gate valve on a hydroelectric plant releases a little more water, or a regulator or a turbine asks for a little more heat—and your electric meter counts it off, and you pay for it.

The window is a question of averages. Your south window salesman can speak with the same confidence as the electric company if you will spare him talking about tomorrow and let him talk about an average day:

On a clear day in January there are about 1,900 Btus per square foot incident on a south wall in New Mexico.

An average January day in Gallup receives about 60 percent of this amount or 1,140 Btus per square foot and, of this, about 75 percent or 855 Btus pass through two layers of glass.

In Las Cruces an average January day receives about 70 percent of the possible sun and we have 998 Btus per square foot admitted through the window.

What about the loss through the window? Again during an average January day in Gallup the temperature outside is about 29°F and we lose (65 - 29)(24)(½) = 432 Btus, and at 42°F in Las Cruces, (65 - 42)(24)(½) = 276 Btus, so when we subtract the losses from the gains we have a net gain of 423 Btus per square foot for Gallup during an average January day, and 998 - 276 = 722 Btus per square foot in Las Cruces.

Similar calculations for single glazing give:

	Gain	Loss	Net Gain
Gallup:	992 Btus	864 Btus	128 Btus
Las Cruces:	1,157 Btus	550 Btus	607 Btus

If we extend our calculations over an entire year, for double glazing we find a net gain during a 220-day heating

season in Gallup of about 100,000 Btus per square foot, and during a 180-day heating season in Las Cruces we find a gain of about 140,000 Btus per square foot.

How much is this energy worth? If you are using electric resistance heating at $.035/kWh the savings are:

100,000 Btus = $1.03

140,000 Btus = $1.44

If you are burning natural gas at our present low prices the savings will be only one quarter that great.

What about the windows? North windows do nothing but lose heat when you need it, and gain heat during spring and summer when you don't want it. Double-glazed east and west facing windows about hold their own during the winter in most parts of the state, but the west facing window admits the afternoon summer sun. The south facing window obligingly turns itself out of the high summer sun.

It is fascinating how individuals and groups are able to announce themselves as solar energy workers and energy conservationists to receive federal grants, to give press conferences—but never notice the sun or its effects on buildings. As an example, I would like to call attention to the Project Conserve questionnaire recently mailed to New Mexico house owners from the FEA and the New Mexico Energy Resources Board: it is quite well known in our state that south facing windows can provide a great majority of the heat needed during a New Mexico winter. The questionnaire, although it asks numerous questions about the windows in one's house, never asks in which direction they face or whether or not they receive sun. The questionnaire goes on for seven pages about installing insulation and weather stripping with no mention of using the sun. It appears likely that this questionnaire was written for people who live in another part of the world, for it also omits wood as a possible fuel for heating systems. Evidently the authors misunderstand the word "free." Sunlight is free, but it is ignored while the cover of the questionnaire announces that the questionnaire itself is free. Actually, the Project Conserve questionnaire has cost us a great deal of tax money.

How can something as marvellous as the different natures of the walls of a building be overlooked and ignored? The answer to this question might suggest answers to many more questions of how we have ignored nature. Total ignorance is often a more comfortable state than some slight knowledge that perhaps leads to brooding and worry. "Boss, go easy on 'em. They didn't know nothin'." This might make the sight of our houses, windows stuck on their sides with no more attention to orientation than thrown dice, kind of stupidly cheerful.

Thermal mass

Large windows or skylights are most useful in houses that have large thermal mass, houses that can absorb the day's sunshine without overheating. The thermal mass of a house can be computed by multiplying the weight of everything in the house by its specific heat. For instance, an eight-inch block wall filled with dirt has a weight of 65 pounds per square foot that, when multiplied by .21, its specific heat, gives a thermal mass of 13.6 Btus per degree, per square foot of wall. A 10 inch adobe wall is about the same. A slab floor is more difficult to judge for it touches the earth below it. It seems reasonable to treat its thermal mass as 15 Btus per degree, per square foot. Such items as wood also enter into the sum. If after computing the thermal mass of your house you find it to be 50,000 Btus per degree, there is still the question of how rapidly it can absorb and give off this heat. When concrete is exposed to direct sunlight the surface temperature rises while the interior may remain cool.

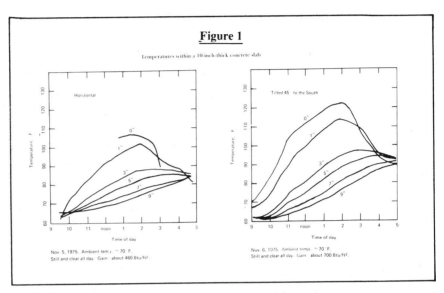

Figure 1

Temperatures within a 10-inch-thick concrete slab

Horizontal
Temperature, °F

Tilted 45° to the South
Temperature, °F

Time of day

Nov. 5, 1975. Ambient temp., ~70°F.
Still and clear all day. Gain: about 460 Btu/ft².

Nov. 6, 1975. Ambient temp., ~70°F.
Still and clear all day. Gain: about 700 Btu/ft².

No building can afford to have large internal surfaces get as hot as the surface of our block. One square foot of sunlight must be spread over several square feet of masonry if you want to be able to remain in the room while it is being charged with heat. Some methods:

1. Have the wall at a very oblique angle to the sun. (Notice the effect in our graph, Figure 1.)

2. Bounce the sun off a white wall first and spread it out over a wider area.

3. Admit the sun in patches that sweep the floor and walls as the sun moves—no one area is then exposed for any great length of time.

Of course, one needn't take these precautions too seriously. There may well be parts of a room where it is fine to let a slab floor climb in temperature to 110°F during a winter day.

If we consider that a square foot of south window may admit some 1,500 Btus during a day, we see we may need 10 square feet of concrete or masonry comfortably to absorb the heat delivered by one square foot of window. The arrangements of windows and skylights with floors and walls is an interesting problem: Do you ride out a hot Janu-

ary afternoon as you charge the room or do you relent and open a window? Fortunately not all materials must be treated so gingerly when it comes to charging with heat. Masonry and concrete may need the sun diffused, but water containers can absorb sun, even when it is concentrated by reflectors.

How valuable can a 55-gallon drum be? A house with large south windows has been overheating during the day and is too cold at night. In an effort to even out the fluctuations, we bring in a number of drums of water and place them behind the south window. The room is now comfortable during the day and the night. A drum standing behind a south window, on an average Albuquerque winter day, picks up 10° during the day, then loses it to the house at night.

The drum—although it doesn't create heat by itself—can take credit for the 500 lbs. x 10°F = 5,000 Btus it stores because it delivers this heat when you want it instead of when you don't want it. If the drum performs this service 150 times during the winter, it has delivered 750,000 Btus to make us comfortable. In addition to this enormous

Continued

"ASK ME ABOUT SOLAR ENERGY"
ALTHOUGH NOTHING CAN REPLACE AN ACTUAL
NUCLEAR EXPLOSION - WE HAVE FOUND WE CAN
DO ALL OF OUR SOLAR ENERGY WORK BY
COMPUTER SIMULATION

"ASK ME ABOUT SOLAR ENERGY"
IN THIS AGE SOLAR TECHNOLOGY MUST
BE ECOLOGICALLY CONSCIOUS, FOR EXAMPLE
OUR DRUMS ARE 100% RECYCLED

"ASK ME ABOUT SOLAR ENERGY"
I DON'T KNOW AMIGO, BUT MY DADDY
TAUGHT ME HOW AND WE'RE ALWAYS
WARM IN WINTER AND COOL IN THE
SUMMER

Drawings are by Peter Voorhees, for the Consumer Conference on Solar Energy Development held in New Mexico, October 1976.

SMALL BUILDINGS *continued*

amount of heat—worth from perhaps $1.50 to $8—we also find the drum ready to provide cool during summer days. It absorbs heat during the day from a warm house and then gives it off at night to cooler air blowing through the house.

It isn't long before the can or drum of water has quietly paid for itself by doing no more than accepting heat in the day and giving it off at night.

When placing a drum wall you must remember that as the drums increase they begin to loaf—not cycling through such extreme temperature ranges—simply because the windows supply a limited amount of heat to the room. Keep adding more thermal mass until you are comfortable, then stop. Such adjustments can be done after the building is completed.

Ceiling Fans

In a direct-gain house where the thermal mass is too small and the temperatures rise too high, a ceiling fan can help. In the hot December afternoons, the sweating inhabitant can turn on a ceiling fan as other people do in July. This accomplishes two things: he can withstand the high temperature without discomfort and deliver the heat via the swirling air to remote parts of the house where walls, floors and furniture will be charged.

The ASHRAE handbook tells us that a lightly clothed, sedentary person is comfortable at 78°F if the air is still and that he is also comfortable at 83°F if the air is moving at a gentle 150 feet per minute or 1.7 miles per hour.

The Reflector Shade

If the ground immediately below a south window is white or reflective, the window above gains not only its own sun, but a great deal of sun reflected from below it. In some applications this front reflector can be a door that doubles as a summer time shade and, if people are willing to operate it, as nighttime insulation. Similar devices can be installed on skylights.

Comfort

The ASHRAE *Handbook of Fundamentals*, ASHRAE Comfort Standard 55-66, defines thermal comfort "as that state of mind which expresses satisfaction with the thermal environment." What an excellent, if sly, definition!

Much of the work needed for the use of solar energy in heating houses is in making people less fussy about temperature fluctuations. Someday a scientist will raise rats in cages subject to different temperatures. I'll wager that the rats that see a fairly wide temperature fluctuation will outlive the steady 70°F rats. In the meantime, how does the person whose house is not 70°F defend himself against his neighbors who understand this to be a deviation in more than just temperature? It is a sign, to many, of a system out of order, broken down, a failure, a symptom of slovenliness on the part of the owner, or a moral failure on the part of the owner to live up to an unwritten contract between himself, the consumer, and the producers of modern equipment.

I think it is a real shame to live in a house or work in a building that is never allowed to tell you, even gently, what time of year it is. If during the winter your house is somewhat chilly and during the summer it is warm you at least have one less thing to remember when you get up in the morning.

New Mexico Research, Inc.

New Mexico, despite its small population, has been unusually successful in using the sun as compared to other states, maybe because the AEC brought so many scientists and engineers to our state. But I don't think this is the reason. I would suggest that our state's achievements are more the result of a certain backwardness. Dirt instead of paved roads, loose building codes, poor surveillance by inspectors, many of the buildings being made by "primitive" instead of "advanced" methods. Where else could people escape to build what they wanted to.

If we are ahead, do we owe this lead to the atomic bomb and the scientists it attracted to New Mexico or do we owe it to the adobe brick? It is important to know because we have many problems to solve. If we have found a style or method that leads to answers it shouldn't be buried; it should be encouraged. I think most of the creative work done on the solar heating of houses has been done on a low budget. This is why it is a double shame now for enormous grants to be given to scientists to devise solar heating systems: It is a waste of money, and the results tend to obscure the successful low budget work.

The successes of the amateurs, although adopted by friends and neighbors and finally copied by the scientists, do not seem to lead the scientists or those in power to inquire about the working methods and attitudes that helped the idea along. The innovators, rather than being considered a group who have perhaps struck a rich vein, are considered strange aberrants whose quota of good ideas is now most surely exhausted.

Here is an idea a friend suggested, for replacing the complicated pollution control regulations on passenger vehicles with a test that strikes at the heart of the problem and is easy to administer and easy to understand.

The vehicle is stopped—whatever it is—the Santa Fe Chief, a car, a bicycle. The engine is turned off and the passengers try to push it themselves. If they can't get it up to 2 miles per hour, they fail and must continue on foot. Their vehicle is now public property and is recycled by a group who wait around the test like ants around a dying beetle.

Money

"Money, money, money" is entirely the wrong way to think about the sun. The sun doesn't notice money; money doesn't notice the sun. It is hard for our economists and hard-headed businessmen to consider the sun because they trade in shortages. In a way, they are better equipped to think of shadows. Their unpleasant preoccupation has left them unfitted to view the opportunities of life. Our "realistic executive" arrives in dark glasses, a flunky shading him with a parasol. He carries a briefcase filled with shadows.

Listen to the words of a recent ERDA report—COO/2688 - 76/1, page 60. "Likewise, long-term installation costs should be low, so hand-crafted systems, constructed on the roof by tradesmen, should be discouraged even if this is not the least expensive way of building solar systems today."

Solar systems are commonly divided into active systems and passive systems.

The passive system runs without external power.

The active system has fans or pumps powered by central generating stations to manage the collection and distribution of energy. The price of this electricity can be raised. The power can also be turned off.

There is a financial counterpart to this division. Passive financing requires the homeowner to work on the project himself on weekends, or perhaps to reduce the size of the house or make some other local adjustment between his time, his money and the project. The financial counterpart to the active system finds the builder looking to the state or federal government and ignoring consideration of local costs. If his project is paid for by a grant, it arrives through the mail—like an electric wire. If he can forget that the grant is no more than his and his neighbors' tax money minus handling charges it seems to be a good deal.

I would say that the elegant designs we strive for should minimize, along with friction and heat loss, the necessity for taxes and those who handle them.

Water Heaters

After the house is well insulated, filled with thermal mass, and fitted with proper windows and skylights the problem of heating water remains.

The first thing to do is to install shower heads that constrict the flow of hot water. Three gallons a minute can be just as pleasant as six gallons a minute.

A number of solar hot water heaters are on the market. Most use collectors, pumps, and thermostats. They are quite expensive. The collector placed below a heat exchanger and circulating antifreeze by natural convection cuts the cost, but demands a special location. There are other cheaper ways to heat the relatively small amounts of hot water used in a household. Direct-gain water heaters were used years ago in California, today are used in South Africa and New Zealand, and are beginning to be used again here in the United States. The water is heated in a black tank exposed to the sun behind a south window or beneath a skylight. Again, the reflective surfaces, bright aluminum or white, can be used to great advantage by bouncing additional radiation onto the tank. A cylindrical tank enclosed in a clear housing and standing upright above a white ground and in front of a south facing white wall receives enormous glare and is able to raise water 70°F above outside temperatures. The tank absorbs as much as 3,000 Btus per square foot of the tank cross section, daily. If the owner of such a solar heater is guarding his heat carefully he can easily insulate such a tank by placing a large version of the familiar tea cozy over the tank after the sun goes down.

Another inexpensive water heater uses an air collector with a small fan or large thermosiphon ducts to pass the hot air directly over the tank, which becomes the heat exchanger. The heat transfer can be increased by gluing fins to the tank or adding a gravity-circulated heat exchanger as Peter van Dresser did years ago in Santa Fe.

One great benefit from using the sun to heat your house and hot water is that it makes you feel better. Although untested scientifically, it is possible that exposure to the sun enriches the water or the house with orgone energy (or some other scientifically disreputable substance) improving the health, energy and spirits of the users. ☼

The second law of THERMODYNAMICS

By Jan F. Kreider

Energy accounting that ignores the quality of energy while restricting itself to keeping track of its quantity—which the first law of thermodynamics assures us will not change in non-relativistic processes—is an unwise practice. To use both solar and non-solar energy sources effectively, we need a system of accounting based upon the second law of thermodynamics, which asserts that work is the highest-quality/lowest-entropy form of energy.

Nearly all spoken and written works on the energy question, by technical experts, politicians, and laymen, deal with energy conservation. This approach to insuring adequate energy for future use is based on an assumption that energy is a substance that flows or is somehow transportable from place to place by means not specified. It is assumed that it can be "saved" by not using it. The basic thermodynamic error in this pseudo-econometric way of thinking is that all attention is focused on a single property of a system—energy—and not the interactions within or between systems. But it is those interactions—tasks to be performed—that transform energy into its two useful types, work and heat. The only fundamental method of conserving, while effectively using, energy reserves will therefore evolve from paying strict attention to whole processes and system interactions.

A unique and technically favorable thermodynamic attribute of solar energy is its suitability for a wide variety of tasks that require widely varying temperatures. Analysis of this factor has been universally neglected, but it will have the greatest long-term impact on the optimum use of solar energy as well as the use of other energy sources supplemented by solar. Solar radiation can be and has been converted to useful heat at temperatures ranging from ambient to 3,000°C.

The first law of efficiency—η_1—has been widely used to evaluate alternative energy use. This law can be defined as the energy transfer achieved by a device, divided by the energy input to achieve the effect. This efficiency is far from adequate because it may have numerical values greater than one, equal to one, or con-

strained to be less than one by other physical principles. For example, an oil furnace may have a 75 percent efficiency. This implies that a 100 percent-efficient furnace is perfect, an incorrect conclusion because the oil could be used instead as fuel for a turbine powering a heat pump, a system that could provide more heat to a building than the total energy content in the raw fuel.

First-law efficiency also is dependent on the system in use to achieve a desired effect. It completely neglects energy quality and the second law. And it lacks generality when applied to complex systems with outputs that may include a combination of mass transfer, heat production, and work.

The fundamental difficulty with the first-law efficiency is its base energy. Energy is a property that cannot be consumed and is therefore an ambiguous and inadequate measure of its own effective use. The second law of thermodynamics affords a method of ranking energy uses by quality of energy used and by the approach to the thermodynamic limit of a given process.

Energy Quality - The Second Law of Thermodynamics

Since work is the highest quality or lowest entropy form of energy, it is the most valuable form of energy and provides an index that can be used to rank energy conversion processes. The second law permits definition of a property called available energy, defined as *the maximum amount of work that can be produced from thermal energy*. Unlike energy, *available energy* is consumed in a process. Whereas the first law leads to the conclusion that energy should be hoarded (not used for any purpose) to minimize consumption, the second law provides a quantitative measure that can be used to minimize the consumption of available energy in any process. The goal of energy process optimization is to minimize the consumption of energy.

Second law efficiency, η_2, is defined as *the ratio of the minimum amount of available energy required to perform a task, A_{min}, to the available energy, A, actually consumed by use of a given system.* Maximizing η_2 necessarily lessens consumption of a fuel or other energy source because availability, an extensive property,

is proportional to the mass of fuel. A waste of availability is a waste of mass of fuel and hence a waste of energy.

In the case of solar thermal systems, no mass of fuel is consumed, so the maximum η_2 will result in the minimum capital investment in power producing equipment. Note that maximum η_2 is related to a process or task to be achieved, not to a specific device.

Available energy or maximum work was first defined by Gibbs as

$$A \equiv E - T_0 S + p_0 V - D$$

where

E = internal energy of fuel
S = entropy of fuel
V = volume of fuel
T_0, p_0 = ambient or "dead-state" temperature and pressure
D = maximum useful work from diffusion processes

The diffusion term D is significant only for combustion and chemical processes and need not be considered in solar systems. Availability can be defined in words as the maximum work that can be provided by any system or energy source as it proceeds to a final equilibrium state with a specified atmosphere or dead state.

Second law efficiency in equation form is

$$\eta_2 = A_{min}/A$$

The Kelvin-Planck statement of the second law implies that the availability (maximum work) of a process transferring a quantity of heat Q at temperature T is simply the product of the Carnot efficiency and the heat transferred,

$$A_{min} = \left(1 - \frac{T_0}{T}\right) Q$$

The entries in Table 1 follow from this basic equation.

Example

Calculate the second-law efficiency of a gas-fired furnace providing heat at 50°C (121°F) if the outside temperature is 0°C (32°F). Assume that the available energy of gas A is equal to its lower heating valve ΔH_L (accurate for most hydrocarbons within a few percent) and that the first-law efficiency for the furnace is 65%.

Solution

The second law efficiency is given by

$$\eta_2 = A_{min}/A$$

from Table 1.

$$\eta_2 = Q_a\left(1 - \frac{T_0}{T_a}\right) \Big/ \Delta H_L$$

Recognizing that

$$\eta_1 = Q_a/\Delta H_L,$$

η_2 is given by

$$\eta_2 = \eta_1(1 - T_0/T_a)$$

Inserting numerical values—

$$\eta_2 = 0.65\left(1 - \frac{273}{323}\right)$$

$$\eta_2 = 0.10$$

Dr. Kreider is president of the Environmental Consulting Service, based in Boulder, Colo.

The low value of η_2 shows the potential for large inprovements in space heating. If the task had been defined as delivery of air at 21°C (70°F) to a space—the real purpose of a heater—the second-law efficiency would have been 0.046, indicating still greater possibility for improvement. Low second-law efficiencies indicate a fundamental mismatch of task and energy supply. In this case, low-entropy fossil fuel is used to provide high-entropy hot water for space heating.

Matching Solar Energy to its Task

The preceding analysis affords a quantitative methodology for matching solar conversion systems to their tasks. For low-entropy uses such as space heating, water heating, and crop drying, low-temperature low-entropy solar energy collected by flat-plate collectors provides the best match between energy source and task. The operation of a solar-thermal generating station, on the other hand, would require low-entropy (high-temperature) thermal energy to match the task of producing shaft work (low-entropy).

Solar energy is unique among energy sources in that its entropy level can be manipulated to provide the best task-to-entropy-level match. This is usually accomplished by varying degrees of solar concentration. A familiar example of such concentration is the damming of rivers to produce hydroelectric power. Falling rain produced by the solar-powered evapora-tion-precipitation cycle of the atmosphere falls widely over the earth's surface. This high-entropy source is converted to a concentrated low-entropy source by gathering the flow from entire watersheds into narrow river valleys. The rivers in turn are capable of producing shaft work (low-entropy energy) from the "concentrated" rainfall. Likewise, in direct solar-to-electric conversion methods, sunlight is usually focused by mirrors or lenses onto a high-temperature receiver. The high-temperature working fluid can then effectively be used to create shaft work in a Rankine-cycle turbine or other heat engine.

Example

Using the second law of thermodynamics, evaluate two proposed systems for heating hot water at 50°C with solar energy. The first system is a flat-plate solar collector capable of delivering water to the water heater at 60°C. The second system is a focusing solar collector capable of delivering water to the heater at 200°C. Heat losses in the exchange process between the solar collector fluid and the water heater are 10 percent of the heat collected and the environmental temperature is 0°C.

Solution

The second-law efficiency of a solar water heater is given by (see Table 1)

$$\eta_2 = n_1 \left[\frac{1 - T_o/T_a}{1 - T_o/T_r} \right]$$

For the low-temperature system

$$\eta_1 = 0.9$$

$T_o = 273°K$
$T_a = 323°K$
$T_r = 333°K$ from which
$\eta_2 = 0.77$

For the high temperature system

$T_o = 273°K$
$T_a = 323°K$
$T_r = 473°K$ from which
$\eta_2 = 0.33$

The flat-plate system is more than twice as effective as the concentrator in producing hot water at 50°C. This phenomenon is reflected in the larger capital investment required to concentrate sunlight. As a result, second-law analysis has reached the same conclusion as an economic analysis based on entirely different principles.

If natural gas (methane) were used to heat water in the preceding analysis with a first-law efficiency of 65 percent, second-law efficiency would be significantly lower. For a flame temperature of 1400°K it is shown that

$$\eta_2 = 0.65 \left[\frac{(1 - 273/323)}{(1 - 273/1400)} \right]$$
$$\eta_2 = 0.12$$

The thermodynamic effectiveness of a gas water heater is less than one-fifth that of a solar water heater. The poor gas heater performance is a direct consequence of the poor match between task and energy source.

Summary

The second law of thermodynamics is a powerful tool in analyzing energy systems. Specifically

1. The second-law of thermodynamics of a process is a proper measure of the effectiveness of energy utilization in that process.

2. The second law provides a method for maximizing the effectiveness of solar energy collection.

References

Berg, C.A.: Potential for energy conservation in industry, *Annual Review of Energy*, vol. *1*, 1976, pp. 519-534.

Berg, C.A.: A technical basis for energy conservation, *Technology Review*, vol. , 1974, pp.15-23.

Berg, C.A.: A technical basis for energy conservation, *Mechanical Engineering*, vol. 96, 1974, pp. 30-42.

Ford, K.W., et al,: *Efficient Use of Energy*, American Institute of Physics, Report No. 25, New York, 1975.

Gibbs, J.W.: *The Collected Works of J. Willard Gibbs*, vol. 1, New Haven, Yale University Press, 1948, p. 77.

Keenen, J.H.: *Thermodynamics*, New York John Wiley and Sons, 1948.

Reistad, G.M.: *Available Energy Conversion and Utilization in the United States*, ASME Paper No. 74-WA/Pwr-1, 1974.

Reistad, G.M.: *Availability: Concepts and Applications*, Ph.D. dissertation, University of Wisconsin, 1970.

Table 1.
Availabilities, First, and Second Law Efficiencies for Energy Conversion Systems.*

TASK	ENERGY INPUT	
	Input Shaft Work W_i	Q_r from reservoir at T_r
Produce Work W_o	$A = W_i$ $A_{min} = W_o$ $n_1 = W_o/W_i$ $n_2 = n_1$ (electric motor)	$A = Q_r(1 - T_o/T_r)$ $A_{min} = W_o$ $n_1 = W_o/Q_r$ $n_2 = n_1(1 - T_o/T_r)^{-1}$ (solar, Rankine cycle)
Add heat Q_a to recover at T_a	$A = W_i$ $A_{min} = Q_a(1 - \frac{T_o}{T_a})$ $n_1 = Q_a/W_i$ $n_2 = n_1(1 - T_o/T_a)$ (heat pump)	$A = Q_r(1 - T_o/T_r)$ $A_{min} = Q_a(1 - T_o/T_a)$ $n_1 = Q_a/Q_r$ $n_2 = n_1\left[\frac{(1-T_o/T_a)}{(1-T_o/T_r)}\right]$ (solar water heater)
Extract heat Q_c from cool reservoir at T_c (below ambient)	$A = W_i$ $A_{min} = Q_c(\frac{T_o}{T_c} - 1)$ $n_1 = Q_c/W_i$ $n_2 = n_2(\frac{T_o}{T_c} - 1)$ (compression air conditioner)	$A = Q_r(1 - T_o/T_r)$ $A_{min} = Q_c(\frac{T_o}{T_c} - 1)$ $n_1 = Q_c/Q_r$ $n_2 = n_1\left[\frac{(T_o/T_c-1)}{(1-T_o/T_r)}\right]$ (absorption air conditioner)

*The heat reservoirs are considered to be isothermal; T_o is the environmental temperature and processes are reversible; $T_r = T_a = T_o = T_c$.

I'LL TAKE THE HYBRID

By Dan Scully

Is the sun shining? Is the wind blowing? Is it raining? Is the food growing? These used to be fundamental questions; in fact, some were the foundations of religions. Today, does it really matter? Hell, our food magically appears, even if it is from 2,000 miles away, so who cares? Our houses, in effect our skins, consume energy, like gluttons, make little use of what energy is directly available to them and ignore what happens to both our organic and inorganic wastes. While our technology has freed us from jobs of drudgery to pursue more worthwhile endeavors, it has also altered our attitude toward the environment from one of dependence to one of abuse.

Conservation

In nature, waste is unknown. Elements recycle themselves continually. Conservation of all things, including energy, is fundamental. Nature's attitude needs to be a part of the design of our buildings, especially so when using solar energy is contemplated.

Solar energy is diffuse by nature, and not very intense at that. It may take 1 square foot of collector an entire heating season to accumulate as much energy as is contained in 1 gallon of oil. Several things become immediately clear:

• We need to reduce our needs before going after what the sun has to offer us, and then to be as sparing in its use as possible.

• We need a lot of collection area to get significantly usable amounts of energy, and therefore must use every means possible to collect it efficiently.

Each of these is as important as the other, and our buildings can do more than they do now to make both things happen.

Dan Scully is an associate with Total Environmental Action, Inc.

As well as insulation, the design of a building can reduce the heat demand upon it. This is clear from the heat-loss formula, in which all factors are of equal importance, or have equal capacity to reduce heat loss: HEAT LOSS = (AREA x U x ΔT) + INFILTRATION. Architecturally we can deal with all these factors, not just the U value of the wall. We can build thermal/buffer zones around the living spaces to reduce the Delta T—the difference in temperature—between the inside and outside with foyers, greenhouses, unheated garages, thermal zoning of rooms for time-of-day use and temperature requirements, and so forth. On the outside, we can site the building so that winter winds do not sweep across exposed walls, and we can minimize windows on the cooler north side. Each significant factor becomes of greater importance as improvement is made in the others. The greater the wall insulation, the more important window heat loss becomes. In really well-insulated houses, infiltration can account for over 50 percent of the total heat loss. Fix it. But do not cut down on southern windows and their size just because the heat-loss formula tells you so. Instead, change the nature of those windows. In most regions of the country, a double-glazed south-facing window will gain significantly more heat than it loses over the course of a heating season. This does not fully redeem the window. It *does* put it on our side, but still it is BAD at night, according to that basic formula. So at night turn it from a heat-gainer to a heat-saver; insulate it. The nighttime insulating shutter is an important piece of any house that aspires to be a complete solar machine; house as Collection/Storage/Distribution.

Okay, energy conservation comes first, and without it you have missed the point and will be faced with a monumental collector and accompanying system. By decreasing heat-loss rates, you decrease the required collector area to provide the same percentage of heat requirements of the house. The operational efficiency of the collector has much the same effect upon sizing the system. A collector that starts picking up useful heat at 90°F will be smaller than one designed to pick up heat from 120°F degrees, for the same total heat gain. These are all factors that determine how large an architectural element the collector will be to deal with. Actually, total collector area is usually the result of building design requirements, and not of the way the numbers for the system work out. Operational efficiency also influences the size of the storage container that will be required. A low-temperature system requires a proportionally larger storage volume than a high temperature system does.

Insulation can influence the means by which heat can be distributed throughout the house. With an appropriate plan, a very well-insulated house could conceivably have only one point of heat supply and that could be on the floor immediately above the thermal storage—because the insulation will eliminate cool and drafty sections of the house, providing a much more even thermal balance.

One problem with many solar buildings, besides whatever technical problems they might have, is that the solar collector often has been used as the principal determinant of the house form, rather than one of the many inputs that should go into the design of a building that meets the need of those using it, and is buildable as well. Many solar house designers have taken a standard house, drawn back to the south, and thrown a collector at it from 57 degrees. "BONK! Take that, house!" Unfortunately, this cosmic

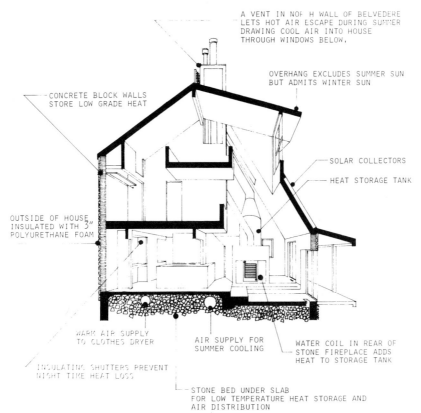

A VENT IN NORTH WALL OF BELVEDERE
LETS HOT AIR ESCAPE DURING SUMMER
DRAWING COOL AIR INTO HOUSE
THROUGH WINDOWS BELOW.

OVERHANG EXCLUDES SUMMER SUN
BUT ADMITS WINTER SUN

CONCRETE BLOCK WALLS
STORE LOW GRADE HEAT

SOLAR COLLECTORS

HEAT STORAGE TANK

OUTSIDE OF HOUSE
INSULATED WITH 3"
POLYURETHANE FOAM

WARM AIR SUPPLY
TO CLOTHES DRYER

AIR SUPPLY FOR
SUMMER COOLING

WATER COIL IN REAR OF
STONE FIREPLACE ADDS
HEAT TO STORAGE TANK

INSULATING SHUTTERS PREVENT
NIGHT TIME HEAT LOSS

STONE BED UNDER SLAB
FOR LOW TEMPERATURE HEAT STORAGE AND
AIR DISTRIBUTION

The Barber house, designed by Charles W. Moore Associates with Richard Oliver; Environmental systems by Sunworks, Inc. It's got a good tilt.

angle looks a little oppressive on the regular house, not to mention the resulting heavy feeling of the wall in the small bedrooms usually stuck behind the collector.

Tilt Angles

In Everett Barber's house, by Moore and Associates with Barber, the optimal tilt angle is used successfully. The collector tilt angle is an integral part of the entire design conception, allowing the interior an open ventilation system, and windows are used on the slope, relieving some of its visual weight. Heated air enters the living space around the floor slab and meanders up through the open house, to be reclaimed at the top and brought back down past thermal storage and recirculated. This scheme uses solar heating components, such as the vertical water storage tank, as elements to be dealt with, and even liked, rather than as dirt shoved under the rug or into the basement. The design also includes such architecturally useful elements as window overhangs to block out summer sun, but let winter sun in. The insulation for the house is put on the outside of the masonry block wall in order to use the thermal mass of the house as part of the heat storage, as a damper to temperature swings.

Buildings need to be cooled in the summer, as well as heated in the winter. At the top of the open space, a ventilating belvedere can be opened to induce air motion up, through, and out of the house. To deal with problems of dumping summer heat from the liquid collectors, an outside-mounted heat rejection coil is placed in the air stream from the belvedere. This rejected heat in effect super-charges the rise of air through the belvedere.

Optimum tilt angles may be just that—optimum—but they are not absolute laws. There is plenty of usable energy at roof slopes from say 45° to 90°. The angle to be used is the result not only of the solar system demands, but also of all the other constraints that go into building design: The ease of construction, the cost, the effects on the rest of the building design, etc. Anybody want to try and build the Trombe concrete wall at a 57° tilt? In some situations a vertical wall is not a bad alternative; windows have already proven that. In northern regions, the sun is low in the heating season. In January, 90 percent of the solar insolation that strikes the optimum tilt at 42° N. Lat., also strikes a vertical south-facing wall. To that can be added a 20 percent increase in insolation due to surface reflection off snow and the like. This vertical wall is, therefore, most effective in December and January and proportionally reduces summer build-up of heat in the collector.

Reflectors

Reflectors used to direct additional solar radiation to the collectors can produce significant gains, at a cost below that of the collectors themselves. A house by Robert Shannon explores the use of large-scale solar reflectors, hinged from the outside, and of insulating shutters, hinged from the inside. The basic shape of the house is cubic, a compromise between minimal surface area and the desire for a wide solar exposure. The solar system is an air-cooled collector site-built into the framing of the south wall. Great aluminum-faced reflectors, pivoted out in space, reflect additional solar insolation onto the solar collectors. The cables on these reflectors theoretically can be adjusted several times throughout the year, to maintain optimal reflective angles. Inside, huge insulating shutters hang vertically against the windows and can be winched up and out of the way, to be tied against the ceiling. It is a sailor's dream for winter living.

Windows

There is just as much energy available to a square foot of window as there is to a square foot of fancy collector. The only difference is in how it is collected/

Continued

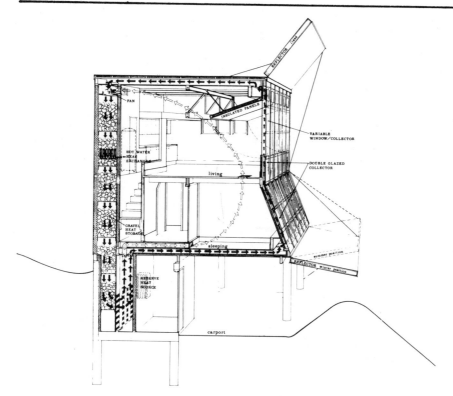

through oversized skylights to the sun, and in turn radiate heat directly into the rooms where the heat is needed, not to the outside. The house design also provides thermal buffers for the main living spaces—the entry, greenhouse, and bedroom spaces.

The Existential Pleasure

One of the excitements of good solar building design is that it can be done many ways, depending on many things: The people using the building, the climate and site, the budget, the idea, available materials—all the things, functional or symbolic, that go into the design of a building. In the end it is really a combination of solar collection methods that is appropriate, not all windows, and not all collectors, but a hybrid combination. And do not forget domestic hot water heating. An appropriate balance, cost effectively and otherwise, needs to be found between the many ways to collect/store/distribute the available energy within a building. You know they missed the point when you see a solar house with a big collector and only a few windows to the south. Instead of Either/Or, how about Both/And? ☼

stored/distributed. If there is enough solar energy out there to warrant using solar collectors, there is certainly enough to make a major contribution toward heating the house during the day with the windows, and let the collectors put their heat into storage for the night. With the appropriate thermal mass behind the windows, the interior space can effectively avoid daily overheating while simultaneously storing some direct solar gain for the nighttime.

A Vermont ski house designed by Robert Shannon of the People/Space Co. uses reflectors to help out the snow in winter.

Thermal Mass

The more windows are used to advantage, the more importance must be placed on interior thermal mass. It is in this area, the thermal mass performance in our buildings, that we will see some of the biggest changes from the standard methods of construction. The trick is to use standard materials and methods, but in some non-standard ways—like insulating on the outside to keep the mass on the inside. Most often, mass is built into buildings as a part of the structural system, whether it be concrete, adobe, or whatever. The Cook house, which I designed at T.E.A., Inc., is a light frame building with water tubes supplying the thermal mass in its center. These are exposed

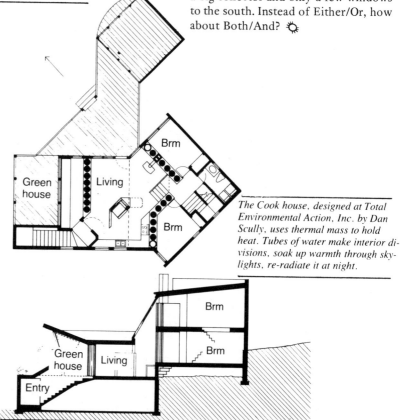

The Cook house, designed at Total Environmental Action, Inc. by Dan Scully, uses thermal mass to hold heat. Tubes of water make interior divisions, soak up warmth through skylights, re-radiate it at night.

59 WAYS TO SAVE ENERGY

*originally entitled "Fifty Ways to Save Energy"
**or "A Fictitious Day in the Life of Bruce Anderson"

By Bruce Anderson

Here I am, lounging comfortably in a nicely overstuffed chair[1], fur-lined shoes[2] discarded, wool-socked feet[3] waving uncertainly in the air, playing with words and ideas and paper[4] and ink[5]: "Fifty Ways to Save Energy." What can be said that hasn't already been beaten to death? I conclude, "Nothing."

I stare pensively into space. Silent clashes of battle draw my attention to the window: fierce combat as Btus muster thousands of successful charges out through the leak around the cold steel[6] sash. With equal vigor, millions of photons of sunlight bounce pathetically off the dirty[7] chicken wire glass[8] trying to get in.

Drowsy, in discomfort at the gory battle scene, my eyes shift back to the sterile whitewash of light drowning every nook and cranny of the library—enough light to grow a fine crop of wheat, I dare say. Not much rainfall, though. I look up to the sky of this brightly lit field to rows and rows of fluorescent[9] light fixtures. Ah—there's the rub. Let's remove two of the four bulbs[10] and make the whole room much more pleasant—and save energy.

But what?! It's already been done! There are only half as many bulbs lighting the room as was called for in the original design and those that remain are still sufficient to imitate a terrific greenhouse[11].

I glance over at the only other endangered species in the room—will she give a knowing glance? But only a dull stare at her table top—a fluorescent hypnosis at her papers that absorb a mere fraction of the more than 5,000 watts we share together in the room, about 4,980 more than we need. Small 20 watt bulbs shining on our respective sheets of paper would be more than adequate[12].

The warm room air seems to thicken. My wool socks[3] feel soggy—my long-johned[13] pits cry for Arid—a few other places need Johnson & Johnson—I wipe my brow and look around for the source of heat. I spot the floor register under the window[14] and rush over to close it. Bending over, my left hand supports me against the cold wall—obviously poorly insulated[15]. A rush of cool air explodes in my face.

A knowing shock penetrates my bones—air conditioning in the middle of the winter[16]! But of course—the excessive heat from the lights can't get out fast enough, even when it's cold outside. I am jolted in the realization that for each unit of electrical energy used by each bulb, three to four units

Continued

Cartoons by George Ellis

Bruce Anderson, author of The Solar Home Book *(Cheshire Books, Harrisville, N.H. 03450; $7.95), just built a solar-heated house in Harrisville, N.H., with a two-story garage for the south wall's insulating shutter during the day.*

1 Heavy overstuffed furniture is usually warmer and cozier than plastic and metal furniture.

2 Wear insulated or fur-lined shoes in the winter.

3 Wear thicker socks (or two pair) in the winter.

4 Use paper at least twice. In particular, use back sides of paper already used before.

5 Use non-disposable pens with ink refills. Remember that every item discarded is discarding resources and energy that went into its production.

6 Avoid metal window sash; it conducts heat (and cold) too well. Use wood instead.

7 Keep windows clean for maximum solar heat gain.

8 Use two, even three layers of glass, not just one.

9 Fluorescent light fixtures are more efficient than incandescent bulbs but natural lighting uses *no* energy.

10 Very often in large buildings, half the fluorescent bulbs can be removed without significantly reducing lighting levels.

11 Many (most!) commercial and academic environments have far more light than necessary.

12 Light is most efficiently used when it shines directly on the item needing the light, not on every single object in a room.

59 WAYS TO SAVE ENERGY *Continued*

of fossil fuel energy are burned at the power plant and only a fraction is actually converted into light, the rest being released as heat[17]!

I run to the front desk. "I'm sorry, sir, the thermostat is upstairs. It controls the whole building[18]—and it's locked[19]." And the light switches? There's only *one* switch for the entire room[20]. As far as I know, it's left on all the time, day and night. Seems they use the light to heat the place[21]. In desperation I race to the window. It's locked too[22]! Looking carefully, I notice it hadn't been opened in years.

"Thud." My compatriot's head has fallen to the desk—a dazed unconsciousness replaces her dull hypnotic stare. I grab my shoes and papers and flee for the door, my coat trailing on the gray carpet behind me. Lunging for the "fresh" air of the softly lit hallway[23], I give a glance to the elevator, but head for the stairs instead[24]. Surely the three

flights of stairs are more easily negotiable for me than was the transformation of sunlight into fossil fuel millions of years ago to be exploited today in our power plants to produce the electricity to hoist the elevator! Bursting through the fire escape, the din of New York City traffic[25] seems all the more oppressive in the pollution-muted sunlight[26]. Little wonder I had chosen to be inside[27]. A direct relationship: less traffic and less heating and cooling and lighting produces less pollution and more sunshine.

I head toward Port Authority—there's bound to be a bus[28] home.

"Hoonnnk" "Screeeecchh" A blur of yellow descends down upon me[29]—I jump—but too late. . .

I wake up in a sweat. I've been undressed down to my long-johns[13]. Next to me, Angie's warm naked body[30] undulates peacefully[31]. I slip out from under the down

---

13 Long underwear significantly improves comfort in cooler environments, enabling room temperatures to be kept lower.

14 Most buildings are so poorly designed that warm air must be introduced at the windows, counteracting the cold drafts falling off the windows as a means of improving comfort.
But warm air against cold windows means greater heat loss.

15 Insulate walls (and roofs and floors)!

16 Many buildings waste energy on excessive lighting and then require air conditioning energy to rid themselves of the resulting heat.

17 Electricity is usually an extremely wasteful form of energy use. Avoid its use whenever possible. Again, fluorescent bulbs are more efficient than incandescents.

18 Individual spaces within buildings, even houses, should be zone controlled. The heat or air conditioning for many spaces can then be turned off when not needed.

19 So many buildings are designed to be machine controlled. People comfort should be of primary concern when designing buildings.

20 There should be numerous switches strategically located so that individual lights can easily turn on (and off!) according to need.

21 A once common myth—leaving lights on saves energy. Rarely true. Turn them off!

22 Again, too many buildings are hermetically sealed, ostensibly in the name of keeping mechanical control of the sake of saving energy. Very often, if people can open windows, energy can be saved (and people will be more

comfortable!).

23 I should have helped her, and I feel awful now that I didn't. Sorry. I would have turned the lights off as I left—but no switch. At least the hall was sensibly lighted. And it looked and felt so much more comfortable.

24 Avoid elevators; use the stairs and save energy and improve your health.

25 Too much needless driving! Walk, bus, bike, or train.

26 Let's eliminate pollution and let the sunshine help keep us warm.

27 Working and playing outside means less energy use inside for heat, air conditioning, and lighting.

28 Ride a bus and leave the driving, and the saved energy (and less pollution) to someone else.

29 The taxi was driving much too fast. Slow down and live—and save energy too.

30 Although I strongly advocate sleeping naked, a few nightclothes save energy or. . .

31 Sleep with someone—saves energy.

32 Tired of heavy layers of blankets? Try down quilts. Hmmm, warm.

33 Flannel sheets (often called "sheet blankets") are the greatest!

34 Turn your thermostat back as far as possible at night. This is one of the best possible energy saving efforts you can make! Because this house is so energy efficient, it drops only one or two degrees in temperature at night. Turning thermostats back to 45° or 50°F in ordinary houses is not

quilt[32] and the flannel sheets[33] to the freshness of 66°F pitch black air[34] and stumble to the window. The Styrofoam insulating shutter wrapped in beige terrycloth pulls away easily from the window[35], and I set it against the warm wall[36] filled with the equivalent of 8 inches of fiberglass insulation[37]. Starlight streams through the Thermopane[8] wood-frame[6] windows, protected on the outside from the cold by yet another pane of glass, an ordinary storm window[38]. The deciduous trees have lost their leaves for the winter, and through the scraggly hug of the sugar maple branches the dim horizon forebodes the eternally endless day/night cycle. During the winter we have the benefit of that early morning heat from the sun, but only finely filtered light manages to squeak through the dense leaves during the summer, keeping us cool[39].

Continued

uncommon and saves considerable energy.

35 Movable insulation, like Styrofoam covered with cloth, can reduce heat loss through windows to about one-fifth or one-tenth.

36 A well-insulated wall is warm instead of cold. This saves energy by reducing heat loss, and the warmer wall surfaces permit comfort at lower temperatures, also saving energy.

37 Many houses today are being built with much thicker walls. Where 3½ inches of fiberglass insulation was once the "standard," today, insulation having an insulating value equivalent to eight inches is often the standard, or should be.

38 Thermopane windows in combination with storm windows are fast becoming popular. In combination with movable insulating shutters, you can have an extremely energy efficient window!

39 Trees remain the best shading devices known to man. They let the sun in during cold winter months and keep it out during hot summers. The transpiration of their leaves adds to their cooling effect around the house.

40 An excellent energy conserving house eliminates or greatly reduces the need for a large furnace. The sound of the furnace blower is then also eliminated or greatly reduced.

41 There are many clock designs that require no mechanical energy. Clocks don't use much but it all makes a difference.

42 On the average, showers use less than half the hot water of a bath. An energy-conserving shower uses even less.

43 A solar water heater makes good economic sense everywhere, especially if your water is presently being heated by electricity.

44 Using less hot water saves energy, obviously. Using less cold water also saves energy by reducing the pumping energy required to deliver it to your house. Using less also conserves that very valuable resource, water.

45 Taking a shower together may not necessarily save energy—although it certainly *can*. But it sure is fun, and that in itself is enough of a reason to do it.

46 Warm liquid always warms the body, if not the soul.

47 Light a lantern instead of flipping a switch and notice how much more you enjoy light—*and* save energy!

48 Wine can help keep you warm too—or make you insensitive to cold. And buy local wines—saves on energy transportation costs.

49 A locally-made mug saves industrial and transportation energy, and supports the local economy.

50 Tiled floors absorb and hold heat from the sun.

51 Again, movable insulation for the windows. This clever and very effective device is known as Beadwall[TM], developed by Zomeworks Corporation in Albuquerque, New Mexico.

52 All the heat coming through the windows must be stored by heavy material (such as concrete and brick and water) in the house to keep the house from overheating.

53 Sometimes the heavy materials are not sufficient or are too expensive. An insulated bed of rock or gravel under the floor is ideal for storing the excess heat. Warm room air is blown through the gravel and at night is blown back out to heat the house.

59 WAYS TO SAVE ENERGY *Continued*

The stillness[40] of the night is interrupted by Dad's grandfather clock downstairs; it chimes six times[41]. Shall I shower and dress now or wait for Angie? Although our water is heated by a solar collector[43] we just had installed in March*, and although we have changed to water-saving faucets[44], Angie and I still shower together[45].

Perhaps a hot cup of tea while I wait[46]. I strike a match, and take the refurbished Cape Cod lantern[47] by the hand. As the warm glow and I slip through the den together, Tasha, our fickle seal-point Siamese, snuggles on my over-stuffed chair, barely mustering a raised eyebrow. I give a knowing glance to the few drops of wine[48] remaining in the cobalt-glazed mug[49] on the ceramic tiles of the floor[50] next to the neat stack of still-untouched paper[4]. So that's how I got to bed last night!

Without warning, a dull hum of distant motors and suddenly the entire south wall of the den begins to shed its insulating veils, and the early morning rays of the sun leap across the floor[50]. Small blowers—vacuum cleaner size—

have been told by a simple sensor that it's light outside—time to draw the millions of tiny polystyrene beads out from between the two layers of glass spaced 4 inches apart. During the day, the house is lighted and heated by the sun coming through the glass. But at night, the beads fill the space between the two layers and transform the heat-losing window into a well-insulated wall[51]. The thick walls and floors of the house absorb most of the heat[52]; excess heat is circulated by a small fan through the gravel under the floor, preventing the room from overheating and storing it for when the sun doesn't shine for several days [53]. Within a minute, the blowers have finished their chore and the winter-wooded mural covers the entire south wall.

It's been three days since the sun shone. The house had gotten up to 73°F, but since then it's been down in the 20's most of the time outside, and it's dropped down to 66°F inside. Another day without sun and I would have been able to fire up our wood stove[54]. These damned energy-efficient houses, anyway—hardly ever have the opportunity to experience the penetrating pleasure of wood heat!

As I snuff out the unneeded lantern, and reach the kitchen, I smile pleasurably at the ripening tomatoes[55] in the greenhouse[56] just beyond the round oak table. With tomatoes at 85 cents per pound, our greenhouse is doing more than saving energy!

I pour a cup of cold water into a mug[49], plug in the electric heating coil[57] and drop it in the water. Outside, the Jacobs[58] is quiet now, but I know with confidence that we have plenty of electricity stored up in our batteries.

The water soon starts bubbling. English Breakfast tea always gets me going in the morning, and I retreat to the den with the hot cup. Tasha's lying in the sun now[59]. I reclaim my chair and settle in again for another try. Let's see now. What can I say about saving energy that hasn't already been said fifty times before?

---

54 Wood-burning stoves and furnaces are great alternatives to oil, gas, coal, and electricity in many parts of the country. In energy conserving houses, a good, easy-to-use, efficient wood stove is a pleasure to use.

55 Grow your own food. It saves a whole lot of growing, harvesting, processing, and transportation energy—and is healthier to eat.

56 A well designed greenhouse not only provides solar energy, provides fresh vegetables, and provides a pleasant avocation, it also reduces heat loss by acting as a buffer zone between the house and the cold outdoors.

57 A simple electric coil that dips into the water heats primarily the water instead of the stove burner, the pot, the air, the cup, *and* the water. If you must use a tea kettle to heat water, heat no more than what you intend to use. Use as lightweight a teakettle as possible.

58 A "Jacobs" is a windmill with a long history (from the early 1930s) of use. Wind power is becoming an increasingly viable means of producing electricity.

59 Lie in the sun. A great way to stay warm—and healthy!

---

* The federal government will probably pass legislation this year to provide tax rebates to everyone who purchases solar equipment. On a solar water heater purchase of $1,500, the rebate (refund) could be as much as $525.

Passive: when applied to solar energy, a method of collection that uses building design and materials to gather and store solar heat (or cool), rather than special hardware, machinery, or energy-consuming equipment.

The state—and state-of-mind—of
THE PASSIVE ART

By David Wright and Barbara Wright

Man has devised ingenious systems using stored solar energy in the form of coal, oil, gas, uranium and the like to assist his life support techniques. We are discovering that these systems, while logical for a short term, are not sufficient for the long range scheme of things.

Nature manages to accomplish the miracles of life on a low temperature scale. Virtually all of our natural earthly processes are sustained by diffused energy. Photosynthesis, body metabolism, and weather cycles make use of low grade conversion of energy to produce the forces which power our earth's systems. The main power input, that great fusion reactor called the sun, is the only necessary driving force. Man, on the other hand, has created high energy techniques of coping with his environment, and we call this progressive. This quest for progress has driven us to overdependency on high temperature and relatively inefficient thermal processes to accomplish a myriad of tasks that at this time we consider necessary for our quality of life.

In many cases, we are finding that high energy active methods are not the most efficient means of using our technology. Certainly space conditioning, food production, water heating, desalinization, and crop drying, among other things, are easily accomplished using primarily the incoming solar resource. Imagine a design that would make efficient use of all energy and materials. This involves thinking on both major and minor scales. Passive systems that do not require high temperature thermal processes to function are probably our most readily available and even our best long range solution to many of today's energy problems. In order to succeed in the implementation of passive processes in our houses, offices, and factories, we must reevaluate all of our systems, and draw from history, nature and technology at many different levels.

The first step toward this goal is to ascertain what forms of energy are being wasted. We need to know how many units of heat energy are required to do a job, and then use our knowledge to do that job in the most efficient practical manner. Once the task has been singled out and objectively reviewed, we can get on with creating the mechanisms for implementation.

Today the state of the art of passive systems is evolving rapidly. We now are aware that many tasks formerly requiring the combustion of fuels can be accomplished by much lower temperature means. In the forefront are the heating, cooling, and ventilation of the buildings in which we live, learn and work. This is just the tip of the iceberg. Many creative minds and hands are exploring avenues of solving both short and long range problems through passive techniques. The simple reminder that enough usable solar energy falls on the surface of the earth to provide all of our energy requirements as long as the

Mr. Wright is an environmental architect based at the Sea Ranch, Calif., and a member of the Solar Age *advisory board. Barbara Wright translates.*

THE PASSIVE ART *continued*

Phase change material: a substance that changes from one phase (for instance, a solid) to another (for instance, a liquid) with a change in temperature, and that with each change either stores or releases heat.

sun will shine, is enough to stimulate this long overdue research and development.

Basically, the approach today falls into three categories. The first and most evident is to use what we have and know about already to the best advantage. The use of commercially made or homemade materials and devices for new tasks involves looking around in our environment for things that can be used in different and better ways. Junkyards, hardware stores, the yellow pages, and many other places hold untapped reserves. Steve Baer's discovery that steel drums, once used to transport oil or chemicals, could be filled with water and used to collect and store energy to heat and cool without electricity or any fuels, was a remarkable reevaluation of a material and man's needs. The Harold Hay experiment—an insulated ice chest, filled with water and covered and uncovered systematically—taking advantage of direct solar gain, deep space radiation, and evaporation to heat and cool the contents, led to the "Skytherm," another example of how the laws of physics and existing materials can effect man's

needs. Both have led to further innovative speculation on how and why we do things with what we've got.

The building materials industry currently markets numerous materials and devices that can be adapted to passive solar designs. Large transparent and sloped solar collection surfaces can be made using sliding glass door replacement units. These factory sealed, double tempered, altitude adjustable glass units, which come in four basic sizes, are exceptionally durable and high quality transmission surfaces. They usually can be purchased quite reasonably in orders of ten or more. Other industrial materials such as fireproof wood fiber insulations with "R" values up to 7.0 per inch, neoprene windshield glazing gaskets, ultraviolet resistant greenhouse fiberglass, flexible high temperature hose, and other such items can aid solar designers with a ready-made pallet of suitable materials.

The other approach to passive systems is to create new tools and materials to accomplish tasks that appear feasible. Here, we step beyond the adaptation of existing ideas and products to reach for the means to solve specific problems with new ideas and products. Take, for example, the work of Suntek (Day Chahroudi and others)—the sandwiching of clear gel, which becomes opaque at higher temperatures, between transparent sheets to create a self governing solar gain surface and prevent excess absorption, is an instance of finding a low-key solu-

tion to a problem that exists in greenhouses and direct solar gain structures. They and other groups are working on the capacity to store passive or ambient heat in new and existing structures, another problem. Today development is being done on encapsulating phase change materials in concrete masonry units, wall and ceiling tiles, plasterboard, etc. This will allow the fabric of our buildings to use the latent heat of phase change in order to maintain equalized temperature in spaces at all times. Within a temperature range of a couple of degrees, a building will have enough heat capacity to accept or give off energy to sustain comfort, equalizing the normal daily and seasonal loss/gain demands.

Now our focus on passive applications will intensify. With a sustained effort, we will undoubtedly become aware of the solutions which now elude us. The products and systems contained in this catalog are the starting point.

From now on, it is eminently important to search out and use all of the methods available to better use the energy of the sun. We must, however, at the same time beware of half baked solutions, fast buck artists whether corporate or individual, and even the entrepreneur hiding under alternate energy slogans who is marketing a good product at an outrageous price.

May we all become more actively passive in our attitudes towards energy consumption. ☼

State of the vocabulary

A PASSIVE SOLAR GLOSSARY

By Charles Michal

Arguments abound concerning the most appropriate definition of the term "passive." I like to use the term, when referring to heating or cooling systems, when either the heat transfer processes are entirely unaided by mechanical power or when the heat collecting and storing functions are served by the architectural elements of the building, elements like windows, floor slabs, and heavy-mass walls.

●**Attached greenhouses** figure prominently in passive solar building designs. An attached greenhouse (or any greenhouse for that matter) can be solar heated itself with the addition of **thermal mass** and perhaps **movable insulation;** the attached greenhouse may also serve simply as a **buffer** zone, windbreak, or extra glazing layer for the major structure, with the accompanying energy benefits of a lower **overall heat loss factor.** Whether the attached greenhouse is considered a thermosiphoning space-heater or collector in which one happens to grow green things, or a food and flower producer that provides a little solar heat for some other space, it nevertheless is an appropriate tool for the designer of passive solar.

●**Beadwall** is the proprietary name of a concept for **movable insulation** developed by Dave Harrison and Steve Baer of Zomeworks Corp. By means of vacuum-type electric blowers, PVC piping, and special valves, millions of tiny polystyrene beads can be blown into the cavity between the two sheets of material in a double-glazed window, window wall, or skylight, transforming the **overall heat loss** factor of a building instantly. During times of potential solar heat gain and whenever views and natural light are desired, the

Mr. Michal is a design associate with Total Environmental Action, Inc.

beads are sucked from between the double-glazing by a return vacuum line. The process is repeated as desired.

●**Conduction** refers to heat flow through solid objects and is a process found in almost every application of solar heating. Insulating materials conduct very little heat: in the **Trombe wall** and the radiant floor slab, high thermal capacity combines with the conducting process to produce a delayed response to temperature changes, a characteristic of importance in passive design.

●**Drumwall** is the name given to a passive system configuration used by Steve Baer in 1971. Very similar to systems studied at MIT in 1947, the Drumwall system uses stacked 55-gallon steel drums filled with water for thermal storage capacity, and places this thermal mass against south-facing glass walls for maximum solar absorption. As in most successful passive systems, space heat losses to the outdoors during periods of no solar heat gain are prevented by **movable-insulation** devices over the glass walls. Heat is delivered to the space day and night by **radiation and natural convection** from the drums.

●**Free or natural convection** combines with **radiant heat** as the means by which heat is most often transferred in passive solar heating. Like radiant heat exchange, **natural convection** (sometimes known as gravity convection) is not well understood in a practical design and engineering sense. Driven by localized temperature difference that causes their volumetric weight (density) to change, air, water and other fluids develop internal circulation patterns that move primarily up and down. Under the influence of gravity, the heat energy is displaced, or transferred, along with the moving fluid. This type of convection is most often random and uncontrolled, but the same principle provides for chimney effects and thermosiphoning, useful passive design concepts.

●**Glare** is a second most common concern in passive solar space heating design (see **overheating**). The larger glass areas seen in many designs provide not only solar energy for heating purposes but an abundance of daylight that may cause vision problems due to the angle of the light, the contrast between bright windows and dark walls or work surfaces, specular reflections of the sun's image, or other lighting and vision related factors. Many passive concepts, like **Trombe walls** and **thermosiphoning air collectors,** involve in part the idea of letting sun in but keeping light out.

●The **greenhouse effect** provides the basis for useful solar collection, whether in active or passive systems. Because light energy will, while heat energy will not, pass through certain otherwise transparent materials like glass, trapping solar energy is accomplished when light strikes opaque objects beneath glass. (The light is partially absorbed into the opaque object, and re-radiated as heat that can't escape **except** by **convection** and **conduction.**) For this reason glassed-in spaces that are insufferably cold unless heated when the sun isn't shining often overheat when the sun does shine, regardless of outdoor temperature. (See also **transparent insulation**).

●**Movable insulation** is one of the most useful concepts in passive solar use. A variety of means exist (see **Beadwall**), and others can be developed, to insulate glass or other transparent portions of passively heated structures against heat losses when the sun isn't shining. Paradoxically, "opening up" a building to admit solar energy can increase the **overall heat loss factor.** The **greenhouse effect** helps prevent radiant heat loss, but during long cold nights, much energy is lost by **conduction** through the collection area. Insulating devices can be used to transform such transparent solar-energy-admitting wall or roof areas to heat-retaining elements at appropriate times, thereby increasing several fold the effectiveness of the passive solar heating.

●**MRT (Mean Radiant Temperature)** is an approximate indication of the effect that the surface temperatures of surrounding objects have on human comfort. (See **radiant heat transfer**). Whenever surface temperatures are different from air temperature, comfort cannot be judged by air temperature alone. As the MRT of one's surroundings goes up, air temperature can come down and comfort be maintained. Likewise a cold interior surface will cause discomfort even when air temperatures are high—a situation guaranteed to be energy consumptive.

●An **overall heat loss factor** is the numerical indication of a building's thermal energy needs as determined by outdoor temperatures. In many cases thermal energy needs are controlled by factors other than outdoor air temperature; nevertheless the overall heat loss factor provides an easy measure of the extent to which energy conservation efforts have been made, efforts that must be made prior to, or at least simultaneously with efforts to solar-heat buildngs. The heat loss factor is most usefully expressed in Btus per °F per hour, or Btus per °F per day, or the metric equivilent. Of particular interest to some passive system designers is the ratio of overall heat loss to the passive collection area (windows, **Trombe walls**, etc.); the lower the ratio the better.

●**Overheating**, or a temperature rise above a comfortable level in passive designs is the most common concern people have; legitimately so, since the experience of the 1950s showed that large glass areas alone did not make for a comfortable solar house. Overheating is simply the effect of more heat energy available than can be used **or** stored. Good passive design calls for reducing the need for heat energy (reducing the **overall heat loss** factor) while increasing the structure's ability to store energy (additional **thermal capacity**). Overheating created by energy input greater than the passive system's ability to absorb it usefully can be cured by venting.

● **Radiant heat transfer** between objects is the means by which energy is often released or exchanged within passive solar buildings. Like the other forms of heat transfer (conduction and convection), radiant heat exchange is driven by a temperature difference between the objects affected. Heat flow is always towards the cooler body. Unlike the others, radiant heating requires no contact, either directly or through fluids, between bodies. Radiant heat, like light, travels only in straight lines and, most important, casts "cold shadows." The actual rate of radiant heat exchange is enormously complex, affected by temperature differences, shapes, and surface properties (emittance and reflectance), and the distances between objects, an important aspect in energy conservation space design (see MRT).

●**Solar heat gain** is a general term with obvious meaning, but it is often used specifically to refer to the numerical quantity of heat energy received *inside* a passive solar-heated structure during a given time period, before any losses or energy transfers are taken into account. Solar heat gain factors as established by professional societies (ASHRAE) and manufacturers of conventional cooling equipment (Carrier Corporation) provide passive solar system designers with a working equivalent to the clear-day insolation values used in active solar system design.

●**Solar ponds** should not be confused with the **Skytherm** concept. The solar pond is a shallow body of water that has such high salt concentrations that the normal direc-

tion of gravity or **free convection** within the pond is reversed. Solar-heated water at the surface of the pond is displaced by cooler water from below, proving that heat doesn't always rise. This reversal is possible because the cooler water is unable to sustain the same level of salt concentration as the warm water and is therefore *lighter* and rises to the top of the pond. The heated water tends to stay at the bottom and does not cool off readily. There are natural solar ponds but most interesting to passive solar design are man-made versions that seek to use this remarkable phenomenon for space heating by drawing the collected, stored solar energy off as heated water from the bottom of the pond.

●**Skytherm** is the proprietary name given to a passive system configuration developed by Harold Hay in 1967. His roof pond concept is as subject to variation by other designers as is the **Trombe wall** idea. In its simplest form the roof pond incorporates a shallow pond of water in thermal contact with a strong but highly conductive flat roof or ceiling structure. The ceiling heats the space below, primarily by radiation. The pond, which serves as thermal mass, is exposed to solar heat gain and protected against wasteful heat loss at appropriate times by activating some system of **movable insulation** mounted above the pond. Originally developed for flat roofs in southern latitudes, and also effectively used for radiational cooling to summer night skies, the pond concept has been used as far north as Concord, New Hampshire.

●**Stratification** refers to the tendency of the **natural convection** process to establish different temperature layers (strata) within a contained fluid such as air or water. Stratification is a delicate thermal and gravitational balance easily upset by outside forces (breezes, drafts, currents, fan or pump activity). Stratification cannot be maintaned over time without additional heat energy input (remember entropy?). Normally stratification results in hotter temperatures above and cooler below (but see **solar ponds**), an ordered layering that is good in a thermosiphoning domestic hot water heater and bad in high ceilinged rooms during the winter.

●**Thermal mass or thermal capacitance:** refers to the varying abilities of different materials to store heat energy. The stored energy (expressed in Btus in the English system of measurement) is manifest in the temperature rise of the material. Materials of high thermal mass store more energy per unit weight for a given temperature rise than materials with low thermal mass. Thermal mass provides both heat storage and protects against overheating in passive solar systems. Thermal capacity per unit **volume dollar** spent are useful ways to express this material property. Concrete, masonry materials, and water are most often used to provide the thermal mass in passive solar applications.

●**Thermic Diode** is a name coined by MIT researchers for a fairly sophisticated version of a thermosiphoning hot water solar system for space heating. A clever oil-filled one-way valve (to prevent reverse thermosiphoning) provides a one-way thermal circuit between a thin film of fluid (water) for solar collection and a thicker, larger layer for thermal storage separated by insulation—all of which is contained specially—fabricated modular building component. The panels form either exterior walls or roof/ceiling assemblies to collect and store solar energy which then radiates as heat to the interior of the building. Some versions of the thermic diode use a force or free **convection** loop to deliver stored heat to the building spaces.

●**Thermosiphoning air collectors** are passive subcomponents especially suited to low-cost adaptation of existing structures to partial solar space heating, and also to various hybrid systems where both active and passive systems are employed. Thermosiphoning, a form of **free convection** affected by the restrictions to air flow and the vertical heights employed in the collector design, is the driving force that moves cooler room air through a simple flat-plate collector to be heated and then discharged back to the room. To accomplish this without fans and mechanical controls, the space to be heated has to be at the same level or above the collector, a fact with significant architectural implications.

● **Transparent insulation** is a concept for new building materials that would combine the best features of **movable insulation** and the **greenhouse affect.** Transparent to light, yet highly resistant to conductive heat flow and opaque to radiant heat loss, such materials will, when developed for general construction use, significantly alter the architectural and engineering aspects of solar energy use. Work is currently underway in California to test prototypes of such materials developed at MIT.

●**Trombe walls, or Trombe-Michel walls,** are passive system configurations in which the **thermal mass** and the collection area are provided by a modified concrete or heavy masonry exterior wall (preferably south-facing) that releases stored solar energy to the building interior both by radiant and naturally convective heat exchange. lThe configuration is named after Felix Trombe and Jacques Michel of Odeillo, France, who first used such walls in 1956. A glass or other transparent outer layer is always used with the concrete walls; beyond this, designs vary with wall thickness and inner & outer heat transfer characteristics. A Trombe wall, modified with **Beadwall** as the outer glazing (first used by TEA, Inc. in 1974,) does for a building's south walls what Harold Hay's Skytherm does for roofs as far as passive space heating is concerned. ☼

Passive Products

ENERGY SHELTER™
manufactured by:
A.C.M. Industries, Inc.
Box 185
Clifton Park, N.Y. 12065
George Keleshian; phone: (518) 371-2140
A structure using little or no external energy source to maintain normal design temperature and humidity levels in nearly any climatic/geographical condition, while providing competitive building cost and architectural flexibility. **Features and options:** buildings not dependent on solar assistance, however Solar X Thermosyphon and SO$_2$ Inwall collector/exchangers systems are available. **Installation requirements/considerations:** buildings erected by competent local contractor and building trades, supervision by ACM technician suggested. Foamlock™ process to be performed by ACM technician or competent operator. **Guarantee/warranty:** five year limited. **Maintenance requirements:** normal building maintenance. **Manufacturer's technical services:** architectural and engineering, field supervision by ACM personnel. **Regional applicability:** Northeast, including S.E. Canada, Ohio and Indiana. **Availability:** contact factory. **Price:** available on per job basis; square foot estimates given for ballpark price.

KALWALL SOLAR FURNACE
manufactured by:
Kalwall Corp.
1111 Candia Road
Manchester, N.H. 03103
Drew Gillett; phone: (603) 668-8186

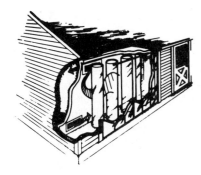

The Solar Furnace consists of Sunwall, Sun-Lite storage tubes, and an insulating curtain. Sun-Lite collector/storage tubes are filled with water and set vertically next to the Sunwall to serve the combined function of solar energy absorption and thermal storage. **Features:** most solar systems separate the collection from storage; this type of system simplifies the entire operation into one unit. **Installation requirements/considerations:** must be installed on a south-facing wall for satisfactory results. **Guarantee/warranty:** none

expressed or implied. **Maintenance requirements:** At five year intervals the panels may be refinished with Kalwall Weatherable Surface to provide surface erosion protection. **Manufacturer's technical services:** sales and technical staff are available to discuss customer's requirements at the above phone number. **Regional applicability:** nationwide. **Availability:** from Kalwall Solar Components Div., P.O. Box 237, Manchester, N.H. **Price:** $10 to $12 per square foot.

SKYLIDS ^{r} **/U.S. PAT. # 3,884,414**
manufactured by:
Zomeworks Corp.
P.O. 712
Albuquerque, N.M. 87103
Steve Baer; phone: (505) 242-5354
Insulating louvers that fit beneath skylights or glazed roofs. They open and close in response to the sun by shifting the weight of a Freon gas. They open when the sun shines and close in the evening, automatically **Features and options:** in addition to being good insulators they also serve as effective sun screens and can keep a space cool in the

summer. **Installation requirements/considerations:** preframed, ready to install into a rough opening, much like a pre-hung door. No special tools are required. **Guarantee/warranty:** one year on materials and labor. **Maintenance requirements:** the polyester tube seals may eventually need replacement. **Manufacturer's technical services:** contact Zomeworks for any problems. **Regional applicability:** anywhere the sun shines. **Availability:** F.O.B. Albuquerque, N.M. **Price:** $241 to $425.

DRUM WALL (PLANS)
manufactured by:
Zomeworks Corp.
P.O. Box 712
Albuquerque, N.M. 87103
Steve Baer; phone: (505) 242-5354
Two 24 by 36 inch pages of blueprints describing the general and specific features of the drum wall heating and cooling system. The drum wall is a passive solar heating system that can be used in south walls. Also included are tips for instructive experiments with solar energy and instructions on how to build drum clips. **Price:** $5.

1977 SOLAR ENERGY & RESEARCH DIRECTORY

By a Special Ann Arbor Science Task Group

Here is a new Directory to give you bona fide manufacturers and researchers in the rapidly moving solar energy field—and tell you their respective products and research areas. You get complete, new listings for some 700 manufacturers, design and construction firms, researchers, government-sponsored R&D groups, energy conservationists, and distributors—all currently involved in solar energy development. Includes hundreds of names and addresses of responsible persons active in solar, and myriad literature sources—for real, currently available products and ongoing research.

The 1977 Directory was prepared by a special Ann Arbor Science task group from questionnaires mailed to a carefully researched list of companies, institutions and organizations active in the solar field—emphasizing research and manufacturing firms. Each questionnaire was carefully screened and evaluated. Because the total compilation was done in 1977, the entire content is current and applicable *now*.

Not a collection of pages prepared by the individual organizations represented, but a complete cross-referenced, and convenient-to-use book—organized so you can quickly find the new, up-to-date information you want—when you want it.

Everyone interested in or working with this promising power source will find this new Directory a valuable document. Of especial use to builders, scientists, architects, environmentalists, energy specialists, politicians, attorneys, consulting engineers, government, industry, universities, energy conservation groups—anyone who wants to intelligently explore the prospects and/or capabilities of solar power.

FIND OUT WHO'S DOING WHAT IN:

Bio-Conversion • Wind Conversion • Ocean Thermal Energy Conversion • New Construction and Retrofitting • Solar Water Heating • Solar Heating & Cooling • Solar Electric Power Generation • Solar Refrigeration • Solar Powered Transportation • Solar Ponds • Solar Furnaces • Solar Engines • Design and Construction • Heat Pumps • Turbines • Solar Energy Storage • Photovoltaics • Total Systems • Components • Homes • Commercial Buildings PLUS Geographical index.

1977   **Cat. No. 40171**   **$22.50**

How It Was—Background Value
HERE IS THE ORIGINAL 624-PAGE...

SOLAR DIRECTORY

Edited by Carolyn Pesko

Gives information on work in this rapidly growing field, of interest to:

Government • Industry • Universities • Architects • Engineers • Scientists • Builders • Lawyers • Politicians • many others interested in Energy Conservation.

SOLAR DIRECTORY CONTENTS

• Solar radiation measurements • Ocean thermal gradient systems • Bioconversion • Wind conversion • Inorganic semiconductors • High and medium temperature collector systems • New construction technologies • Water heating • Space heating and cooling • Power generation • Total systems • Lay & technical periodicals • Reprints • Books • Solar engines • Transportation • Cooking • Greenhouses • Passive systems • Retrofitting • Homes • Condominiums • Academic buildings • Solar projects information • Foundations providing funds • Incentives for energy conservation • Bibliographies • Information services • Original researched information & reprints
 Scientific, financial, economic, social, and legislative considerations are also included.

1975   **624 pp.**   **Cat. No. 40109**   **$20.00**

An Inquiry Into the Future
of Our Industrial Society
FUELS, MINERALS, AND HUMAN SURVIVAL

By **C. B. Reed**, Geological Consultant, Austin, Texas

What alternate energy systems are available to ease present problems? What others for the future? Can we cope with the pollution some of them create?

These are important questions for everyone particularly if you are involved with energy needs, if you work in pollution control (the two problems are inseparable), involved with environmental health in any way, or perhaps just interested in this major issue.

CONTENTS
1. THE NUCLEAR STORY / The nuclear power plant—Safety and pollution—Breeder reactors and plutonium cycle—Waste disposal is imperative—Our vested interests—The peaceful atom—Questions and answers—13 graphs and diagrams.
2. ALTERNATE ENERGY SYSTEM / Power from fusion—Sun-power systems—Interim power sources—Coal.
3. CONSERVATION AND DEPLETION / Our diminishing resource base—A challenge to Cornucopia—Opinions and conclusions—30 resource-depletion curves.
4. APPENDICES / Radioactivity—Resources—Ice Salt—tables of available resources and projected requirements. 48 graphs, diagrams, 43 figures, 5 tables, 198 pages.
1975   **Cat. No. 40083**   **$12.50**

Building your own SOLAR GREENHOUSE

By Bill Yanda and Rick Fisher

The site you have chosen for a greenhouse may demand attention before you can begin foundation work. In certain cases, a site that is not level can work to your advantage. If the terrain slopes away from the existing structure, for instance, you might consider "sinking" the floor level of the greenhouse (Figure 1). A good depth for attached solar greenhouses is the same depth as the home foundation. One of the main reasons for excavating is to lower the profile of the unit so that it fits beneath existing eaves of the house. It will require more excavation than simply leveling the site, but it can also result in more useable vertical space.

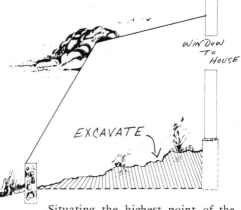

Situating the highest point of the greenhouse interior directly adjacent to a house window or doorway will also supply more useable heat to the house. If the ground slopes laterally to the side of the house, you may wish to design a split-level floor plan rather than level the entire site. Whether split-level, sunken or used in its existing state, the site should be relatively level (side to side, front to back) before beginning foundation work.

The excavation depth for a green-

Bill Yanda and Rick Fisher are authors of The Solar-Heated Greenhouse *(John Muir Press, Santa Fe, N.M., $6.50) from which this article has been excerpted, with permission.*

house is determined by several factors. Many people have the misconception that if you dig a little way into the earth, the below-grade soil will be thermal storage. Actually, one must go down a considerable distance below the frost line to reach earth that would constitute a heat gain for a winter greenhouse.

"Pit greenhouses" or "grow-holes" are based on this principle. They are dug out several feet below the frost line to enjoy the benefits of the earth's thermal storage. We have observed that grow-holes perform slightly better than solar greenhouses *only* in extremely cold weather (below -20°F in the New Mexico region).

To achieve increased performance in a pit type of greenhouse adjacent to the house, you might have to dig down several feet below the foundation of the dwelling. This is *not* advised. If you have an existing deep cellar or basement with strong walls and good drainage away from it, an attached grow-hole might do quite well. The hot air in the apex of the greenhouse would enter low in the home and the cool air in the basement would be circulated into the lower part of the greenhouse. The problem is that any design of this nature would require extensive excavation, landscaping and thorough knowledge of strength, condition of existing walls.

The foundation

If the foundation is built properly, many future problems will be avoided. Careful measurements are essential. We will assume for the purpose of these construction steps that you are building a greenhouse with a rectangular floor plan and that you are attaching it to the house.

To determine the 90° corners of the structure, place one edge of your framing square against the existing wall and extend the other edge with string Stake

the string at the distance you have determined for the outer boundary of the greenhouse. After repeating this procedure for the other end wall, measure to see that the two strings are parallel ("A" to "B" equals "C" to "D"). Connecting the outer perimeter stakes should produce a rectangle ("A" to "C" equals "B" to "D"). To double check your 90° corner angles, see that the diagonal measurements are equal ("A" to "D" equals "B" to "C").

We will describe the poured concrete/rock type of foundation because it is widely used and easily understood and constructed by the home builder.

Along the perimeter of the greenhouse excavate a trench to the desired width and depth. Make it at least 4 inches wider than the walls of the greenhouse and at least 16 inches deep (if a massive wall is planned). Drive stakes into the trench at 6 to 8 foot intervals, leaving exactly 6 inches of the stakes exposed above the bottom of the trench. Check that the trench is level by laying a flat board on the stakes and taking a reading with the level. Do this around the perimeter of the trench. Fill when necessary; then smooth out the sides and bottom with a flat-nosed shovel. To double check the level, we recommend:

The Old Carpenter's Water Trick

You want to check one end of the foundation trench with the other. You don't have a transit and the 2x4 won't bend. Get a friend, then take a regular garden hose and lay it in the trench. Drive stakes in the end corners; each must be exactly the same height from the bottom of the trench. Holding the ends of the hose flush with the top of the stakes, fill it with water. If the extremes are level, the water level at each end will be equal.

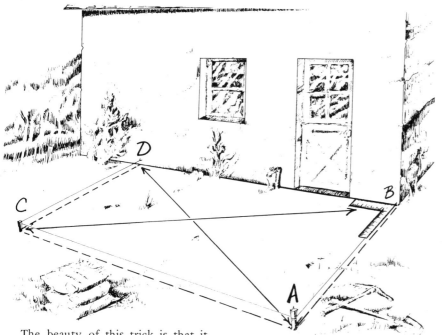

The beauty of this trick is that it will work for any length and over rough terrain (with a couple of people and plenty of garden hose). Of course, the hose ends have to be held higher than any point in between.

After the trench is dug, leveled and cleaned out, keep all interested gawkers away from the edges so they don't cave in the sides.

Before the foundation is poured, the outside and bottom of the trench should be insulated with 1 inch or more of rigid styrofoam. Cut the panels to size and fit them into the trench. They can be temporarily propped in place until the concrete is poured. Another method of insulating the perimeter is to wait until the foundation has been poured and the concrete has hardened; then dig a trench around the outside of it. Line the trench with sheet plastic and fill it with sawdust, dry pumice or styrofoam beads. Enclose the loose insulating material with the plastic to keep it waterproof and cover the trench with dirt.

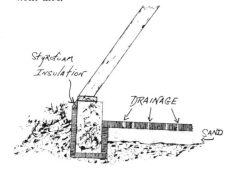

Another prepouring step is to insert reinforcing material in the foundation trench. If you have to meet stringent building code requirements, this may be mandatory. "Re-bar" or "re-rod," as it's called, can be used in 1/2 inch or 3/8 inch diameter. It can be bought and cut to length at any building supply store. Two lengths of re-bar are laid along the bottom of the trench about 8 to 10 inches apart, supported 4 to 5 inches off the bottom by rocks. Ends that meet are lashed together with baling wire.

I've poured foundations with and without re-bar. I often throw as many river rocks as I can find (5 to 8 inches in diameter) into the bottom of the trench and forget about the re-bar. I haven't noticed any settling or cracking in the foundations I've built this way.

It's a good idea to bring the foundation above grade (3 or 4 inches). This automatically eliminates some drainage problems and is definitely necessary if frame, adobe, or other water-soluble materials are going to be used to build the walls.

Old lumber can be used for the forms to restrain any concrete that is above grade. Almost anything will do to secure them in place; large rocks, blocks, stakes, wire. Be sure the forms are the right distance apart and *well braced* so they don't spread with the weight of the concrete. Anyone who has worked with concrete can testify

to its weight. Once you've had the terrifying experience of seeing a large mass of wet concrete start moving toward you, you'll always over-brace forms and wire the braces together.

On the inside of the forms, mark a level line for the top of the foundation (a chalk line works well) about 3 inches above the actual level to which you are going to build the foundation so that it doesn't smear when the wet concrete is being poured. Another way is to chalk line the exact level and height, and drive nails halfway in along the line, for an accurate guide.

The fast way to pour the foundation is to have the ready-mix concrete truck back up to the site and dump it on you. But often the concrete companies won't deliver in small quantities, or the site is impossible to reach. In that case you have the option of buying premixed dry bags or making your own mix from cement and sand. If mixing your own (much cheaper), a standard concrete recipe is five parts sand and 3/4 inch gravel (mixed equally) to two parts dry Portland, and water. This is a heavy job so line up a few friends. The entire foundation should be poured at one time. You don't want to have seams from two or more separate pourings.

When you are ready—sand and gravel in place, Portland bags stacked, shovels in the ready position, wheelbarrow greased, beer iced down—consistency is what you want in the mix. It should not have dry clumps or an overabundance of any ingredients. The mix should be wet without being runny. If you pull a hoe through it, it should make nasty noises. When the mix is just right, it reacts like Jello when patted with a trowel. Nice stuff.

Start at one end of the trench and work around. After a load, usually a full wheelbarrow, is dumped, spread the

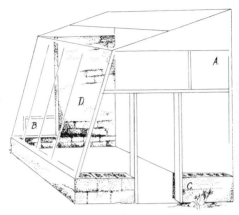

full wheelbarrow, is dumped, spread the concrete along the trench. Push and work the mix down into the trench with a trowel. Don't be gentle. You want to avoid holes or pockets in the foundation. Keep adding loads of concrete until you've nearly reached the level line or marker established as the top of the foundation.

As you work around the trench, pat and smooth out the top of finished areas. After the cement begins to set up, insert an anchor bolt (screw threads exposed) for sections that will be framed above the foundation. With a square, make sure that these bolts are perpendicular and in line with where the plate will be and that you've left about 1½ inches extending above the poured foundation. The plate, which

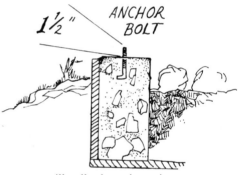

we will talk about in various contexts throughout this chapter, is a piece of lumber, in this case a 2x4. The foundation plate provides a base for the frame walls of your greenhouse. The top plates give vertical studs and roof rafters something to hang on to. In general plates serve as weight supporting members of any frame structure.

Note: clean your tools immediately after use or they'll never be the same. If for some reason you have to leave a load in the wheelbarrow or mixer for a short time, pour a small amount of water on top of it and cover as tightly as possible. This also applies to mortar and plaster mixes.

After the pouring is done, check to see if any areas have sunk and make sure that the above-grade forms are secure. When the concrete has set up or hardened (usually within three or four hours), spray it with a light mist of

water or cover with wet hay or straw. This prohibits rapid evaporation that might crack or weaken the foundation. Spray it every few hours for the next day or two (don't bother at night).

When the foundation has a feeling of permanence, the forms can be removed. Clean well and recycle them into shelves, tables, or bed frames for the greenhouse interior.

There is a way to avoid laying a foundation under the frame portions. I'll give it to you as an option. It might be useful if you are in a big hurry or plan a temporary structure. Level the ground where the frame walls will be. (Don't dig a trench; just level the earth.) Wood plates (I've used railroad ties) are laid directly on two inches of sand and staked in at 3 to 4 foot intervals. The stakes can be metal or wood but should be at least 36 inches long. They can be screwed, nailed or bolted to the ground plate. The wood should be treated with copper naphthenate as a preservative. Don't use fresh creosote or pentachlorophenol ("penta"), as these chemicals give off fumes that are noxious to plants.

Whatever foundation method you choose, the most important considerations are:
• Is the weight evenly distributed?
• Is the foundation level?
• Will the water drain away from it?

When the concrete in the foundation has cured, you can begin work on the massive masonry walls. Different types of masonry construction call for different techniques. Let's use hollow pumice blocks for our example. (Basically the same technique applies to building brick or adobe walls, except that you can use mud for mortar in the latter case.) The standard size pumice block is 15½ inches long x 7½ inches high x 7½ inches wide. For estimating the amount needed, use the dimensions 16 x 8 x 8 inches because the added space will be filled by mortar. Before making your estimation, determine the exact size and location of any vents or doors in the walls. They must have jambs or frames built around them, and

that lumber, usually one and a half inches wide, should be included in your calculations.

I like to avoid openings in masonry walls whenever possible. They involve a precision and degree of patience that I often lack. It's usually easier to locate vents and doors in frame areas.

As the diagram shows, there's a high vent in the eastern frame wall (A). The low southwest vent sits on the masonry wall (B). The lowest part of the door is set in the east masonry wall (C), the upper four-fifths in a frame section.

When the size of openings in masonry walls is determined, the estimate of the total number of blocks needed can be made. Determine the square footage of the walls and estimate one block per square foot plus 10 percent for cutting. For our example I'd buy eighty full blocks and twenty-five half blocks. You may want to order more than we've estimated. Masonry blocks look rather formidable, but actually are quite fragile and easily broken until they're in the wall.

The mortar used in laying blocks is made with masonry cement and screened sand. The standard proportions are 5½ parts screened sand : 2 parts masonry : 1 part Portland. Mixing equipment is the same as for the foundation. Small amounts are made as needed. A large triangular trowel is used to spread the mortar.

To lay a perfectly straight and level block wall, poles can be erected precisely at the outside edge of each corner of the structure. Marks are made up the poles at 8 inch vertical intervals. Poles and marks are checked against each other with heavy twine and a string level. They should all be at exactly the same level for each course of blocks, and they indicate the top of each layer. That's the precise way to do it. Another method is to simply begin laying the blocks, checking for level, plumb, and straightness as you go. In an area as small as our 10 x 16 foot example, this should be sufficient. Lay the first course of blocks all the way around. Check for square on the corners. Now, using the level, build up several courses of block at each corner. String is strung between corners and straight runs laid down to the string. Checks for vertical can also be made with a carpenter's level or plumb line. If you're new at this kind of work, don't trust your eye too much.

Have you ever watched an experienced mason building a wall and seen the beautiful fluid movements he makes? Don't expect to match this degree of skill. Try to put a uniform thickness of mortar (3/8 to 1/2 inch) on all seams. Any way you can get it to stick to the blocks is cricket. Try putting a small amount on the trowel and scraping it off with that downward and outward motion. If you can't get the knack of this, put a larger amount on the trowel and shake it over the edges. I resort to my hands occasionally. "Seat" (firmly tap down) the block with the handle of the trowel. It should be evenly supported by mortar, and level when it's in place. See that the seams are staggered, not only directly over another.

A method used to "tie-in" a new masonry wall to an existing structure is to bend a small (8 x 10 inch) piece of metal lath to form a 90° angle and fit this between the new wall and the home wall. Tack the lath securely to both walls. Lay the mortar and firmly seat the next course on top of the lath, pushing the block tightly to the existing wall. This should be done three or four times in a wall as tall as the one in our example, the 8 foot west wall.

When the block walls are up to their final height, fill all center holes in the blocks with concrete (the same mixture you used for pouring the foundation). This strengthens the walls and adds mass—heat storage capability—to the structure. Before the concrete dries, insert anchor bolts where the framing plates will be applied to the top and sides of the walls.

When all the masonry walls are up, relatively straight, level and plumb, it's a good time to have a celebration. You can finally see and feel the results of your brain and muscle work. The hardest part is over. Enjoy it!

Frame walls (clear, opaque)

A frame wall does not have any appreciable mass; therefore it cannot store heat for a structure. However, when properly insulated, it will keep the heat in. The vertical framing members (studs) in the *clear-wall* sections are erected at 47 inch intervals. The fiberglass that we recommend comes in 48 inch widths. The 47 inch center allows for overlapping the panels. Studs in the *solid frame* wall are on 24 inch centers, as are the roof rafters. All corner studs and upper plates are double-width for added strength. A scale drawing of your greenhouse will help you to estimate how many 2x4s of various lengths will be needed for the rough framing. Add the total linear footage of doors, vents and horizontal nailers (firestops) to the estimate. Add another 20 percent.

You'll save money by having as few leftover scraps as possible. This is accomplished by making all principal framing members slightly shorter than an even number. Hence, if the stud in the south wall is 7 feet 10 inches high, it can be cut from an 8 foot piece. On the other hand, if the stud is 8 feet 1 inch high, a 10 foot 2x4 usually must be purchased, and you're left with a 23 inch scrap. At current lumber prices, that's sinful. In your scale diagram, *make* the lengths come out economically by slightly changing angles and dimensions in the drawing until it works. If you just can't make it come out right, then plan to use the scrap lumber for tables, shelves, boxes, bed frames or other things. A 23 inch scrap, for in-

Building the front wall: working flat on the ground is one way to make sure it's square.

stance, could be used for a firestop in the sheathed (solid frame) wall.

It's important to get a proper dollar value for your lumber. As with most things in life, you do this by choosing it yourself. Occasionally it will be "company policy" not to let the customer look through the stock. Don't do business with a company like that.

Let's begin by framing up the front face. First, you have the option of framing the wall in place or building it "prefab" on the ground. This decision is mostly a matter of personal preference. We're going to prefab the front face in our example, and frame the side walls in place.

In the diagram, the length of the roof and foundation plates (top and bottom framing members) is exactly 16 feet (1). The top plate should be double to prevent bowing under the weight of the roof. So the first step is to nail the two 2x4s together with 16-penny nails, at an angle so that they don't protrude from the wood.

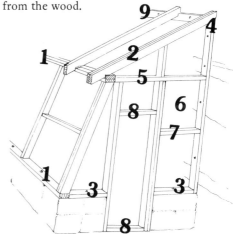

Lay the plates (bottom and top) on the ground and mark them where the vertical studs (47 inch centers) will be attached. Corners should be double-width.

Continued

GREENHOUSE *continued*

The next step is to cut the lumber to the appropriate length. In our example, we have seven studs cut to 7 feet 7½ inches. With the plates, this makes an 8 foot high front face.

If the south wall is to be vertical, the cuts are square on each end of the studs. In our example, we have tilted the front face 20° to make a 70° slope. When the studs are cut, nail them at the marks to the top and bottom plates with 16-penny nails. Note: always buy high quality nails.

The south panel can now be lifted into place on the low front wall. This will take several folks working in unison. In order precisely to mark the junctions of the anchor bolts in the low wall with the bottom plate, gently lower the prefab frame structure into its permanent position. While several people hold the framework in place, one particularly aggressive type can pound on the plate with a hammer directly over the protruding anchor bolts to mark their position. Take the face off again and drill holes at the indicated marks. The front face can then be installed on the wall, temporarily braced, and bolted down.

At this point put up a few rafters (2) for braces, the ones on the corners and a couple of others. Nail them down to the top plate of the front face and to the plate or rafters on the house side. When you have finished, take a break. Stand back and admire your work.

The next step is to cut and place the lower plates on the east and west walls (3). Bolt the plates down. To connect the frame walls of the greenhouse to the house, a stud is securely tied-in to the house wall. It should fit snugly from the bottom plate to directly under the rafter plate (4).

You now have a funny-shaped box: the perimeter of the east and west walls to fill in with framing lumber. We're going to divide that box with a plate that will also serve as a header for the lower frame walls (5). Hold a piece of lumber level across the span and mark its intersection with the front face. On the east the plate can also be the top of the door frame. On the west, it sits on the

massive wall for the majority of the span. When the pieces are cut, nail them into place.

The east and west walls can now be framed in place. The insulated walls will be framed on 24 inch centers (6). The easiest way to do this is to mark the bottom plate at the intervals where the upright studs are to be nailed. Take a carpenter's level and hold it against the side of the first stud. Keep the base of the lumber on the bottom mark and get the stud exactly plumb. Then mark the point of intersection of the stud with the header on both the stud and the plate. Cut the stud about 1/16 inch longer than the mark you've made. Toenail (drive nails in at an angle) the studs in as you go. When the studs are cut, fitted, and nailed, install horizontal spacers (7), and the vent and door plates (8). Make certain they're level, and nail them in with 16-penny nails.

Roof rafters

With luck, you will be able to tie the roof rafters of the greenhouse into the existing rafters of your house. If not, attach a 2x6 inch or 2x8 inch plate to the wall of the house as a base for the rafters (9). Expansion bolts or large wood screws are used to get a *secure*

tie-in to the wall. Don't scrimp here. The greenhouse roof must bear its own weight plus, in many areas of the country, snowloads *and* the additional loads of snow sliding off the roof of your house.

The rafters for a short span (under 10 feet) can be 2x4s or 2x6s set on 24 inch centers. For longer spans use heavier lumber or put the rafters on closer centers.

Shallow notches are cut in the wall plate or joist hangers can be used instead. The rafters are toenailed to it at these points. At the intersection of the rafters and the south face top plate, very shallow notches (called "bird's mouth" notches) are cut in the bottom of the rafters so that they will rest snugly on the plate. Again, toenail them in.

Putting up the rafters (and the roof) can force a person into some pretty strange acrobatic contortions. I would caution you that 8 to 9 feet up in the air is higher than you might imagine, especially from an aerial view. Get in the habit of *not* leaving any tools or materials lying about on the rafters or the roof, even for a moment or two.

When all the rafters are in place, mark the intersection of the clear roof

Assembling the pre-constructed walls requires a little help from your friends, some long lumber for temporary braces, and quite possibly a sense of humor.

with the insulated roof areas. Use a chalk line to do this. To determine that junction in your unit, use a sun movement chart. It will usually be about halfway down the rafter span.

Rafters and framing lumber can be treated in more humid climates. Copper naphthenate is recommended. Some paints have a preservative in them. After it's been treated, all framing lumber that will be visible within the clear walls and ceiling area should be painted with a glossy white enamel or latex. This will help reflect light into the greenhouse and also enhance its appearance.

Choosing the glazing

Traditionally, glass has been used for the clear surfaces in greenhouses. Due to its resistance to high temperatures, glass is also used extensively for covering solar collectors. Prefabricated sheets of double-layered glass are available from major manufacturers and may be purchased in various sizes. Many types of sliding glass doors are also double glazed and can be used in the greenhouse. The obvious advantages of glass are that it allows a view to the outside and is highly resistant to the harmful effects of weathering. It is, however, easily broken and demands extreme care and technical skill in installation.

Technological advances in the design and manufacture of plastics have produced important alternatives for the greenhouse builder. Some types of semi-rigid fiberglass/acrylic sheeting are guaranteed for twenty years to transmit enough light for photosynthesis. Fiberglass transmits nearly the same amount of light as glass, even though it's translucent rather than clear. However, fiberglass can become cloudy or brown as a result of ultraviolet ray damage; Tedlar coating, for instance, has an ultraviolet retarding characteristic that helps greatly to preserve the clarity of the plastic to which it is applied. Corrugated plastic is recommended for clear roof areas; it is easily installed and resistant to hail damage. Flat fiberglass is readily attached to clear-wall frames with rubber-gasketed nails or lath strips. I recommend the

use of flat fiberglass on all vertical and near vertical surfaces. Flat material has a 20 percent smaller surface area than corrugated; therefore far less area for heat loss. The quality of light transmitted through fiberglass is diffuse. It doesn't give the sharp, clearly defined shadow areas of glass. This is beneficial to plant growth. The cost per square foot of new fiberglass is considerably less than the cost of new glass.

There is an ecological question concerning the use of plastics in general. Plastic is a petrochemical product and is not biodegradable. The supply of petrochemicals is dwindling rapidly, and the atmosphere is becoming polluted by petrochemical wastes. Nondegradable products also constitute a form of pollution on the earth.

I feel that our only hope for maintaining an ecological balance on the earth depends on a thoughtful, positive use of modern technology and its products. I suggest, therefore, that the intelligent and careful use of plastics in the greenhouse will serve as a valuable example of how to put our fossil fuels to beneficial use . . . before they are exhausted.

Having built and painted the frame walls, you are now ready to install the outer layer of fiberglass. Most of the products (Lascolite and Filon, for example, brands we've used and can recommend highly) are sold in 4-foot-wide rolls, up to 50 feet long. For this reason, we suggest that you install one 4-foot panel at a time. Beginning at one end of the frame wall, measure the height of the section carefully. After measuring and marking the material with a felt-tipped pen or razor knife, cut the section with sharp wire-cutting shears. Apply an even bead of sealant, silicone or butyl rubber to the wood frame you are going to cover. With two or three people holding the edges of the plastic, align it in front of the frame. It is handy to make corresponding marks on the fiberglass and the wood frame. Nail the top edge to the upper frame member in the center. Use a gasketed nail; do *not* nail it in completely. Check to see that the sides and bottom will fit

Two pairs of hands help when placing the translucent plastic glazing, one to hold and pull, one to nail.

properly. Nail the outside edge in the center (again, halfway in). Pulling slightly on the remaining two edges (bottom and inside), nail them temporarily in the centers of the frame. You can use flat-headed 1 inch nails for the inside or leading edge or simply lay it on with no nails. The next section of plastic will overlay it and must fit snugly. The rubber-gasketed nail heads protrude about 1/8 inch from the wood and would cause bumps in the overlaid plastic. Nail the material down to the horizontal braces first. Then pulling diagonally on the four corners of the sheet. Work the bulges out of the plastic from the center to the corners. If a major bulge has developed, try to detect it early; remove the temporary nails and realign the sheet.

There are two very expensive items used in this method, silicone sealant and gasketed aluminum nails. Lately I've used a cheaper method that I believe is just as effective. Check and mount as before. Don't use any silicone sealant. Get small galvanized nails and drive them into the fiberglass about every six inches. After all panels are up, cover edges and overlaps with thin (1/4 x 1½ inch) wood lath. If there are any bulges or leaks, they can be sealed with regular caulk.

After cutting the second panel, lay another bead of adhesive over the leading edge and the top and bottom. Proceed as above, overlapping the second panel by 1 inch on the edge previously flat-nailed. Use nails to attach the overlapping section. Follow the steps given

Continued

for the first panel, again using 1 inch flat-headed nails in the leading edge to be overlaid by the next panel. For vents and removable panels, install the fiberglass on the ground.

The irregular shaped panels in the corners of the east and west wall are sized by cutting a rectangular piece to fit the length of the frame. Hold the piece up to the panel and mark the shape with a chalk line. Save the scraps for other irregular shapes and vents.

Roof installation

Spanners (7) like those used between the vertical wall studs will be nailed between the roof rafters to receive the corrugated plastic and solid roofing material. Be sure to toenail-in a straight row of these spanners along the line separating the clear from the solid roof. In all other roof areas, the spanners can be staggered for ease of nailing. They should be spaced up and down the rafters at 3 to 4 foot intervals.

You are now ready to install the corrugated clear roofing. One sheet of corrugated is held in place at a corner. Align it carefully along the bottom and side edges. The roof can overhang the side by two corrugations and the bottom by about 6 inches unsupported. Get it "true," that is, perfectly in line with the rafters and plates. An out-of-line first piece will throw the roof whacky. When you've got it right, tack it down temporarily at the top. On the bottom plate, under the fiberglass, insert a corrugated sealing strip. I recommend the foam type, but redwood or rubber will do fine.

Now nail down the fiberglass every three to four corrugations with gasketed nails to the bottom plate and rafters. Nail into the high ridge rather than the valley. This will prevent water seepage through the nail hole. Overlap the sheets at least one and, if possible, two corrugations and continue across the clear area in this manner. Don't nail the top edge (to be overlapped by the opaque roof) yet. The only real trick to nailing into fiberglass is to have a steady hand and a good eye. A missed hammer

blow can splinter the material, leaving an ugly opaque mark (to say nothing of the damage to your thumb).

The solid roof is installed next. If plywood and composition roofing material is to be used, lay a strip of corrugated molding across the top edge of the clear/solid roof junction. Nail the plywood sheets onto the rafters, overlapping the foam strip and the clear roof area by 3 to 4 inches. The foam molding above the corrugated plastic will create a tight seal against heat loss. You can now apply your composition shingles or other roofing material over the plywood. Vents in the roof are *not* recommended for any but the most experienced builder. It is extremely difficult to seal them against air and water leaks.

In New Mexico a common roofing material is corrugated galvanized steel. It is relatively cheap, easily installed and maintenance free. If you use metal, no plywood is needed. Simply align the corrugations of the galvanized with those of the plastic (they match), and nail it down to the rafters and spanners. Again, the solid roof should overlap the clear areas. For nailing, hold lead-headed roofing nails with a pair of

pliers over the designated spot (on a ridge and over a rafter or spanner) and strike solidly with a hammer. Wear eye protection.

Six-millimeter polyethylene (with an ultraviolet inhibitor) is suggested for the interior clear walls and roof areas. It is susceptible to weathering but will be protected by the fiberglass outside layer; it should last three to five years before having to be replaced. I've used both Monsanto 602 and Tedlar.

The thin plastic is more easily installed than the outside fiber acrylic. Large sections of clear area can be covered at once. Using scissors or a razor knife, cut out the sections to be attached. Make your cuts at least 3 inches longer than measured. With three or four helpers holding the extremities of the sheet, begin stapling along the centermost stud. Work outward toward the sides and corners, stretching the sheet as staples are driven 8 to 10 inches apart. Wood lath (1/4 to 1½ inches) is nailed over the plastic-sheeted studs with finishing nails to produce the final seal. You can stain or paint the stripping to make it more attractive.

Inside

Solid frame walls can be insulated with a wide variety of materials. Rockwool, Fiberglas, Styrofoam, polyurethane, pumice, and cork are all excellent. Foil-backed insulation stops a lot of radiant heat loss through the walls and acts as a vapor barrier, but it can be prohibitively expensive. If you do use it, install it with the foil backing facing the interior of the greenhouse. Recently I've been stapling heavy-duty tin foil on the interior side of Fiberglas batt. That's cheap and effective.

The amount of insulation applied should be as much or more than is used in the walls and roof of a well-built house in your area. In New Mexico, I usually use at least 4 inches of Fiberglas batt in the walls and 6 inches in the roof.

Always wear a long-sleeved shirt, gloves and button your collar when installing Fiberglas insulation. Wear safety glasses for rafter work, especially if you have sensitive eyes, and try not to breathe too much.

Insulated frame walls can be paneled with any material that you find attractive. Paneling materials call for various adhering techniques. Wood products are usually simply nailed with finishing nails. As the insulated wall will not store heat, the inside should be a light color to produce a reflective surface. Water-sealing will help to protect the interior against deterioration due to high humidity.

Doors and vents must be tight fitting and weather-stripped around all joining surfaces. If you are using fiberglass, you may choose to use glass for the vents and/or doors. Clear views to the outside can offer attractive accents to opaque and translucent walls. When glass is used, it should be double glazed to reduce heat loss. If possible, windows and doors should slide or hinge away from the greenhouse interior. Hinges are mounted on top of vents to open out. Remember, the high vent that you put in one wall (east or west, depending on the direction of the prevailing winds—should be on the downwind side of the prevailing air flow) needs to be about one third larger in square footage than the low vent in the facing wall.

In determining the dimensions for an outside door, remember that you will be moving large quantities of soil into the greenhouse. Make the door wide enough to accommodate a wheelbarrow (plus your knuckles). Thirty-two inches is good. It is also advisable to make the height of the door a standard measurement for convenient access. In very cold climates an air lock over the exterior door will save large losses.

Sealing the greenhouse

All massive walls should be insulated on the outside. An effective insulating material for this is Styrofoam or styrene panels (1 or 2 inches). They can be stuck to the walls with a heavy-duty construction adhesive. Use it liberally. If the wall is to be plastered, cover the

---

Materials

1 bag masonry
5 bags Portland
1 yard sand
1 yard 3/4 inch aggregate
2 yards dry pumice or pea gravel
80 full pumice blocks
25 half pumice blocks
8 6-inch anchor bolts
72 feet rebar
30 8-foot 2x4s
3 10-foot 2x4s
2 16-foot 2x4s
2 8-foot 1x4s
2 10-foot 1x4s
40 feet of 1x12 (for floor beds and shelving)
400 feet of wood lattice moulding (for trim and tables)
1/2 lb. concrete nails
300 to 44 aluminum nails or 3 lbs. small galvanized nailx
10 lbs. No. 16 common nails
5 lbs No. 8 common nails
2 lbs. No. 8 finishing nails
3 lbs. blue sheetrock nails (for sheetrock)
6 sets hooks and eyes
1 set 3½ or 4 inch butt hinges (for door)
2 sets 2 inch butt hinges (for vents)

7 2x4 joist hangers
2 4x4 joist hangers
3 pulls
8 corner braces (to reinforce door and larger vent)
2 2 feet by 8 feet by 2 inches styrofoam panels
150 square feet of 4 or 6 inch Fiberglas insulation, 24 inches wide
2 packages 3/8 inch foam strip (weatherstripping)
32 feet corrugated stripping (foam or redwood)
4 pieces 1/4 or 3/8 inch sheetrock
2 pieces 3/4 inch Celotex or equivalent exterior sheathing or paneling, i.e., 64 square feet rough lumber
4 pieces 8 feet corrugated roofing material
2 tubes silicone caulk — clear
1 tube regular caulk
1 gallon good quality white latex paint
1 gallon good quality dark color latex
1 pint dark stain (for lattice moulding)
200 square feet flat fiberglass/acrylic (greenhouse quality)
70 square feet corrugated fiberglass/ acrylic (greenhouse quality)
250 square feet polyethylene (greenhouse quality)
6 55-gallon drums with tops (water tight) and/or a number of smaller water tight containers

---

Styrofoam with tar paper. Then use firring nails and chicken wire over that. The wall can now be plastered with a hard coat (5 parts sand : 3 parts Portland : 1 part lime).

Exterior covering of the frame walls can be any material that suits your aesthetic and economic criteria. I've used old lumber, plywood, Celotex (exterior fiber sheathing) and metal siding. The most important thing is to make sure there are no leaks that allow water into the frame walls. Remember—higher panels overlap lower ones for waterproofing. Caulk anything that looks suspicious.

If there is one most critical factor in the construction of your solar greenhouse, it is that *all joining surfaces fit tightly together.* Tightness is, in fact, one of the higehst goals you can attain, both as designer-builder and proud owner. The primary reason to strive for tightness is to reduce air leaks. But an-

other important reason is to produce an attractive environment that reflects the care and thoughtful work that went into the design and construction.

During each phase of building, ask yourself where heat loss is likely to occur. Wherever large surfaces are joined (such as the foundation plate to the foundation), a layer of Fiberglas or foam insulation will help solve the problem. Sealants such as silicone and caulking materials (wood filler, spackling compound) should be used on any crack or opening that could transmit air through the structure. For larger openings, such as might occur at the house/greenhouse junction, use metal lath and plaster to build air-tight walls (insulate between them). As we mentioned earlier, vents and doors must be completely weather-tight. The importance of sealing and insulation cannot be overemphasized, as it can make the success or failure of your greenhouse. ☀

Green-
houses

SOLERA
manufactured by:
Solar Technology Corp.
2160 Clay Street
Denver, Colo. 80211
Richard S. Speed; phone: (303) 455-3309

Solar building system collects and stores solar energy to provide most of its own heating and supplemental heat to other structures. Wide variety of applications (heating, greenhouse, patio enclosure, home addition, solariums, crop dryer, shelter). **Features and options:** highly insulated wall panels and thermal windows. Optional inner windows, pop-in panels, and hot-air collectors reduce heat losses at night and during cold weather. **Installation requirements/considerations:** may be surface mounted to new or existing homes in free-standing, attached, or abutted positions; modest building experience required. **Guarantee/warranty:** one year on materials and workmanship. **Manufacturer's technical services:** will install at additional charge. **Regional applicability:** unlimited. **Availability:** from factory.
Price: for basic structure per square foot of floor area — $8; installed — $17.

SOLAR ROOM (PLANS)
manufactured by:
Garden Way Publishing Co.
Dept. 130ZZ
Charlotte, Vt. 05445
Douglas Taff; (802) 425-2147
Extensive set of plans for the construction of solar greenhouse added to an existing house. Designed to supplement space heating requirements by as much as 33 percent.
Price: $9.95.

WE CAN'T GROW ON LIKE THIS.

We've always operated on the assumption that bigger is better. But is it?

Like the dinosaurs, societies and economies can grow too big for their own good.

America is fast approaching that point. The natural resources we need to live — clean air, water, land fuels, metals — are getting scarcer. Some are on the verge of extinction. Others are becoming prohibitively expensive.

At the same time we're wasting tremendous amounts of these precious resources. And our wastes pollute our communities, our nation, our world.

We need to learn to use our resources efficiently and economically and to share them better so that everyone gets a piece of the pie.

We need to conserve the raw materials that jobs depend on, because if we deplete our resources now, things will be that much tougher later.

We need to put people to work *doing* things instead of just making things. The things we *do* make have to save resources instead of wasting them. We can build mass transit instead of freeways, rebuild our cities instead of spawning new suburban sprawl, put people to work cleaning up our environment instead of despoiling it. Harsh prescriptions? Maybe. But ones that will assure a more prosperous future.

For a better tomorrow, let's stop using resources like there's no tomorrow.

Not Man Apart
the complete environmental newsmagazine

A Friends of the Earth publication — one year memberships: 20.00 (24 issues)

124 SPEAR SAN FRANCISCO CALIFORNIA 94105

Insulation:
A MOVING TALE

By Sandra Oddo

This is not a scientific essay. This is the story of a living room with 280 square feet of window—240 square feet of which face north and northwest. In the spring, birds flash past and the buds nod. In the summer, inside can be outside—just open a casement—and the late afternoon sunlight lies lazy on the floor. In the fall, three sides of the room are a fiery mural as the leaves turn. But, oh, the winter. Oh, especially, those cold black nights with icy fingers, that leach the warmth and comfort of the hearth away.

The windows have storm windows, of course, and the fixed parts are double-glazed. Not enough. Insulating curtains helped for awhile but, oh, the ever-rising heating bills. Last fall, insulating shutters in mind, I took a tour of the local building supplies places.

"Insulating shutters" is still a phrase that draws blank stares from insulation salespeople. Here's what I found: Fiberglas, foil or kraft paper faced, 3½ or 6 inches thick, for walls and ceilings. Unfaced glass fiber 6 inches thick, for unrolling in your attic over the insulation you've already got. Packages of polystyrene beadboard, 3/4-inch by 13½-inch by 4 feet, for insulating between the firring strips when you put up panelling in your basement playroom. Styrofoam, 4 by 8 feet, 1 inch thick. Ditto, 2 by 8 feet—but not tongue-and-groove, which exists. Four by 8 foot by 3/4 inch sheets of beadboard, mostly already crumbling around the edges. And in one place, stacked toward the back of a warehouse, 2 by 8 foot by 2 inch beadboard, on sale.

The windows come in eight sections,

Sandra Oddo is editor of Solar Age *magazine.*

each 8 feet long by 5 feet high, and are set along the top of a waist-high stone wall (yes, it's insulated inside) with a 12-inch sill. Shutters can't swing on hinges, first because they'd knock over most of the furniture, second because there's no place to fasten the hinges. They can't flap from the ceiling because the windows turn two corners—and what do you do where the raised shutters overlap? My esthetic sense revolted. But it didn't object too badly to the idea of six fixed panels (at the four junctures of the windows and the two ends), each about 3 feet wide, sized to cooperate with the mullions—and removable in summer.

So I bought nine sheets of 2-inch beadboard and some 1 by 2-inch firring strips, three packages of ten 8-footers for $5 each. When you buy firring strips for uses where their shapes might matter, by the way, pick them out yourself. Firring tends to be junk wood, and can develop all sorts of peculiar kinks and crotchets. The firring got cut into 5-foot sill-to-ceiling lengths, fastened to 3-foot cross pieces top and bottom, and screwed edgewise into the heavier mullions, two screws per section, just to keep them in place and tight against the glass. The beadboard was friction-fitted into the rough frame, a 2-foot section butted to a 1-foot strip. It would have been better, of course, if it had been seamless, but the building supply supplier hadn't cooperated. The scraps were used to block up basement windows and, backed by Masonite, to make insulating panels to screw to the outsides of outside doors, between door and storm door. The insulating curtains—we still liked the warm orange corduroy—were taken

apart and the material stapled in pleasing pleats to the tops and bottoms of the panels. Ninety square feet of window, removed from the critical list.

That left only 190 square feet to go, and these shutters would have to move. The idea is to stack them during the day in front of the fixed panels, and because they were to stack three-deep I decided, for reasons that seem incomprehensible to me now, that they should be only 1 inch thick. The edges would take a beating, so they needed a frame. More 1 by 2, this time flat, nailed to hardboard (seven 4 by 8 foot sheets, on sale, $14.84) cut to fit the 5 by 6 foot sections between the fixed panels, two shutters to a window. The hardboard would go against the glass. Beadboard (seven sheets, $16.80) was cut to fit inside the frame and pasted down. White glue, by the way, doesn't work too well; some mastic left over from attaching a bathtub enclosure to the wall worked fine. Corduroy was pleated, stretched tight over top and bottom, and stapled there and along the sides.

It was down to 10°F the first night we friction-fitted the shutters into place. We slept warmer. But the next day when I took them off and stacked them against the fixed panels, where the orange corduroy caused them to blend, visually, into one nice 4 to 5-inch thick insulated block, they demonstrated an alarming tendency to fall over into the room with resounding crashes, scattering furniture and cats. Fortunately, they had been cut to fit within a half-inch of the ceiling, so some strips of 1-inch quarterround ($4.50) solved the problem. It was nailed like molding to the ceiling in

Resisting Heat Loss

The number of kinds of insulation is finite, and fairly small. With rising stress on energy conservation we might expect some new materials for new applications. Meantime, the main types of insulation available are:

Glass fiber, made by melting glass, extruding it in fibers, and packing the fibers together with resins. Batts 3½ inches thick have an R-value of 11, or about R-3 for each inch of thickness.

Mineral wool, made from melted rock or slag by the same process. It has a similar R-value, and like glass fiber is relatively inexpensive and easy to apply.

Cellulose fiber, made from wood pulp or recycled paper chemically treated to resist fire and water. It comes in batts and blankets but is most used as loose fill. Some settling and deterioration may be expected with time. R-value is similar to mineral wool.

Rigid plastic boards, polystyrene and polyurethane, either molded or extruded. Extruded form has an R-value of 4-6 per inch of thickness; molded form has a slightly lower R-value. Polystyrene is quite water-resistant; polyurethane has a higher R-value but may pick up a little moisture and swell with time. Both will burn quickly, with toxic fumes, so should be protected from fire (3/8-inch gypsum board will do it, inside; even better, put the rigid insulation outside so that your walls can be part of the thermal mass of the house).

Polyurethane foam, used to spray between existing walls. It is effective, but somewhat unpredictable, depending greatly on the skill of the spray operator. It gives off toxic fumes if it should burn, and can swell after it is installed, but it has a value of more than R-5 per inch.

Urea-formaldehyde foam, patented by Rapperswill Corp. as Rapco-Foam, a sponge-like insulation for spraying between existing walls. It is non-flammable, has an R-value of about 5.5 per inch, and is unlikely either to shrink or to absorb moisture—but in the hands of an unskilled operator it can develop a lingering formaldehyde smell.

Particle board, made from miscellaneous wood fibers. It's good for soundproofing, has an R-value of 2.6 per inch, is inexpensive, but like wood is susceptible to termites.

Expanded mineral materials like vermiculite or perlite. These are verminproof and fireproof, available as loose granules (R-2 or less per inch), or as board (R-3).

One note: before you go buying massive quantities of insulation, check local building codes. Chicago, for one, prohibits the use of polystyrene.

---

front of the windows, about 5 inches out, flat side toward the glass, where it acts as a stop to tipping shutters and as a track to control them as we slide them into place. Painted the same color as the ceiling, it disappears.

The same time it takes to put the shutters in place is only seconds longer than it took to pull the curtains. A gentle push along the sill (the corduroy slides easily), and a settling into place against each other between the fixed panels. On very cold days, they're simply left in place. We haven't kept accurate track of the shutters' effect on heating, but the bill for December and January, each of which set low-temperature records, was a hair less than it had been the warmer year before. It's possible that before the winter was out, the shutters already had saved the $71.30 they cost to build. We certainly were more comfortable.

Other Possibilities

There can be many approaches to the challenge of movable insulation. Given an existing house and fairly rigid limitations, I could have used a product that spared the hassle and the loss of insulating area caused by frames and hardboard, something that came thicker, in more flexible sizes. Without the 12-inch sill, the shutters would have had to be ceiling-hung, and the hardware for that is hardly imagined yet.

Forethought in designing a new house opens up realms of possibilities.

Pockets in the wall, perhaps, into which shutters can slide? Mirrors on the inside, to bounce back light and heat at night? Beadwall, Steve Baer's styrene-bead snowstorm between two panes of glass, an R-8.5 insulating value, blown in and out night and morning by vacuum motors? Sunshine Design's 1/4-inch foam roll-up shades that stick tight to the frame with Velcro? Barn-door affairs that can slide across the *outside* of windows at night, maybe activated by photo sensors? Kalwall's Mylar curtain that rolls down like a window shade for insulating values from R-4 to R-10? As far as insulating shutter design goes, it's a fluid situation—in movement, one might almost say.

Continued

INSULATION
continued

Costs

Ninie sheets 2 by 8 feet by 2
  inch polystyrene beadboard $20.16
Thirty 8-foot 1 by 2s 15.00
Seven sheets 4 by 8-foot
  by 1 inch polystyrene
  beadboard 16.80
40 feet of 1-inch quarterround <u>4.50</u>
  $71.30

Plus ½ gallon of leftover mastic, about 37 yards of cover material that would have cost something if it hadn't already been there, and various nails and staples.

Some manufacturers

Dow Chemical Corp.,
Styrofoam and Thurane (urethane).
marketed by Amspec, Inc.
1880 Mackenzie Drive
Columbus, Ohio 43220

E.I. duPont de Nemours & Co., Inc.
1007 Market Street
Wilmington, Del. 19898

Johns-Manville
Insulation Center, Drawer 17-L
Denver, Colo. 80217

Monsanto Co.
800 N. Lindbergh Blvd.
St. Louis, Mo. 63166

Olin Corp.
120 Long Ridge Road
Stamford, Conn. 06904

Owens-Corning Fiberglas Corp.
Fiberglas Tower
Toledo, Ohio 43601

PPG Industries, Inc.
One Gateway Center
Pittsburgh, Pa. 15222

Rapperswill Corp.
305 East 40th Street
New York, N.Y. 10016

R-VALUES OF INSULATING MATERIALS

Material and Description		Density (lb/ft^3)	R-value* per inch thickness	R-value* for listed thickness
Blankets and Batts:				
Mineral wool, fibrous form (from rock, slag or glass)		0.5	3.12	—
		1.5-4.0	3.70	—
Wood fiber		3.2-3.6	4.00	—
Boards and Slabs:				
Cellular glass	90°F	9	2.44	—
	60°F		2.56	—
	30°F		2.70	—
	0°F		2.86	—
Corkboard	90°F	6.5-8.0	3.57	—
	60°F		3.70	—
	30°F		3.85	—
	0°F		4.00	—
	90°F	12	3.22	—
	60°F		3.33	—
	30°F		3.45	—
	0°F		3.57	—
Glass fiber	90°F	4.0-9.0	3.85	—
	60°F		4.17	—
	30°F		4.55	—
	0°F		4.76	—
Expanded rubber (rigid)	75°F	4.5	4.55	—
Expanded polyurethane (R-11 blown; 1" thickness or more)	100°F	1.5-2.5	5.56	—
	75°F		5.88	—
	50°F		6.25	—
	25°F		5.88	—
	0°F		5.88	—
Expanded polystyrene, extruded	75°F	1.9	3.85	—
	60°F		4.00	—
	30°F		4.17	—
	0°F		4.55	—
Expanded polystyrene, molded beads	75°F	1.0	3.57	—
	30°F		3.85	—
	0°F		4.17	—
Mineral fiberboard, felted core or roof insulation		16-17	2.94	—
acoustical tile[1]		18	2.86	—
acoustical tile[1]		21	2.73	—
Mineral fiberboard, molded acoustical tile[1]		23	2.38	—
Wood or cane fiberboard				
acoustical tile	1/2"	—	—	1.19
acoustical tile	3/4"	—	—	1.78
interior finish		15	2.86	—
Insulating roof deck[2]	1"	—	—	2.78
	2"	—	—	5.56
	3"	—	—	8.33
Shredded wood (cemented, preformed slabs)		22	1.67	—
Loose Fills:				
Macerated paper or pulp		2.5-3.5	3.57	—

Source: ASHRAE Handbook of Fundamentals, 1972.

Insulation

NIGHTWALL
manufactured by:
Zomeworks Corp.
P.O. Box 712
Albuquerque, N.M. 87103
Insulating panels of beadboard that attach directly to the window pane by means of Zomeworks' magnetic clips at 18 inch intervals. Zomeworks will sell clips to those wishing to purchase beadboard separately. **Price:** app. 20¢ per linear foot of window perimeter plus app. 20¢ per square foot of beadboard. Magnetic clips are app. 30¢ each on a minimum order of twenty clips.

SR6421 PIPE INSULATION
manufactured by:
Urethane Molding, Inc.
RFD #3, Rt.11
Laconia, N.H. 03246
James M. Annis; phone: (603)524-7577
A molded urethane jacket, with a durable exterior designed to insulate supply and return lines. **Installation requirements/ considerations:** place jacket over water lines, gluing the butt ends and taping until the adhesive sets. **Guarantee/warranty:** unlimited. **Maintenance requirements:** none once installed. **Manufacturer's technical services:** call or write at any time for information. **Regional applicability:** no limitations. **Availability:** direct from man-

ufacturer. **Price:** $2.85 per foot, minimum of 100 feet.

BEADWALL ʳ MOVEABLE INSUL—ATION SYSTEM
manufactured by:
Zomeworks Corp.
P.O. Box 712
Albuquerque, N.M. 87103
Steve Baer; phone: (505)242-5354
Polystyrene beads are blown into a cavity between double translucent glazings to minimize heat loss at night. During the day the beads are withdrawn from the cavity so that heat and light may enter a space. **Features and options:** costs are reduced as the size of the installation is increased, **Installation requirements/considerations:** panels are most economically installed in pairs using standard materials. **Guarantee/warranty:** one year on material and labor when customer pays freight. **Maintenance requirements:** anti-static agent must be added every six months or so. The beads will have to be replaced, probably after ten years. **Manufacturer's technical services:** phone David Harrison at Zome-

works for problem trouble-shooting. **Regional applicability:** any climat 4,500 Degree Days or colder. **Price:** plans are $15 and include a licence to build 144 square feet of beadwall; additional square feet ate 10¢ a square foot.

Fiberglass Reinforced Polyester:

Performance tests for FRP COLLECTOR COVERS

By James White

More and more, solar collector covers are being made of fiberglass reinforced polyester. As a cover material it has many advantages, and it also has some problems. Proper evaluation of weatherability is important. Here, Mr. White assesses three basic types: Kalwall Sun-Lite Premium (#1), a proprietary material; Sun-Lite regular (#2), an acrylic-modified, highly light-stabilized polyester; and standard grade fiberglass reinforced polyester sheet (#3) of various types. Solar energy transmission, ultraviolet and thermal degradation, surface erosion, impact resistance, and thermal shock are considered.

When considering fiberglass reinforced polyester (FRP) for a solar collector cover, one of the most important properties is solar energy transmission. Accurate measurement has caused many researchers problems because of the light-scattering phenomena of diffuse materials such as FRP. The preferred method, according to the American Society for Testing and Materials, is ASTM E 424 (Test for Solar Energy Transmittance and Reflectance of Sheet Materials) Method B. This method requires a 28 inch square sample to minimize the effect of light scattering. Initial solar energy transmission for a "super class" FRP can run between 80 and 90 percent.

Method A of ASTM E 424 has been used with widely varying degrees of success. Small samples, and the location of the sensing device relative to the integrating sphere surface have caused problems. However, with proper care, solar energy transmission by wavelength curves can be generated. See figure 1. Fiberglass reinforced polyester has the desirable properties of very high transmission over the typical solar spectrum and near opacity in the longwave range for excellent heat-trapping properties.

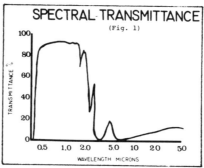

SPECTRAL TRANSMITTANCE
(Fig. 1)

Utraviolet degradation

Degradation due to ultraviolet radiation has long been of great concern to people designing or using products exposed to sunlight. Researchers in the FRP industry have come a long way in retarding ultraviolet degradation. A typical non-light-stabilized general purpose polyester can lose more than 15 percent transmission in just fifty hours of exposure to a sun lamp. One of the earliest attempts to improve the UV resistance added ordinary aspirin for light stabilization. After fifty hours' exposure to a sun lamp, a general purpose polyester with aspirin added will lost only 5 percent transmission.

Obviously, today's researchers have gone much beyond aspirin in the field of light stabilization. Altering the polyester backbone (modifying the glycols and acids that make up polyester), adding acrylic, adding sophisticated light stabilizers, and applying special coatings or films are necessary to produce a quality solar collector cover. In order to facilitate research, several different weatherometers were developed and are in general use today. The most common are the fluorescent, carbon arc, and xenon weatherometers. The fluorescent weatherometer has a high concentration of UV and causes more severe changes than the carbon arc, or the xenon.

Although it is extremely difficult to correlate weatherometer hours with real time outdoors, many researchers use 250-400 weatherometer hours as approximately one year of actual weathering. Two thousand hours equal approximately five years of real time. Samples were exposed to a fluorescent weatherometer for 2,000 hours. Color change and light transmission readings were taken at 500-hour intervals.

Color change measurements were taken in accordance with ASTM D 1929. Material #1 had a color change of 3.5 after 2,000 hours; material #2 had a color change of 10. Depending on formulation, a standard-grade FRP sheet could have a total color change of around 28. (Figure 2). In order to verify the weatherometer results, color-change measurements were taken on a sample of material #1 weathered in South Florida, considered to be the most severe natural environment in the U.S. because of large quantities of sunlight, heat, and moisture, for five years. It was found to be 4.4. (A specially-coated piece of material #1 had a color change of only 1.1!)

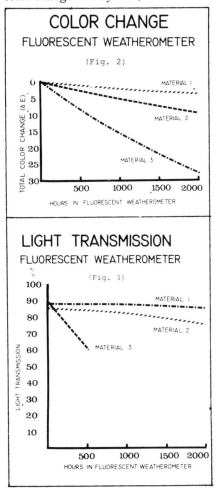

COLOR CHANGE
FLUORESCENT WEATHEROMETER
(Fig. 2)

LIGHT TRANSMISSION
FLUORESCENT WEATHEROMETER
(Fig. 3)

Light transmission measurements were taken on the same weatherometer specimens.

Material #1 lost only 3 percent transmission after 2,000 hours, while material #2 lost 11 percent, and material #3 lost up to 20 percent in only 500 hours of exposure time. See figure 3.

It is apparent from this data that the grade of fiberglass reinforced polyester is extremely important to decreasing ultraviolet degradation.

Thermal degradation

Another extremely important area for solar collector covers is thermal degradation. In most efficiently operating flat plate collectors, cover temperature will not be above 200 degrees F; therefore, tests were conducted on samples aged continuously in a 200 degree F oven for one year. The drop in solar energy transmission for both materials #1 and #2 was approximately equal, about 10 percent. The standard grade sheet, #3, lost more than 50 percent of transmission. See figure 4.

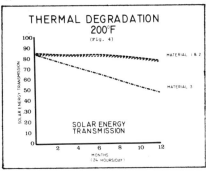

THERMAL DEGRADATION 200°F
(Fig. 4)

Cover plate temperatures higher than 200 degrees F may occur during stagnation, in collectors with improperly designed venting systems. For example, with 300 Btus per square foot, per hour, of insolation, if no fluid were flowing through the collector, the absorber plate could reach 380 degrees F and the inner cover of a double cover could reach 260 degrees, if the outside temperature were 60 degrees F. For this reason, short-term tests were conducted at 300 degrees F. (Figure 5). After 300 hours (equal to ten hours a day for thirty days), material #1 lost only 2 percent of

its solar energy transmission, while material #2 lost 4 percent. Extending the test to 5,000 hours, material #1 lost approximately 10 percent of transmission at 300 degrees F, material #2 lost 22 percent, and material #3 lost 40 percent.

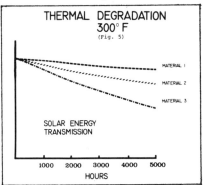

THERMAL DEGRADATION 300° F
(Fig. 5)

Surface erosion

Surface erosion is a weathering factor that should be considered for maintenance of long-term performance. Surface erosion is the actual physical wearing-away and oxidation of the surface. The resulting exposure of fibers on the surface is sometimes called "fiber bloom." Measurements of both average roughness and peak-to-valley roughness were taken with a Clevite 1200 Surfanalyser, in order to test surface erosion.

The surface erosion for material #1 is not noticeable for the first three years of outdoor weathering in South Florida. At the end of four years, however, some surface roughness was noticeable, and after five years there were about 55 micro-inches (1/1-millionth of an inch) of average erosion. A standard grade of fiberglass can have more than 105 micro-inches of average erosion after only two years of South Florida exposure. See figure 6. In order to halt this kind of surface erosion, a proprietary high-temperature coat-
Continued

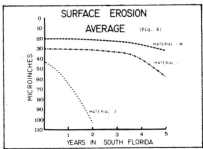

SURFACE EROSION AVERAGE (Fig. 6)

FRP COLLECTOR COVERS

continued

ing manufactured by Kalwall Corp. called Kalwall Weatherable Surface can be applied. After five years of weathering, only 14 micro-inches (hardly noticeable to the human eye) of average erosion was measured on material #1 with this coating.

A more dramatic measurement is of peak-to-valley roughness (figure 7). Material #1 showed a maximum roughness of approximately 300 micro-inches change, while the coated sample showed only 100 micro-inches. Both samples have been weathered for five years in South Florida. A standard grade sheet can have more than 1,000 micro-inches of erosion after only two years, with the same exposure.

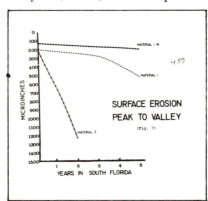

SURFACE EROSION
PEAK TO VALLEY
(Fig. 7)

Impact resistance

One of the major reasons fiberglass reinforced polyester is used for solar collector covers is its remarkable impact strength and shatter resistance. Unlike glass, which can easily be broken into dangerously sharp pieces, FRP is completely shatter resistant.

The best way to measure impact resistance for solar collector covers is the falling ball method. To prove that fiberglass does not lose its remarkable impact strength after many years of outdoor exposure, a fourteen-year-old sample was taken from a building and tested. The

control (unweathered) sample required 25 foot-pounds (a 6.4 lb. steel ball dropped from 4 feet) to cause a rupture of the material, while the fourteen-year-old sample required 32 foot-pounds (the same ball dropped from 5 feet) to cause rupture. See figure 8.

Low temperature impact does not cause a problem. Tests have been conducted at -40 degrees F, and the results showed almost a 50 percent increase in the dynamic load required to cause failure.

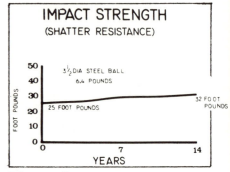

Thermal shock

Many times during the life of a solar collector a rain storm or other rapid change in temperature may cause severe thermal shock to a heated collector. To test FRP resistance to thermal shock, a sample was heated to 300 degrees F and then quickly submerged in cold water. The thermal shock did not cause any harmful effects or noticeable degradation.

In summary, it has been shown that high grades of fiberglass reinforced polyester exhibit excellent weatherability. Critical properties for solar collector covers have been examined and shown to be highly acceptable for safe and efficient use in the solar industry. Special thanks are extended to the American Cyanamid Company and to Owens-Corning Fiberglas Corporation, for their generous technical assistance and testing.

James White is a former product development manager for Kalwall, and an active member of professional societies like the Society of the PLastics Industries, the Society of Plastics Engineers, ASTM, SEIA, and ASHRAE.

Glazing

TEFLON FEP FLUOROCARBON FILM
manufactured by:
Plastic and Resins Dept.
DuPont Co.
Wilmington, Del. 19898
R.C. Ribbans; phone: (302) 999-3456

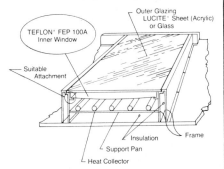

High light transmission, transparent fluorocarbon film for use as inner glazing in solar heat collection devices. **Features:** 96 percent solar transmission; transmits more ultraviolet, visible light and infrared radiation than does ordinary window glass, making collector plate efficiency correspondingly higher. **Installation requirements/considerations:** highly flexible, stretchable and responsive to thermal expansion and electrostatic forces. Stapling and contact adhesives have been used as a temporary measure until final assembly in the collector. Secure fastening must be obtained through mechanical clamping. **Regional applicability:** worldwide. **Availability:** direct from DuPont Co., Plastic Products and Resins Dept. **Price:** 18¢ per square foot/1 mil thickness.

SUN-LITE PREMIUM
manufactured by:
Kalwall Corp.
1111 Candia Road
Manchester, N.H. 03103
Thomas Minnon; phone: (603) 668-8186
A fiberglass sheet for use as a cover plate on solar collectors. Has high solar transmittance; impact resistant, insensitive to thermal shock; lightweight. **Features and options:** Kalwall Weatherable Surface (KWS) applied at factory to protect the cover sheet from surface erosion. **Installation requirements/considerations:** maximum unsupported span distance of 30 inches. Minimum edge distance to mechanical fastener of ¾ inch. **Guarantee/warranty:** none expressed or implied. **Maintenance requirements:** at five year intervals the panels may be refinished with Kalwall Weatherable Surface to provide surface erosion protection. **Manufactur-

er's technical services:** sales and technical staff are available to discuss customers requirements at the above phone number.

Regional applicability: worldwide. **Availability:** from Kalwall Solar Components Div., P.O. Box 237, Manchester, N.H. **Price:** 33¢ to 50¢ per square foot.

SUN-LITE GLAZING PANELS
manufactured by:
Kalwall
1111 Candia Road
Manchester, N.H. 03103
Thomas Minnon; phone: (603) 668-8186

Covers for individual solar collectors or for large arrays of collectors are pre-fabricated by the Kalwall Corp. using Sun-Lite fiberglass sheeting. All feature fiberglass-aluminum bonding system. **Options:** include custom sizes on both the ½ inch and 1½ inch thick panels. **Guarantee/warranty:** none expressed or implied. **Installation requirements/considerations:** maximum recommended spans on ½ inch and 1½ inch panels, 24 inches and 48 inches respectively. **Maintenance requirements:** at five year intervals, the panels may be refinished with Kalwall Weatherable Surface to provide surface erosion protection. **Manufacturer's technical services:** sales and technical staff are available to discuss customers requirements at the above phone number. **Regional applicability:** nationwide. **Availability:** from Kalwall Solar Components Div., P.O. Box 237, Manchester, N.H. **Price:** $1.95 to $2.25 per square foot.

UX-V POLYESTER FILM
manufactured by:
Martin Processing, Inc.
P.O. Box 5068
Martinsville, Va. 24112
Raymond A. Woody; phone: (703) 629-1711
Ultraviolet inhibiting polyester films in thicknesses from 2 to 7 mils. Light transmission is cut off sharply at 0.4 microns and therefore practically none of the ultra-

violet region is transmitted. Does not exhibit the same type of degradation associated with regular polyester film. **Price:** consult factory.

ACRYLITE SDP
manufactured by:
CY/RO Industries
Wayne. N.J. 07470
Phone: (201) 839-4800
A double-skinned architectural and glazing material available in 16 milimeters-thick acrylic or polycarbonate versions. **Features and options:** clear, white translucent, or solar tints for tailoring of light transmission to application. **Installation requirements/considerations:** Acrylite SDP is a combustible thermoplastic. **Price:** consult manufacturer.

SUNWALL(R)
manufactured by:
Kalwall Corp.
1111 Candia Road
Manchester, N.H. 03103
Bruce Keller; phone: (603) 627-3861

Sunwall is the Kalwall panel system, a "sandwich" panel with face sheets permanently bonded to a supporting aluminum grid core. The panels are highly insulating solar heat transmitting material. System is used as the weatherproof wall or roof as well as collector cover. **Features:** clamptight aluminum installation system. **Installation requirements/considerations:** complete instructions available with each job—consult factory on specific questions. **Maintenance requirements:** At five year intervals the panels may be refinished with Kalwall Weatherable Surface to provide surface erosion protection. **Manufacturer's technical services:** quotation service when inquiry accompanied by architectural plans. Technical services to discuss requirements. **Regional applicability:** worldwide. **Availability:** direct shipment from factory. **Price:** $5.50 to $6.50 per square foot.

TUFFAK, PLASTIC POLYCARBONATE SHEET AND FILM

manufactured by:
Rohm and Hass Co.
Independence Hall West
Philadelphia, Pa. 19105
D.T. Espenshade; phone: (215) 592-3000

Tuffak polycarbonate sheet and film is clear, super-impact resistant, weather-resistant, formable, and machinable material supplied in sheet sizes from 24 x 48 inches to 72 x 96 inches, thicknesses of 1/32 to ½ inch, and in rolls of film of various widths in 200 foot runs and 5 to 30 mil gauge. Solar tinted, translucent, and colored grades are also available. **Installation requirements/considerations:** use of Tuffak sheet in architectural and lighting transmission applications must take into account the combustibility and fire characteristics of the material. These hazards can be kept at an acceptable level by complying with building codes and UL standards and accepted rules of fire safety. Tuffak is rated as a self-extinguishing material. Stress limits for continuously imposed loads should not exceed 2,000 psi under ambient temperatures up to 124°F. **Maintenance requirements:** outdoor installations require little maintenance and resist exposure to weathering and sunlight at least five to seven years. Use nonabrasive cleaners if needed, although abrasion resistant grade is available, keep most organic liquids from contact. **Manufacturer's technical services:** staff of engineers in Plastics Engineering Laboratory at Bristol, Pa., available for consultation and problem solving. Trained technical representatives in principal American cities. **Availability:** marketed through distributors— see Yellow Pages. **Price:** from distributors.

PLEXIGLAS G, PLASTIC ACRYLIC SHEET

manufactured by:
Rohm and Hass Co.
Independence Mall West
Philadelphia, Pa. 19105
D.T. Epenshade; phone: (215) 592-3000

Plexiglas acrylic sheet is a clear, impact-resistant, light-weight, weather-resistant, formable, and machineable material supplied in thicknesses from 1/16 to 4 inches and in sizes up to 10 x 12 feet. Plexiglas sheet is available in colors and translucent colors, as well as shades providing solar control. **Installation requirements/considerations:** use of Plexiglas sheet in architectural and light-transmitting applications must take into account the combustibility and fire characteristics of the material, but these hazards can be kept at an acceptable level by complying with building codes and UL standards and accepted rules of fire safety. Design limits below 1,500 psi are necessary to prevent stress-crazing of surfaces. **Maintenance requirements:** outdoor installations require little maintenance and endure exposure to the sun for ten to twenty years with little change. If cleaning becomes necessary, use nonabrasive cleansers. Keep most organic liquids out of contact. **Manufacturer's technical services:** staff of engineers in Plastics Engineering Laboratory at Bristol, Pa., available for consultation and problem solving. Trained technical representatives in principal American cities. **Availability:** marketed through many distributors—see Yellow Pages. **Price:** from distributors.

FLEXIGARD

manufactured by:
3M Company
3M Center, Bldg. 223-2
St. Paul, Minn. 55101
D.H. Flentje; phone: (612) 733-2184

Designed to be a durable replacement for plastic or glass or for use as a composite with glass. Light transmission and insulating properties similar to glass, but lighter weight, nonbreakable. Flexibility, for handling and storage. **Features and options:** durable, flexible, transparent, weather resistant, lightweight, ease of handling. **Installation requirements/considerations:** Flexigard film was designed for use in greenhouses, as storm window glazing, in solar energy panels. Installed with same ease as glass panels. **Maintenance requirements:** virtually none. **Manufacturer's technical services:** call for assistance. **Regional applicability:** all regions. **Availability:** direct from 3M. **Price:** 35¢/square foot.

SOLAR ABSORBER PLATE MATERIALS

The absorber plate is the warm heart of a solar collector—and like hearts are subject to hardened arteries and sudden seizures unless precautions are taken. This absorbing article, part of a longer discussion of copper, steel and aluminum, is reprinted from Solar Age, Vol. 2, No. 5*

Questions and Answers

By S.H. Butt and James M. Popplewell

We have been asked to provide, in question and answer form, comparisons of the principal materials now used for solar absorber plates. Since we are concerned primarily with water heating and space heating applications, we will leave out the plastics widely used in swimming pool heaters. The principal materials we are concerned with are aluminum, copper, carbon steel, and stainless steel.

Question: How do absorber plates fail?

Answer: The most prevalent causes of failure are corrosion, and mechanical failure induced by elevated temperature occurring under "no flow" or "stagnation" conditions.

Question: How do corrosion failures occur?

Answer: In very nearly all cases, when corrosion failures occur the corrosion is localized. This means that corrosion attack occurs at one or a few spots, resulting in perforation of the wall of the fluid passage. Localized corrosion should be distinguished from general corrosion, that is, the gradual wastage of the material. Some of the forms of localized corrosion are pitting corrosion, crevice corrosion, and erosion-corrosion. Pitting and crevice corrosion are most common with aluminum and stainless steels. Erosion-corrosion is responsible for many failures in copper alloys. In the case of carbon steel, all forms of corrosion are encountered.

Question: Which of the four materials will provide the longest life?

Answer: If conditions are "right" for localized corrosion, it generally is quite rapid and the time until failure is often only weeks or perhaps months. If any of the four materials are exposed to an environment that will cause localized corrosion of the material, its life will be unacceptably short. The general rate of corrosion of aluminum, copper, and stainless steel is so low that, if localized corrosion is avoided, they will last a very long time—100 years or more. Carbon steel is subject to localized and general corrosion, as are the others. If the system is properly protected by the use of inhibitors, general corrosion is usually insignificant and localized corrosion can be prevented with the exception of erosion-corrosion. The key to long life is proper inhibition coupled with maintenance of the inhibitors at a proper level. Erosion-corrosion can only be eliminated by flow velocity control and proper design. This includes the elimination of debris that may cause blockages, resulting in local turbulence. I think we can conclude that all four

Sheldon H. Butt   *James M. Popplewell*

materials can fail rapidly if they are mistreated or can last a very long time if they are treated properly. The belief that, even when properly treated, there will be significant differences in the life of the four materials is a fable.

Question: Which materials can be used with tap water?

Answer: Generally, copper and stainless steel can be used with untreated tap water; although in certain cases, some aggressive waters may cause corrosion. In the case of copper, flow velocities must be controlled to avoid erosion-corrosion. Neither aluminum or carbon steel should be used in this environment under any circumstances. The term "tap water" covers a very broad variety of potable waters from a variety of sources. We also want to point out that the term "stainless steel" covers a very wide variety of alloys of widely varying composition and widely varying characteristics. Most "stainless steel" absorbers we have seen are fabricated by welding-usually in a "waffle" pattern. Many stainless steels are sensitized to corrosion in the weld zone. Many others are chloride ions in closed systems where oxygen levels are low, and the waffle pattern necessarily includes many crevices. Because of these two problems, the great majority of stainless steel alloys on the market are *not* suitable for use in this type of absorber plate. There are some that are satisfactory. The person planning to use stainless steel in an absorber should consult with the corrosion experts at the steel mill before making a choice. As a further warning, the stainless alloys that are suitable are usually rather special. You would be unlikely to find them in stock at a metal warehouse.

*For back issues ($2.50) write Box Z, Post Jervis, N.Y. 12771. Mr Butt is president of the Solar Energy Industries Assn. Mr. Popplewell is associate director of the Olin Brass Metal Research Laboratory.

Continued

Questions and Answers *continued*

Question: Which materials can be used with glycol-base antifreeze solutions?

Answer: Providing that the glycol solution is buffered and inhibited and providing further that it is maintained, all four of the materials can be used—assuming all other necessary precautions are taken. If the glycol is not buffered and inhibited to start with or if it isn't properly maintained, none of the materials is safe to use. The inhibitors used are added in order to keep the water in the mixture from attacking the metals in the system. The buffers are used to maintain an optimum pH where the inhibitors are most effective. At elevated temperature, either ethylene glycol or propylene glycol will degrade, forming glycolic acid and other organic acids. These acids are very corrosive. This degradation is usually accelerated in open systems when the oxygen level is higher. The buffers in the solution prevent the solution from becoming acid. However, they can become used up so that the solution eventually becomes acid, although this may not occur immediately. The rate at which the glycols degrade depends on temperature and oxygen level and therefore, the problem becomes worse when collectors "stagnate." To conclude, glycols should not be used unless the condition of the solution can be monitored so that it can be replaced when it needs to be.

Question: What about using inhibited water?

Answer: Once again, if the proper inhibitors are chosen and if they are maintained and such other precautions as are necessary are taken, "inhibited" water can be used successfully with all four materials. The common inhibitors don't last forever and in service it is generally necessary to renew them periodically. It is very important to remember that, in most cases, the concentration of the inhibitor is very important. If enough is not present, it doesn't do the job. If makeup water must be added to the system, additional inhibitor must also be added to maintain inhibitor concentration. In the case of anodic inhibitors such as chromate, depletion may often accelerate corrosion. Cathodic inhibitors and some of the organic film formers are not usually subject to this problem.

Question: In your answers to the previous questions, you have mentioned twice the necessity of taking other appropriate precautions. What do you mean?

Answer: First of all, if two or more metals are being used in a system that has a water-base heat transfer fluid, the different metals must be isolated from one another with insulating couplings. Otherwise, galvanic corrosion of the less "noble" metal will occur in the area of contact. This is particularly important in the case of aluminum and carbon steel, both more active than either copper or stainless steel.

The ions in solution in most tap water encourage corrosion. Copper and stainless steel are generally more tolerant. To avoid possible problems, inhibited water or water-antifreeze mixtures that are to be used with aluminum or carbon steel should be made up using distilled or deionized water. This will result in a lower level of harmful ions such as chloride being present, providing some additional corrosion protection.

Another worthwhile precaution to take is to use a "getter" column with aluminum or steel when copper parts are used in the system or with aluminum when either copper or steel parts are used elsewhere. Characteristically, the getter column consists of a plastic cartridge containing "sacrificial" metal that will pick up ions of more "noble" metals. If these are allowed to deposit on the inside surface of the aluminum or steel absorber plate, they can cause localized galvanic cell corrosion. You should seek the advice of the corrosion experts at the mill manufacturing the absorber plate material concerning the design of a getter column. It must also be pointed out that a getter column will not completely eliminate corrosion when used alone. It will, however, provide insurance in the event of a breakdown in inhibition.

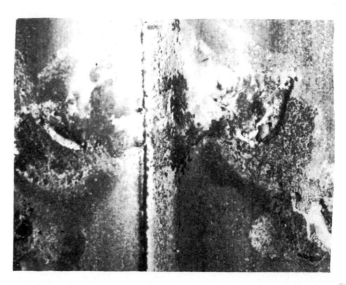

A scanning electron micrograph of a failed aluminum solar panel operated in untreated potable water. The perforation and associated corrosion products are clearly visible. Magnified 100 times.

Failure of aluminum due to erosion-corrosion associated with a blockage. Perforation is marked. Operated with a commercial antifreeze solution for six months. Magnified fifty times.

Question: What is the difference between deionized water and softened water?

Answer: Normally, softened water is simply tap water in which an ion exchange resin is used to replace calcium ions with sodium ions, thus removing the scaling tendency of the water. Everything else in the tap water that might cause corrosion normally remains in the water. Therefore, softened water is *not* a satisfactory alternative to distilled or deionized water. In some cases, in fact, a water may become more corrosive when softened.

Question: Are there other heat transfer fluids that can be used?

Answer: Yes. There are a number of noncorrosive organic heat transfer fluids on the market that can be used and that will not corrode any of the four metals, either alone or in combination with other metals. Things to look out for are the flash and flame points of the organic fluid since some of them could present a serious fire hazard if a leak developed and hot fluid came in contact with the air. Some degrade at high temperatures and the temperature at which this occurs varies from one to another. Some of them are toxic, requiring precautions against contamination of potable water. Others are not. You should explain your application thoroughly to the fluid manufacturer and get his recommendation. *Some* of the non-aqueous organic fluids hold great promise for providing long trouble-free life. They should be seriously considered as an alternative to glycol-water solutions. An effective drier should, however, be installed in the system to prevent ingress of moisture that may result in a form of aqueous corrosion.

Question: Returning to your answer to one of the first questions, you mentioned mechanical failures at "stagnation" temperatures. Will you explain more about this?

Answer: Depending upon the collector design, stagnation temperatures on a hot summer day in bright sunlight can reach 300°F or 400°F—perhaps even more. Since a succession of hot sunny days will fill up almost any thermal storage system, stagnation under these conditions is very likely to occur. The mechanical design of the absorber plate must be such that it will stand exposure to these temperatures.

One of the most frequent causes of failure is the weakening of soft solder joints. Soft solder doesn't have much strength at 300°F—even less at 400°F. As the absorber plate heats and cools, stresses are set up and these often are high enough to cause mechanical failure of soft soldered joints. An example would be the case of an absorber plate in which the tubes were soldered to the plate itself. Unless the assembly is designed so that the plate mechanically holds the tube, using soft solder is risky.

Different metals have different coefficients of thermal expansion. This means that, when they heat up, they expand differently. In a large assembly, such as an absorber plate, the stresses caused by differences in the thermal expansion of two different metals, such as copper and steel, or copper and aluminum, can be enough to rupture joints—not only soft solder joints, which are easy to rupture, but even mechanical joints. For example, depending upon design, the stresses set up because of different coefficients of thermal expansion may cause a copper tube held by spring tension in a plate of another material gradually to loosen. ☼

The scanning electron microscope-microprobe facility at Olin Brass Metals Research Laboratories. This instrument is used extensively in corrosion evaluations.

An Olin Brass solar collector used to evaluate corrosion of different materials.

PLASTICS
for picking up solar heat

By John M. Armstrong

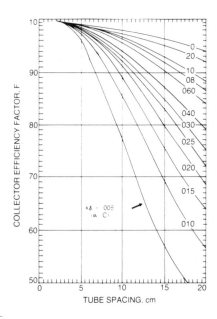

Fig. 1—Collector Efficiency Factor for Plate and Tube Collector

This curve series from "The Hottel, Whillier, Bliss Collector Model" by W.A. Beckman, University of Wisconsin shows that when the tube spacing is under 2.5 cm (1 in.) a variation of 80 to 1 can be made in conductivity without a significant loss in efficiency. (From "Workshop on Solar Collectors for Heating and Cooling of Buildings," NSF-RANN 11-21-74)

"Plastic for heat exchange?!? Impossible!" This normal first reaction to using non-metallic materials for solar collectors quickly changes when the concept is fully understood.

Would you buy an electric blanket made out of aluminum foil? Would you design an ice skating rink with copper plates? The obvious answer is that in specific heat exchange applications there are many other factors that may be more important than conductivity.

Flat plate solar collectors have a heat exchange rate of about one Btu per hour per square inch or less. For such a low rate it just doesn't make sense to pay for copper or aluminum or even steel. Are there, therefore, nonmetallic low cost materials that will withstand all the conditions imposed and collect this low heat flow efficiently?

Because of their high conductivity, aluminum and copper collector plates only require tubes every eight inches or so to draw off heat absorbed from the sun. As the tubes draw off heat from the area of plate right next to them, heat absorbed in areas further from the tubing flows in to replace the heat drawn off. For rubber and plastics the conductivity is lower, so the heat will not flow across the collector surface as well. But this is not an insurmountable problem—the answer is simply to reduce the distance between the coolant used to draw off the heat and any point on the collector surface. In other words, if the heat will not flow across the collector surface to the coolant, take the coolant closer to the heat. See Figure 1.

Manufacturers of non-metallic collectors have devised a number of ingenious ways to do this. The Burke collector, for example, sandwiches a fiberglass screen between two pieces of plastic to spread the coolant out over most of the surface of the collector. Fafco forms liquid passageways in black polyethylene plates. An English company is using a honeycomb plate.

The key development in plastics, though—the one that makes use of nonmetallic substances feasible in glazed collectors—has been the recent availability of an expanded line of heat-resistant materials known as elastomers. Conventional plastics melt or decompose at the 300°F temperatures glazed collectors can reach. Many elastomers can withstand temperatures of 400°F or more. Elastomers are actually synthetic rubber—petroleum-derived hydrocarbon compounds like plastic that have the look, feel, and elasticity of rubber. They are produced using a variety of different formulations, each with unique physical, chemical, and thermal properties. Most can be molded and extruded like plastic.

With the two general objections to non-metallic materials—low conductivity and poor heat resistance—overcome, what then are their advantages over metal? In a word, *many*.

Cost. Plastics and elastomers are considerably less expensive than copper and aluminum. The Burke collector, the Fafco collector, and Calmac's SUNMAT collector all sell for under $3 per square foot in quantity. And this includes the cost of single glazing in the case of the SUNMAT.

Weight. Non-metallic collectors weigh only about one fifth as much as copper or aluminum collectors. This difference in weight reduces installation costs—the collectors are much easier to handle and the structure on which the system is mounted need not be shored up. Shipping costs are also considerably less. The light weight also provides an advantage in terms of performance—because the system has less mass, less energy is required to heat it up to operating temperature from a cooled down state. In periods of intermittent sunshine, lightweight collectors, which have in technical terms *faster transient response*, will produce more usable energy.

Flexibility. Many plastics and elastomers are thermoplastic—they can be bent, rolled and twisted without damage and without losing their original shape. The SUNMAT collector takes advantage of this property by doing away with the conventional rigid panel concept and employing a rollout mat design. The elastomer tubing used in the system is shipped in rolls and can be cut and tailored at the installation site in sizes many times larger than the conventional 3 foot by 7 foot or 4 foot by 8 foot panel. Also flexibility eliminates problems of differential expansion and together with lighter weight reduces shipping and handling costs.

For solar energy to be a major influence in our society installed costs must come down by a factor of three or four. Mass production techniques alone will never accomplish this. Use of low cost materials and new designs is the only answer.

Mr. Armstrong is vice president of the Calmac Manufacturing Corporation.

Absorber plates

ENERSORB
manufactured by:
DeSoto, Inc.
1700 South Mt. Prospect Rd.
Des Plaines, Ill. 60018
Kenneth Lawson; phone: (312) 296-6611
Flat black non-selective solar absorbing coating. When used in conjunction with Super Koropon Epoxy Primer, Enersorb's two-component urethane resin system provides durability over substrates like aluminum, copper, ferrous materials. **Features:** absorptance — 97 percent; emittance — 92 percent. **Installation requirements/considerations:** applied best by spray painting; satisfactory results have been obtained on high speed production lines and on job sites. **Regional applicability:** unlimited. **Availability:** contact factory.
Price: consult factory.

ABSORBER PANELS
manufactured by:
Berry Solar Products
Woodbridge at Main
P.O. Box 327
Edison, N.J. 08817
Calvin C. Beatty; phone: (201) 549-3800
Absorber panels in various dimensions with 24 inches by 96 inches standard. Substrates of copper and stainless steel. Selectively coated with black chrome. Tube-in-sheet design (copper) and two-panel welded sandwich (stainless steel). **Features and options:** selectively coated with black chrome; designed to meet collector manufacturer's specificatons. **Guarantee/warranty:** one year defective materials and workmanship. **Availability:** from manufacturer. **Price:** on request.

SOLARFOIL AND SOLARSTRIP
manufactured by:
Berry Solar Products
Woodbridge at Main
P.O. Box 327
Edison, N.J. 08817
Calvin C. Beatty; (201) 549-3800
Continuous electroplating of selective black chrome on copper, stainless steel, and aluminum in metal thicknesses of .0014 inches to .013 inches and in widths up to 24 inches. Also available with adhesive for direct bonding to absorber panel. **Features and options:** available in coil form for use by collector manufacturers in fabricating flat plate absorbers and as tube wrap in concentrating and evacuated collectors. Formable and solderable without damaging the selective coating: **Installation requirements/considerations:** Solar-Foil with adhesive backing may require furnace curing for proper bonding, depending on no-flow, stagnant-plate temperature levels. **Guarantee/warranty:** selective

coating optical properties guaranteed on shipment to be .92 minimum absorbtivity and .10 maximum emissivity. No guarantee as to performance or adhesive properties. **Maintenance requirements:** none. **Manufacturer's technical services:** data on adhesives and their application. **Regional applicability:** no limitation. **Availability:** from manufacturer. **Price:** consult factory.

ROLL-BOND ABSORBER PLATES
manufactured by:
Olin Brass
ROLL-BOND Products
East Alton, Ill. 62024
John 1. Barton; marketing manager; phone: (618) 258-2443

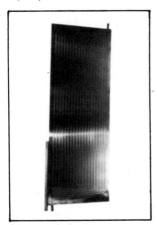

Panels consist of two sheets of metal, either aluminum or copper, bonded together and expanded in certain areas to form channels permitting fluids or refrigerants to be circulated. Flexible design in addition to standard panels. **Options:** connector tubes, alodine pretreat and paint, and other fabrication operations. **Guarantee/warranty:** panels warranted to conform with final specs and/or drawings and will be free from defects in material and workmanship subject to specified conditions and limitations. **Maintenance:** aluminum panels cannot be used with raw water—corrosion information available. **Manufacturer's technical services:** will discuss specific design features and offer suggestions for consideration as may be appropriate. **Regional applicability:** wherever solar panels are used. **Availability:** from manufacturer, no distributors at this time. **Prices:** Copper: 9120-22x96 inches are $64 each plus $25 packing charge per shipment. 9121-34x96 inches are $96 each plus $25 packing charge per shipment. Aluminum: 7610-17x50 inches are $20 each plus $15 packing charge per shipment. 0606-34x96 inches are $39 each plus $25 packing charge per shipment. Reduced prices for production quantity orders are available starting with orders of 100 pieces or more.

COPPER ABSORBER PLATE
manufactured by:
Solar Development, Inc.
4180 Westroads Drive
West Palm Beach, Fla. 33401
Don Kazimir; phone: (305) 842-8935

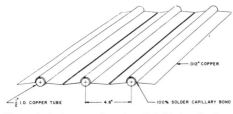

SDI copper absorber plates are normally manufactured in the following configurations:

4' x 10' sinusoidal (45 lbs.)
4' x 10' parallel with ½" headers
4' x 10' parallel with ¾" headers
4' x 10' parallel with 1" headers

Absorber plate (.012 inch copper) is grooved to accept one half of pipe circumference, 100 percent capillary flow solder bond. Piping is ½ inch copper spaced 4.6 inches on centers. Various lengths, widths are available in either series or parallel flow patterns. For parallel flow header sizes are ½, ¾, 1 inch or larger. Stocked sizes are 4 x 10 feet and 2 x 10 feet. **Features and options:** maximum available length is 20 feet. Plates can be coated with flat black paint or black chrome on nickel flash. **Guarantee/warranty:** limited warranty is 5 years for hot water and space heating, 1 year for pool heating. **Manufacturer's technical services:** system design services are provided at no charge when SDI products are used unless the engineering is quite expensive. **Regional applicability:** unlimited. **Price:** orders of 25 or more: 4 x 10 series, $110 unpainted. 4 x 10 feet parallel, ¾ inch headers, $115 unpainted.

ALUMINUM HONEYCOMB
manufactured by:
American Solarize Inc.
19 Vandeventer Ave.
Princeton, N.J. 08540
Joe Beudis; phone: (609) 924-5645
Honeycomb is .004 inch thick aluminum used on absorber plates to increase collector efficiency by cutting reflectivity and increasing absorbing area per square foot. **Features:** sold with permanently black surface for absorbing or polished for reflecting. **Installation requirements/considerations:** sits on plate—no fastening needed. Can be installed when making a collector panel or on site. **Guarantee/warranty:** coating guaranteed to stay on for thirty years without loss in absorptivity when used in recommended operating conditions. **Maintenance requirements:** none. **Manufacturer's technical services:** design services available, field supervision when necessary. **Regional applicability:** wherever high efficiency is needed. **Price:** consult factory.

SOLAR ABSORBER PLATE/3482S

manufactured by:
Tranter, Inc.
735 E. Hazel St.
Lansing, Mich. 48912
Robert S. Rowland; phone: (517) 372-8410

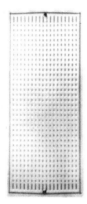

Two sheets of carbon steel (stainless steel available also) seam-welded together, then pressure formed to provide fluid circuitry; 90 percent internally wetted surface for greater heat transfer efficiency. **Features and options:** large headers for flow distribution, sloped headers for drainage, special fitting for threaded or welded joint. **Installation requirements/considerations:** inhibit the internal fluid, do not put excessive strain on the fittings. **Guarantee/warranty:** absorbers are warranted to be free from defects in material or workmanship for one year, excluding internal or external corrosion. **Maintenance requirements:** monitor the internal fluid to be sure suitable corrosion inhibitors are present. Also use a desiccant in the box for humidity control. **Manufacturer's technical services:** various gauges of steel and sizes available for differing operating pressures and collector box designs. **Regional applicability:** may be used anywhere as long as suitable freeze protection and inhibitors are used. **Price:** varies with quantity.

SUNSTRIP

manufactured by:
Burton Industries, Inc.
243 Wyandanch Ave.
North Babylon, N.Y. 11704
Burton Z. Chertol; phone: (516) 643-6660
Modular absorber plate of finned tongue-and-groove aluminum extrusions with a circular central passage, within which a thin walled copper tube is formed and thereby intimately bonded. Corrosion resistance of copper is obtained, while bulk is at cost of aluminum. Lateral assembly of strips produces plate of any desired width that is structurally firm, resistant to thermal stresses and distortion, and adaptable. **Features and options:** lengths from 6 feet to 20 feet available. Width variable in 2 inch increments. Basic surface is natural aluminum and user applies desired coating. Optionally, a black paint coating or anodized finish can be included by manufacturer.

Installation requirements/considerations: may be integrated into roof skin or sidewall structure, as well as within standard collector panels. Headers must be added per requirements of user. **Guarantee/warranty:** one year. **Maintenance requirements:** no special maintenance. **Manufacturer's technical services:** applications and technical consulting. **Regional applicability:** no limitations. **Availability:** from manufacturer. **Price:** $3 to $3.50 per square foot in strip form.

ABSORBER PLATES/C-18 AND C-6

manufactured by:
The Solaray Corp.
2414 Makiki Hgts. Dr.
Honolulu, Hawaii 96822
Lawrence Judd; phone: (808) 533-6464
All copper absorber plate and water ways—type "L". Nickel black chrome selective surface, pipe connections silver solder brazed, pressure tested 300 psi. Plate formed to runner pipes, soldered plus bond coating copper-graphite compound for efficient heat transfer. **Features and options:** custom sizes can be made. **Guarantee/warranty:** 5 year warranty. **Maintenance requirements:** none. **Manufacturer's technical services:** on request. **Regional applicability:** any area. **Price:** app. $3/square foot depending on quantity.

ABSORBER PLATE

manufactured by:
Barker Brothers
207 Cortez Ave.
Davis, Cal. 95616
David Springer; phone: (916) 756-4558
Tubes are brazed into manifolds using silver and phosphorus, with Chem-glaze Urethane coating that has an absorptivity of 95 percent. Some design modifications may be made upon request. **Features and options:** all copper design. Priced competitive with plastic or aluminum finned collector panels. **Installation requirements/considerations:** may be installed with or without glazing. **Guarantee/warranty:** five year guarantee on materials and workmanship. **Maintenance requirements:** none. **Manufacturer's technical services:** collector sizing for specific installations, calculation of pressure drops and pump sizing. **Regional applicability:** national. **Availability:** may be shipped anywhere in the U.S. **Price:** $4.25 per square foot.

COPPER PRODUCTS
manufactured by:
The Anaconda Company, Brass Div.
414 Meadow St.
Waterbury, Conn. 06720
B.W. Erk, Jr.: phone (203) 574-8500
Copper sheet panels and copper tube for solar collectors and water lines to storage tanks. **Price:** consult manufacturer.

SANDWICH PANEL ™
manufactured by:
Ilse Engineering, Inc.
7177 Arrowhead Road
Duluth, Minn. 55811
John F. Ilse; phone: (218) 729-6858

Flat plate liquid heat exchanger panels used in construction of flat plate solar collectors. They consist of two layers of steel sheet, the surface of one sheet embossed with a dimpled pattern and welded at these dimples to second sheet. The space between the two sheets forms flow passage for heat transfer fluid. **Features and options:** nearly 100 percent coverage over internal surface by liquid flow. Panels have been designed specifically for use in a closed system. **Installation requirements/considerations:** ½ inch NPT female flanges provided for plumbing connections. **Guarantee/warranty:** one year against defects in welding or materials. **Maintenance requirements:** no antifreeze or corrosion inhibitors to replace due to "Internal Self-Drain" system. **Manufacturer's technical services:** installation manuals. **Regional applicability:** U.S. **Price:** consult factory.

Associated products

790 BUILDING SEALANT
manufactured by:
Dow Corning Corp.
Midland, Mich. 48640
A one-part silicone material that readily extrudes in any weather and quickly cures to produce a durable and flexible low-modulus silicone rubber building joint seal. **Features:** consistency is relatively un-changed from −34° to 140°F, can be applied in any season. Can be used with most common building materials. Minus 35° to 300°F working range. **Installation requirements/considerations:** ready to use as supplied, no mixing required. **Availability:** 11-fluid-ounce cartridges and 2 gallon pails. **Price:** consult manufacturer.

HIGH TECHNOLOGY PLASTICS
manufactured by:
High Performance Products, Inc.
25 Industrial Park Rd.
Hingham, Mass. 02043
Richard Totlock; phone: (617) 749-5374
Plastic products made of Fiberglass or rotationally moulded urethane. High strength and low weight. **Prices:** consult manufacturer.

SILICONE CAULK
manufactured by:
Silicone Products Dept.
General Electric Co.
Waterford, N.Y. 12188
Silicone caulk withstands temperatures down to −65°F and up to 500°F. Flexible and waterproof. **Features and options:** available in black, metallic, clear, and white. **Price:** consult manufacturer.

ALUMINUM EXTRUSIONS
manufactured by:
New Jersey Aluminum
1007 Jersey Ave.
P.O. Box 73
North Brunswick, N.J. 08902
George E. Olson; phone: (toll free) (800) 631-5856

Aluminum extrusions used in the manufacture of frames, mounting support structures and reflectors for solar panel frames. **Features and options:** frame structures will be fabricated to customer specifications. Design assistance at no obligation. **Price:** consult manufacturer.

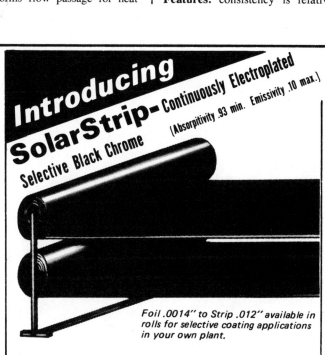

How to recognize a heat exchanger

By Steve Baer

There is no better place to educate yourself in the matter of heat exchangers than a junkyard. Heat exchangers, along with axles, pulleys, and gears, make up a surprisingly large proportion of the metal you find in junkyards. It would take several days to visit all the equipment, installed at work or displayed in a salesroom, that you can see, pick up, and closely examine in a few minutes at a large junkyard. (Most of the pictures in this section were taken at the ACME metal junkyard on North Second Street in Albuquerque.)

Unfortunately, heat exchangers are among the last things to catch the interest of a curious person poking around a junkyard. A broken or rusted-out radiator seems to be more thoroughly a piece of junk than a worn shaft or a sprocket with a bad tooth.

One danger in visiting junkyards is

that it encourages backwards design. The pieces of junk suggest uses for themselves rather than the needs suggesting the design of the equipment. Sometimes an interesting visit to the junkyard proves to be emotionally tiring in the same way that a visit to the pound can exhaust an animal lover.

Axles transfer energy by torque. The rigid shaft, turned at one end, commands the other end to do the same. In heat exchangers it is again a motion transferred through metal. But in this case it is by a molecular trembling, heat passing through the walls and along the fins of the exchanger. It is simple to pass energy by turning a shaft. The forms suitable for exploiting energy transfer by rotation and reciprocation seem monotonous and restricted after having come to appreciate the beauty of heat exchangers.

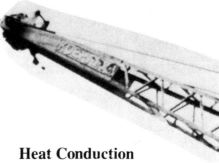

Heat Conduction

Metals vary greatly in their ability to conduct heat. Below are listed various materials and the quantity of heat (in Btu) that will flow through a 1 foot square, 1 inch thick sheet in one hour if the temperature on one side is 1 degree F higher than on the other side.

Copper	2700	Btu
Lead	250	Btu
Gold	2060	Btu
Silver	2900	Btu
Zinc	750	Btu
Aluminum	1500	Btu
Iron	400	Btu
Air (still)	0.163	Btu
Water (still)	3.85	Btu
Oil (still)	1.2	Btu
Glass	5.8	Btu
Styrofoam	0.25	Btu
Glass wool	0.25	Btu

As an example of what these rates of heat flow mean in terms of the strength of sunlight, we can see that a 1 inch thick piece of lead, to pass the heat of the direct sun through itself, would have to be only slightly hotter (about 1°F) on the sunny side than on the shaded side, for the sun gives about 300 Btu per square foot per hour. Lead is a relatively good conductor. For this same quantity of heat to pass through a good insulator such as Styrofoam, the temperature would have to be more than 1,000°F hotter on one side than the other, long before which the Styrofoam would have burned up.

You'll notice in the headings under the pictures of the heat exchangers such titles as liquid-to-air, liquid-to-liquid, etc., but a glance at our list of conductivity of materials shows that air, water, and oil are all very poor conductors. So what sense is there in building a heat exchanger to pass heat to these substances when they can't move it once they have it? The trick is that these substances, gases and liquids, must flow past or through the heat exchanger to make the exchanger function. For instance, radiators on automobiles rely on the radiator fan and the motion of the car to push the air past the cooling fins; in addition there is a water pump to circulate the water within the radiator.

This is usually the difficult part of the heat's journey—into and out of the exchanger, not through it. To accomp-

lish this, pumps, agitators, and fans are used to keep presenting the surfaces of the exchanger with fresh matter to be heated or cooled.

It is quickly apparent after a glance at the liquid-to-gas heat exchanger that the transfer to gas is more difficult than to liquid; for much more surface is exposed to the outside air than to the inner circulating liquid. All in all, the progress of heat into and out of a heat exchanger is like that of a jet traveller who must fight his way through traffic jams arriving at and departing from the rapid leg of his trip.

The Way Heat Moves

Heat is measured in Btus or calories. It is a quantity like pounds or gallons; and, as with other quantities, you can point to something and say, "That contains so much heat." But with heat you have to be fast, for it is always moving from hotter to colder. It refuses to be domesticated to fit our usual idea of a quantity. It is rather like the animals you see at the zoo which still want to get out; and heat succeeds—slowly, if well guarded by insulation—but nevertheless eventually it escapes if there is a cooler place nearby.

You can take comfort in the fact that heat is not clever. It does not invent new ways to move; it cannot leap-frog barriers.

Heat transfer is generally broken into three categories; conduction, convection, and radiation. We have listed values for conductive transfer and mentioned the function of convection in moving matter past hot or cold surfaces. Radiation is another powerful transfer mechanism wherever there are marked differences in temperature between two surfaces. All of our energy from the sun arrives by radiation. There is no material between us and the sun to carry heat by conduction or convection.

Continued

Liquid-to-air. *This is a section of fin tube. Every inch there are four and one-half fins, so that there are 27 square inches of fin per inch of pipe. The relative areas suggest how much harder it is for the heat to get out of the aluminum fins into the air than it is for the heat to enter the copper pipe from the circulating water. It is also obvious that the heat has no trouble passing through the press fit from the copper to the aluminum. Consider the press fit. The coefficient of expansion of aluminum is .0000124 per unit length per degree F. It would seem that as the pair heat up, the aluminum would expand and part from the copper; if it does part from the copper, it will lose its source of heat and then cool and tighten again around the hot pipe. The relative coefficients of expansion of aluminum and copper would seem to leave the aluminum fin always somewhat uncomfortable and undecided about what to do. Actually, the aluminum has enough spring so that even when it is hot it grips the pipe firmly and continuously (even though we know it must have relaxed its hold somewhat). It's possible that if we tried to move heat from the fins into the pipe, the hot fins would expand and never touch the cold pipe. Thus a mechanical heat exchanger could be a one-way heat valve.*

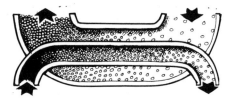

HEAT EXCHANGERS *continued*

Liquid-to-air. *This is a part of a freezer compartment. Cold liquid circulates through the flow channels, cooling the metal plates. This exchanger is made by bonding two formed pieces of sheet steel together. This kind of design is very suitable for solar heat collectors. Notice the larger header which feeds into many small channels.*

Gas-to-liquid. *A boiler. Hot gasses pass through these pipes, from one end of the tank to the other, and heat the tank's contents.*

Gas-to-gas. *This exchanger is composed of hundreds of small (about 3/16-inch) tubes. One stream of gas blows across the tubes while the other gas goes through the tubes from manifold to manifold.*

Gas-to-gas. *This is the head of a piston. The finned head helps to dissipate heat by transferring it to outside air.* ☼

Heat exchangers

TURBOTEC, TURBOFIN
manufactured by:
Spiral Tubing Corp.
533 John Downey Dr.
New Britain, Conn. 06051
Edward H. Cornish; phone: (203) 224-2409
Configured metal tubing increases heat transfer per foot of tube by 2 to 3 times. High bendability permits compact heat exchanger designs. **Features and options:** available in most metals and standard sizes. **Availability:** from manufacturer. **Price:** consult manufacturer.

THRIFTCHANGER r
manufactured by:
Sturges Heat Recovery, Inc.
P.O. Box 397
Stone Ridge, N.Y. 12481
Paul M. Sturges; phone: (914) 687-0281
Heat exchanger designed to recover waste chimney heat from wood fires; 128 tubes, 7/8-inch diameter and 2 feet long with starting collars, instructions, and cleaning brush. **Installation requirements/considerations:** must be installed by mason. **Guarantee/warranty:** five years. **Maintenance requirements:** must be periodically cleaned. **Regional applicability:** wherever heat is needed. **Availability:** from manufacturer. **Price:** $450.

BILOCULINE FTO, PTO, DUPLEX
manufactured by:
Spiral Tubing Corp.
533 John Downey Dr.
New Britain, Conn. 06051
Edward H. Cornish; phone: (203) 224-2409
Double-walled tubing creates highly efficient double-pass heat exchangers, provides two layers of metal between potable water and other media. **Features and options:** available in most metals and combinations, standard tube sizes. **Availability:** from manufacturer. **Price:** consult manufacturer.

Heat transfer fluid

Q2-1132 SILICONE HEAT TRANSFER LIQUID
manufactured by:
Dow Corning Corp.
Solar Energy Div.
Midland, Mich.
Richard H. Montgomery; phone (517) 496-4000
A water-clear liquid having a viscosity of 20 centistokes (77°), fluid from −50°F to 600°F, exhibits heat stability, oxidation resistance, very low vapor pressure and a high flash point. It is essentially noncorrosive to common engineering metals and is virtually nontoxic. This fluid has a high dielectric strength and, as such, will not promote galvanic corrosion. **Features and options:** intended for operation in the temperature range of +50° to 400°F with intermittent exposure to 450°F in systems that are essentially closed. **Installation requirements/considerations:** plastic piping should be avoided. Q2-1132 has a lower specific heat than water and the flow rate must be increased by a factor of 2 to 2.5; this will increase the pumping head. the capacity of the pump will be lowered due to Q2's higher viscosity. **Availability:** from Dow Corning. **Price:** consult manufacturer.

NUTEK-800
manufactured by:
Nuclear Technology Corp.
P.O. Box 1
Amston, Conn. 06231
Thomas F. D'Muhala; phone: (203) 537-2387
Ethylene glycol based heat transfer fluid for use with aluminum, copper, and iron containing systems. Provides corrosion inhibition and protection against freezing to 0°F (−18°C). **Features and options:** protects aluminum and copper in systems containing other aluminum, copper, and iron components. **Installation requirements/considerations:** no special requirements except that the system be clean before use. **Manufacturer's technical services:** consultant; data aquisition, monitoring and analysis; fluid analysis regarding heat transfer fluids. **Regional applicability:** to be used in those areas where freeze protection to 0°F (−18°C) is adequate. **Availability:** in stock; large volume production capability; qualified distributors being set up. **Price:** consult manufacturer.

NUTEK-805
manufactured by:
Nuclear Technology Corp.
P.O. Box 1
Amston, Conn. 06231
Thomas F. D'Muhala; phone: (203) 537-2387
Propylene glycol based heat transfer fluid for use with aluminum, copper, and iron containing systems. Provides corrosion inhibition and protection against freezing to 0°F (−18°C). **Features and options:** protects aluminum and copper panels in systems containing other aluminum, copper, and iron components. **Installation requirements/considerations:** no special requirements except that the system be clean before use. **Maintenance requirements:** periodic checks depending on system design and use. **Manufacturer's technical services:** consultant; data aquisition, monitoring, and analysis; fluid analysis regarding heat transfer fluids. **Regional applicability:** to be used in those areas where freeze protection to 0°F (−17°C) is adequate. **Availability:** in stock; large volume production capability; qualified distributors being set up. **Price:** consult manufacturer.

NUTEK-830
manufactured by:
Nuclear Technology Corp.
P.O. Box 1
Amston, Conn. 06231
Thomas F. D'Muhala; phone: (203) 537-2387
Ethylene glycol based heat transfer fluid for use with aluminum, copper, and iron containing systems. Provides corrosion inhibition and protection against freezing to −30°F (−34°C). **Features and options:** protects aluminum and copper panels in systems containing other aluminum, copper, and iron components. **Installation requirements/considerations:** no special requirements except that the system be clean before use. **Maintenance requirements:** periodic checks depending on system design and use. **Manufacturer's technical services:** consultant; data acquisition, monitoring and analysis; fluid analysis regarding heat transfer fluids. **Regional applicability:** to be used in those areas where freeze protection to −30°F (−34°C) is required. **Availability:** in stock; large volume production capability; qualified distributors being set up. **Price:** consult manufacturer.

NUTEK-835
manufactured by:
Nuclear Technology Corp.
P.O. Box 1
Amston, Conn. 06231
Thomas F. D'Muhala; phone: (203) 537-2387.
Propylene glycol-based heat transfer fluid for use with aluminum, copper, and iron

Continued

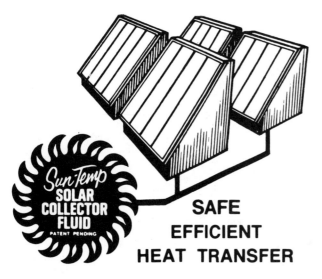

containing systems. Provides corrosion inhibition and protection against freezing to −30°F (−34°C). **Features and options:** protects aluminum and copper panels in systems containing other aluminum, copper, and iron components. **Installation requirements/considerations:** no special requirements except that the system be clean before use. **Maintenance requirements:** periodic checks depending on system design and use. **Maufacturer's technical services:** consultant; data aquisition, monitoring and analysis; fluid analysis regarding heat transfer fluids. **Regional applicability:** to be used in those areas where freeze protection to −30°F (−34°C) is required. **Availability:** in stock; large volume production capability; qualified distributors being set up. **Price:** consult manufacturer.

NUTEK-876

manufactured by:
Nuclear Technology Corp.
P.O. Box 1
Amston, Conn. 06231
Thomas F. D'Muhala; phone: (203) 537-2387

Corrosion inhibitor for solar heating units containing aluminum, copper, and iron components where water is the heat transfer media. **Features and options:** protects aluminum and copper panels in systems containing other aluminum, copper, and iron components. **Installation requirements/considerations:** no special requirements except that the system be clean before use. **Maintenance requirements:** periodic checks depending on system design and use. **Manufacturer's technical services:** consultant; data aquisition, monitoring and analysis; fluid analysis regarding heat transfer fluids. **Regional applicability:** material is to be used with de-ionized or soft water. **Availability:** in stock; large volume production capability; qualified distributors being set up. **Price:** consult manufacturer.

SUN-TEMP

manufactured by:
Resource Technology Corp.
151 John Downey Dr.
New Britain, Conn. 06051
Al Trumbull; phone: (203)224-8155

A non-corrosive, non-toxic, and non-aqueous heat transfer fluid developed for use in solar collectors. **Features and options:** has a boiling point above 500°F and freezing point below −40°F.
Price: 55 gallon drum, $3.50 gallon: $192.50 total
39 gallon drum, $4.00 gallon: $120 total
10 gallon drum, $4.50 gallon: $45 total
5 gallon drum, $5 gallon: $25 total

To Save Solar Heat
KEEP IT COOL

By Gordon Tully

As architects, we at Massdesign have watched the demise of many promising innovations in building technology, either because some unknown risk was discovered during its commercial exploitation, or more often because some obvious problem was "wished away" in order to make the idea more palatable to investors and to users.

Solar heating will likely prosper as an idea, but many of the currently available devices will find early graves in the cemetery of unviable innovations. The problem for the building industry as a whole, and particularly for architects (who are asked at the same time to be very protective of the owner's interests, and to use advanced technology) is to look twenty-five years ahead and figure out which of the current systems will still be there and which will have been removed. It is not surprising that the building industry is so conservative, and that the inexperienced are usually the most adventuresome.

One response to this double bind is tu bury the solar system within the ordinary building construction, and hope that standard techniques can be used to build an essentially new device. This passive approach is an admirable one, but it has many limits, especially for moderate to high-density situations, for retrofit applications, for most home-owners (who need "hands-off" operation), and for highly variable climates. Also, the very existence of passive heating changes the thermodynamic behavior of a building, with often unexpected results.

Another response is to design each element to "Mercedes-Benz standards," and apply fail-safe protection throughout. Unfortunately, each protective device in itself becomes a problem; and

Gordon Tully is president of Massdesign, and Chairman of the Mass, Bay Solar Energy Assn.

the cost of such systems is unsupportable.

A third response, one widely used for institutional and some commercial applications, is to design a low-cost, potentially high-maintenance system, and put it under the care of the existing maintenance personnel of the building. Whether the system will survive repeated maintenance problems depends upon the tolerance of the owner and upon his budget.

A fourth option, and the one we favor, is to design an extremely simple system with few moving parts, to operate at the lowest possible temperatures, using collectors that will survive summertime stagnation with little damage; and then to make the key elements in the system very inexpensive and easily replaced. A designer must project the longevity and potential maintenance problems of the system at every stage of its design, and these considerations must come before either cost or operational advantages, in our opinion. The problem becomes: how cost-effective can you make a long lasting and maintenance free system?

Toward low cost
and low maintenance

We have lived through an amazing period in the history of man, in which the accidental existence of oil and gas has allowed us to completely transform our lives and the earth on which we live. One small effect of this accident is that the plentiful supply of high-temperature fuels has allowed us to use high-temperature fuels in our heating and cooling systems. Large central plants can be built, using steam to drive efficient air-conditioning equipment, to pre-heat large quantities of air with relatively small coils, and to move heat relatively long distances through small pipes.

To use solar responsibly, we need simple, low-cost devices, and that

means low temperatures; but the entire building-mechanical industry is built around the use of medium and high-temperature devices and fluids. Massdesign has tried to deal head on with this problem by exploring the limits of low-temperature systems. Beginning with work formulated by Hugh Adams Russell (who now owns his own firm, Hugh Adams Russell Associates), we had much help from Mark Hyman, who, with his pioneering work towards 100 percent space heating, *had* to work with low temperatures, and led us in that direction. From Will Hapgood at Raytheon and John Bemis at Acorn, who built and monitored a very low temperature system, we continued to learn a great deal.

The first step in designing such a system is to forget about steady-state conditions. One constantly is asked about how much energy a store can hold, or at what temperature a system operates, etc. These are questions that lose their meaning in a properly designed low-temperature system, to be replaced with such questions as:

- How cool did we manage to get the store last night?
- How long does it take the delivery system to deliver a day's-worth of collected energy on a given day?
- How big does the distribution coil have to be to keep the store from overheating?

The efficiency of *any* collector, and most critically that of a simple single-glazed collector (the type we favor because of its low summer stagnation temperature), depends upon the store temperature. As the store temperature goes down, the efficiency of the solar system, considered as an energy source, goes up, mainly because of improved collector efficiency, but partly because of the lower skin loss from the store. It becomes evident that simplicity and durability hinge upon low-temperature storage.

Continued

KEEP IT COOL *continued*

An apparently obvious lesson of thermodynamics is shown by the following example: 2,000 gallons of water at 10°F above room temperature are relatively useless when compared with 500 gallons at 40°F above room temperature, although both contain the same amount of heat. This statement needs partial re-examination. A relatively small fan coil unit operating for twenty-four hours might reduce the temperature of the 2,000 gallon store by 4°F (to 6°F above room temperature), thereby delivering 50,000 Btus to the space. Compare this with a slightly smaller coil reducing the higher temperature store from 40°F to 24°F above room temperature. It would take considerably less time to deliver the same 50,000 Btus, after which the store would be considered "depleted" and the system idled. Upon activating the collectors the next day, the system with the larger store would operate at a Delta-T 18°F lower than the system with the smaller store, and because of the large store would maintain a lower Delta-T throughout the collection period. Clearly the larger store system is the more efficient.

The point illustrated by this comparison is that during mid-winter conditions, when the system typically cannot supply the entire demand, there is plenty of time to deliver whatever energy is collected. By enlarging the store and running the system all the time, a single-glazed collector of simple design can perform at a low enough Delta-T to rival the output of a double-glazed collector, and on a cost-per-Btu basis to rival that of the more complex heat-trap design. The idea is simply to heat the building a little bit all the time rather than to heat it fully for a short time.

As temperatures and insolation rise in the months before and after the peak heating months, there comes a time when even the smallest system can handle the entire heating load, i.e., the building becomes 100 percent solar.

Under what conditions can this occur? As we go from winter to summer results there are a decreasing number of occasions on which auxiliary heat is needed. For any combination of system, house, user, and climate there comes a date after which the auxiliary can be switched off, just as there is a date after which the water can safely be left on in an unheated summer house. This date is empirical—after it the house is almost certain to be 100 percent solar. What can be done to a system to make this date occur earlier?

The answer is the same (fortunately) as that to the question of how to improve collector efficiency in mid-winter—enlarge the store. Since temperature and insolation vary, it was impossible to say how much to enlarge the store without trying out some possibilities on a computer. Massdesign's experience is that the cost-effectiveness of store sizing in an ordinary system is dependent more or less equally upon improvements of collector efficiency in mid-winter, and upon extending the period of 100 percent solar in the spring and the fall. In general, for a single-glazed collector of moderate performance, store sizes of as much as 6 gallons per sqaure foot of collector often prove cost effective.

We have noted that load heat exchangers for these low-temperature systems are not much bigger than those normally used. For fan coil units, the difference is relatively small. For baseboard convectors, an increase in length of from 60 to 100 percent is usually adequate. As the system becomes large relative to the load, system elements are sized more and more for the "spring and fall" conditions, entailing a very large system aiming at 100 percent for the year, the store and the load heat-exchanger both get quite large. Before the 100 percent point is reached, the cost in nuisance of a higher output collector might well be paid, partly to cut down on the roof area, but mainly to allow the store to get warmer and

thereby to cut down on the size of the load heat-exchanger.

Three points can be made here. First, one seldom gets a chance to supply any load completely, because there usually isn't enough room for the collectors. Second, when 100 percent is possible, it might well be worthwhile to size both store and load heat-exchanger for the very low-temperature operation, despite the cost, but few have done any meaningful comparisons along these lines. Finally, some loads such as hot water and process heat, if supplied by solar, require that the store be kept at or above certain temperatures. Note with regard to these higher-temperature loads that, as the percentage supplied by solar goes down, store temperatures can also go down. This makes it possible to do a first-class job of domestic hot water pre-heating with a very cheap collector operating at relatively low temperatures.

What about the cost of large stores? Part of the problem is taken care of by eliminating pressurization. This is a good idea in any case, since tap water drain-down systems without pressurization are much simpler and yield higher collection efficiencies. They work well, too, although you would never know it by asking the legion of mis-informed detractors. Acorn has further reduced the cost of storage by making a tank out of plywood and lined with vinyl and held together with circumferential straps like a barrel. The vinyl gets brittle at 180°F (typical summertime conditions in the systems we design) and probably will need replacing in ten to twenty years. But it si very cheap. Since the plywood merely transfers the load from the vinyl to the straps, it can do its job at these relatively high temperatures. Many other cheap storage designs are possible. Space is more of a problem. At our Massachusetts Audubon Nature Center job, there was only room for 2.4 gallons of storage per square foot of collector, despite which the system has functioned well. It would be more cost-effective with twice the storage.

With un-pressurized storage, care must be taken to avoid air-binding in the distribution circuits if they rise above the water level in the store. The second floor baseboard loop in Mark Hyman's house has operated with no air binding under a negative head of about 12 feet for over a year. But avoid too many loops and constrictions. Naturally, adding pump energy will keep the bubbles moving, but at a great reduction in system C.O.P. Stay with circulators and be careful, and of course do not use air bleed valves.

An added advantage of the tap water system is the possibility of simple hot water pre-heat by means of an immersion coil in the tank. We have a 200 foot coil of 1/2 inch type copper tubing immersed in the tank at the Massachusetts Audubon job, which can heat two or three gallons per minute to 140°F with a store temperature of 165°F or so. We are gathering more exact data on this aspect of the system. Scaled down, such a system might prove to be the best model for domestic hot water heating. A designer of low-temperature systems must deal with the effects of running a distribution system at these low temperatures. Using warm air delivery, some additional investment is required in larger outlet grills and a more massive back-up furnace than is commonly used in cheap warm air systems. This represents a return to an earlier and much more satisfactory practice. In a very small house, forced air delivery at room temperatures may cause some drafts, and require a low-temperature cut-off. But for normal installations, low velocity air can be delivered with no problems at room temperature. Obviously, this is a crucial issue in the design of low temperature systems using warm air delivery.

A new challenge in our attempts to build long-lived active systems with low temperature stores is that of off-peak

storage. Obviously, heating the store at night with electric resistance heat is exactly what we least need. At present, our only answer is yet another storage tank, with a commensurate demand for space. If the European practice of storage radiators is adapted, this additional storage for off-peak use would be in the heated space itself, rather than in the basement, and that would solve the problem. In addition, once off-peak incentive rates are introduced, the valleys between the peaks are filled in fairly quickly and the incentives are discontinued to new users. But for the short term (ten years or so), these rates will be common, and solar will have to find ways to coexist with the new heating demands of off-peak electric heating. Incidentally, as has been pointed out elsewhere, one can never rationally measure a solar system's performance against on-peak electric rates in a world of common off-peak rates. Turning this argument around, the performance degradation of the solar system caused by adding off-peak energy to the store can just as well be subtracted from the off-peak savings! It is a question of which comes first in one's accounting.

The subject of heat pumps in relation to storage is extremely complicated, but one point can be made. The proponents of solar assisted heat pumps have always assumed that one great virtue in operating a heat pump off the solar store is to drive the store temperature down. However, when you compare a solar assisted heat pump's minimum store temperature of 40°F with the kind of system we design which has a minimum temperature of perhaps 65 to 70°F in the winter, it is clear that the difference is not very large. We have found repeatedly that when comparing water source solar assisted heat pumps against a well designed low-temperature system, we can never justify the former.

By a similar argument air systems ought to be measured against low-temperature water systems, rather than

against systems with artificially high low limit cut-offs. Collector inlet temperatures would then be about the same for both systems, but collector efficiencies are much lower in air systems. This fact must be accounted for by a commensurately lower cost in the air systems, which is contrary to current experience.

From its inception, the government's solar program has been intent on developing a new solar industry. In practice, this means encouraging our giant companies to develop or acquire solar divisions. American industry is unfortunately enraptured by a ponderous mythology involving mass extraction, mass production, and mass marketing, which has left it totally unable to deal with the subtle demands of gathering up the sun. By being large, a company loses contact with the long-term maintenance of its product. Industrial response to maintenance and product quality has for many years focused upon replacement with "better" products as soon as possible. Unsuccessful products are habitually discontinued, and replacement parts cease to be available.

These mass marketing strategies cannot work for collectors that are expected to yield a payback only if they operate maintenance free for twenty-five years. One way to solve this problem is to focus upon the development of simpler, understandable designs, perhaps built on site, that can be maintained and constructed by local tradesmen. Another approach is to mass produce extremely cheap, easily replaceable collectors, perhaps made entirely out of plastic. In either case, we are talking about unsophisticated devices with low summer stagnation temperatures, which we must learn to use at low temperatures. We at Massdesign know that it is possible to do a great deal with low-temperature systems, and we hope that our example will encourage others to explore this very fruitful avenue. ☼

Storage

THERMAL STORAGE VAULTS

manufactured by:
American Solarize Inc.
19 Vandeventer Ave.
Princeton, N.J. 08540
Joe Beudis; phone: (609) 924-5645

Vaults are made of lightweight cellular aggregates, vermiculite cement, and other additives. Can be used for rock storage or liquid tubes or be filled directly with liquid. Needs no other insulation. **Features and options:** standard sizes or custom made. **Installation requirements/considerations:** suitable for above or below ground applications. **Guarantee/warranty:** on manufacturing defects and failure to meet warranty specifications. **Manufacturer's technical services:** design service available, field supervision when necessary. **Availability:** f.o.b. from factory. **Price:** $250 to $6,000.

AQUA-COIL HOT WATER STORAGE/SERIES TC

manufactured by:
Ford Products Corp.
Ford Products Road
Valley Cottage, N.Y. 10989
William M. Morrison; phone: (914)358-8282

The Aqua-Coil consists of a stone-lined tank with a pancake coil of finned copper tubing in the bottom. It is equipped with an aquastat that controls operation of a circulator. When the aquastat calls for heat, the heating medium is circulated through the coil to warm domestic hot water. **Features and options:** because the tank is stone lined, it is possible to install a copper coil without creating conditions that could lead to electrolytic corrosion and tank failure.

Optionally available is a unit with an auxiliary electric heating coil and thermostat. **Installation requirements/considerations:** a tempering/mixing valve is recommended at the hot water exit to prevent excessively high water temperatures at the faucet. **Guarantee/warranty:** tank and coil guaranteed for five years; all other parts for one year. **Maintenance requirements:** only non-toxic antifreeze solutions should be used in the exchange fluid (suggest propelene glycol). **Availability:** F.O.B. Valley Cottage, N.Y. **Price:** TC-40 (less circulator)—$144. TC-40C (with circulator)—$187.20. TC-65 (less circulator)—$205. TC-65C (with circulator)—$248. Units with auxiliary electric heating coil: TC-65E (less circulator)—$224.10. TC-65EC (with circulator)—$267.30.

STORAGE TANKS/STJ-30 TO 120

manufactured by:
A.O. Smith Corp.
P.O. Box 28
Kankakee, Ill. 60901
J.A. Cousins; phone: (815)933-8241

Glass-lined storage tanks for use in potable-water systems in capacities from 30 to 120 gallons. Thirty to 80 gallon tanks are insulated to an R factor of 10; 120 gallon tank has R of 7.7 **Features and options:** vertical installation for gravity-flow or forced circulation systems. **Installation requirements/considerations:** two ¾ inch tappings are provided for solar connections: water inlet, outlet, and relief valve openings are at the top of the tank. **Price:** consult manufacturer.

NON-CORRODING FIBERGLAS TANK (INSULATED)

manufactured by:
Owens-Corning Fiberglas Corp.
Tank Marketing Div.
Fiberglas Tower
Toledo, Ohio 43659
Q.V. Meeks; phone:(419) 248-8063

Insulated underground storage tank made of vinyl ester and polyester resins reinforced with Fiberglas, and 3-inch layer of polyurethane foam insulation. **Features and options:** not susceptible to corrosion; nominal capacity of 1,000 gallons; will store water from −40°F to +200°F at 0 to 10 psi. When water is at 200°F with ambient temperature of 0°F the heat loss is 1,300 Btus per hour. **Maintenance requirements:** none, no need for periodic testing or upkeep of cathodic protective systems. **Price:** consult factory.

HOT WATER STORAGE UNITS

manufactured by:
W.L. Jackson Mfg. Co., Inc.
1200-26 E. 40th St.
P.O. Box 11168
Chattanooga, Tenn. 37401
Ralph L. Braly; phone: (615) 867-4700

Glasslined hot water solar storage tanks in 80, 100 and 120 gallon capacities with auxiliary heating element in top third of tank. **Features and options:** fully automatic temperature and pressure relief valve installed as standard equipment. Solar connection is ¾ inch NPT. **Price:** consult factory.

SUN-LITE STORAGE TUBES

manufactured by:
Kalwall Corp.
1111 Candia Road
Manchester, N.H. 03103
Scott Keller; phone: (603) 668-8186

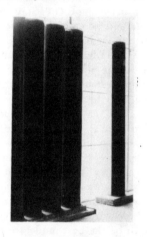

Manufactured from Sun-Lite and designed to contain storage materials in vertical fashion, these tubes are available in standard 12 inch and 18 inch diameter in heights up to 10 feet for use in solar energy storage systems. **Features and options:** other sizes are available on a custom basis, including tanks, can be manufactured with plumbing fittings for interconnections. Tubes are made of solar energy transmitting material for direct heat absorption. **Installation requirements/considerations:** level surface required for base to maintain vertical orientation. **Guarantee/warranty:** none expressed or implied. **Maintenance requirements:** none. **Manufacturer's technical services:** sales and technical staff are available to discuss customer requirements at the above phone number. **Regional applicability:** nationwide. **Availability:** from Kalwall Solar Components Div., P.O. Box 237, Manchester, N.H. **Price:** $32 to $152 each.

NON-CORRODING FIBERGLAS STORAGE TANK

manufactured by:
Owens-Corning Fiberglas Corp.
Tank Marketing Div.
Fiberglas Tower
Toledo, Ohio 43659
Q.V. Meeks; phone (419) 248-8063

Underground storage tanks made of Fiberglas-reinforced plastic—FRP. Available in capacities from 550 to 50,000 gallons; can be used to store water from −40°F to plus 150°F. **Features and options:** not susceptible to corrosion. **Maintenance requirements:** none, no need for periodic testing or upkeep of cathodic protection systems. **Price:** consult manufacturer.

SOLAR POWER CELL/550

manufactured by:
Sunshine Mfg. Co.
4870 SW Main#4
Beaverton, Ore. 97005
David Carson; phone: (503) 643-6172

Thermal power storage cell with a maximum capacity of 2 million Btus and a maximum discharge rate of 1 million Btus

per hour at 400°F. **Features and options:** variable discharge rates and temperatures available as options. Consult factory concerning specific requirements. **Installation requirements/considerations:** insulated and reinforced concrete foundation approximately 6 by 6 feet required along with a vacuum source of 28 inches height. **Guarantee/warranty:** workmanship and operation warrantied against normal operating conditions causing failure for two years from date of installation. **Maintenance requirements:** none. **Manufacturer's technical services:** technical support to distributors. **Availability:** shipped within ninety days. **Price:** $7,500.

STORAGE TANK KIT

manufactured by:
Acorn Structures, Inc.
P.O. Box 250
Concord, Mass. 01742
Stephen V. Santoro; phone: (617)369-4111

Cylindrical tank in kit form constructed of 3/8-inch exterior plywood reinforced with galvaized steel bands. This cylinder sup-

ports and contains the vinyl liner. Three and one-half inch (R 11) fiberglass insulation surrounds sides and cover of the tank. One inch of Styrofoam plus 2½ inches of loose fill vermiculite insulate the base. **Features and options:** tank has 2,000 gallon capacity, can be used for any system that does not require storage temperatures in excess of 160 degrees F. **Installation requirements/considerations:** floor on which tank stands must support a permanent loading of 400 pounds per square foot. All components can be moved through all normal door openings. **Price:** consult manufacturer.

WATER STORAGE TANKS

manufactured by:
The Glass-Lined Water Heater Co.
13000 Athens Ave.
Cleveland, Ohio 44107
Ralph R. Mendelson; phone: (216) 521-1377

Water storage tanks from 40 to 120 gallons. All units can be furnished with or without heat-exchangers, insulation, outer jacket, pumps, controls. **Price:** consult factory.

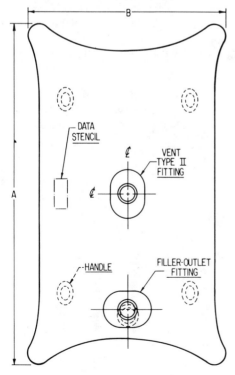

Reflections on a Low Temperature Surface: MORE SUN, MUCH MORE EFFICIENCY

By Jeremy Coleman

Reflectors have been a shining feature in the solar scene ever since the French inventor August Mouchot focused a conical reflector on a steam boiler in the mid-nineteenth century. Today we are familiar with reflectors in virtually all high temperature solar collectors—for only by concentrating the sun's rays is it practical to operate collectors as temperatures become higher than 160°F or so. Most interest in reflectors has been focused on high-temperature, high-tech systems—and yet a reflector can be as simple a device as a field of snow.

Why use reflectors?

Reflectors have a "magical" quality when we look at their effect on solar collection. By increasing the amount of radiation that falls on the collector proper, a reflector increases the effective area of collection and the collector output. Remarkably, collector output can be increased by a greater factor than the same increase in effective collection area would of itself account for.

Consider a collector that puts out 100 Btus per hour at a given set of operating conditions. We add a reflector to this collector in such a way that exactly twice as much solar energy now falls on the collector. We expect that output will now be 200 Btus per hour (double what it was with only one half as much solar energy striking the collector), but upon measurement we discover that output is, instead, 250 Btus per hour: one and one half times greater than expected!

To explain this puzzle we need only remember that collection efficiency

(η_{col}) is a function of the temperature difference (ΔT) between the absorber and the surrounding air divided by the rate of incident radiation (Q_{inc}) on the collector face:

$$\eta_{col} = F\left(\frac{\Delta T}{Q_{inc}}\right)$$

This relation is a negative function—that is, efficiency goes up as $\frac{\Delta T}{Q_{inc}}$ approaches zero, which is just another way of saying that collection efficiency is highest when heat loss from the collector is lowest.

In the example cited, the radiation (Q_{inc}) incident on the collector was increased. Not only was there more energy to be collected, *but it was collected more efficiently*. For example, if efficiency in our example was initially 50 percent, it was increased to 63 percent by adding the reflector. This often-overlooked factor is a compelling reason to look at reflectors in conjunction with all solar installations.

Diffuse and specular reflection

We need to distinguish between two types of reflective surfaces: specular and diffuse. An example of a specular reflector is a mirror, or a flat and highly polished piece of aluminum. Light rays hitting a specular surface are reflected away at exactly the same angle at which they strike the surface. An example of a diffuse reflector is a field of fresh fallen snow. Light rays falling on a diffusing surface are reflected away equally in all directions, so that the reflected light is of even intensity when viewed from any direction, regardless of the direction of the light source. In the Arctic snow fields the even brightness of the snow in all directions is the cause of snow blindness.

The distinction between diffuse and specular surfaces is important to the concept of reflectors for flat plate collectors. To investigate this distinction, a computer program for clear-day radiation was run. To determine the quantity of day-long radiation striking a south-facing vertical surface in Marlboro, Vt. for the eight months of the heating season. Figure 1 illustrates this quantity for three conditions: light

Jeremy Coleman is an associate at Total Environmental Action, Inc.

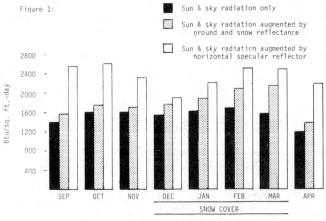

Figure 1:

■ Sun & sky radiation only

▨ Sun & sky radiation augmented by ground and snow reflectance

□ Sun & sky radiation augmented by horizontal specular reflector

Clear-day radiation striking a south-facing vertical surface in Marlboro, Vt.

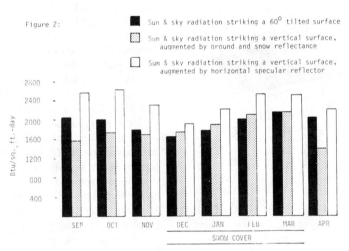

Figure 2:

■ Sun & sky radiation striking a 60° tilted surface

▦ Sun & sky radiation striking a vertical surface, augmented by ground and snow reflectance

□ Sun & sky radiation striking a vertical surface, augmented by horizontal specular reflector

Clear day radiation striking a south-facing 60° tilted surface compared to clear day radiation striking a south-facing vertical surface in Marlboro, Vt.

coming from the sun and sky only; light from the sun and sky augmented by reflection from the snow and ground from an unobstructed field in front of the vertical surface; and light from the sun and sky augmented by a horizontal specular reflector set directly in front of the vertical surface. Assumed, for the purpose of the simulation, were the following: 1) snow cover occurred from the beginning of December through March; 2) reflectivity from the snow was .7; 3) reflectivity from the ground was .2; 4) reflectivity from the reflector was .9; 5) the size of the reflector was equal to that of the vertical surface; 6) the ratio of the height to length of the vertical wall was 1:4 (this has significant impact on the augmentation from the reflector).

The results are startling. Reflection from the specular reflector increases clear-day radiation on the vertical surface from 24 to 87 percent, depending on the month, while reflection from the snow and ground increases this day-long total from only 5 to 37 percent, in spite of the fact that the specular reflector is small in comparison to the unobstructed field (stretching to the horizon).

It can be seen that a finite specular reflector dramatically increases available radiation, with even greater potential for increased collection efficiency. An enhancement of 50 percent in incident radiation might result in a 100 percent increase in collector output. Even reflection from ground and snow becomes more significant in this respect.

When reflected radiation is accounted for, the implications for collector design become apparent. Figure 2 compares the clear-day radiation striking a collector at an optimum 60° tilt angle with the radiation on a vertical surface, including reflection from snow and ground on the one hand and from a specular reflector on the other. It is apparent that radiation on the vertical surface compares favorably with the 60° optimum when snow and ground re-

flectance is accounted for. (Counting this component, optimum is about a 72° tilt angle for space heating. The addition of a reflector causes the vertical surface to surpass its 60° counterpart in available radiation for the entire heating season. Significantly, the quantity of clear-day radiation available to a vertical surface (counting diffuse reflected radiation from ground and snow) surpasses the radiation available to a 60° tilted surface during the coldest months of the heating season—providing a better "fit" between space heating load and collector output for the vertical collector.

Work with reflectors

The potential benefits to be gained from reflective surfaces are great, yet little work with reflectors has been done to date. This is only partially due to lack of imagination by solar designers. Far greater restraints are imposed on reflector use by practical considerations. There is no low-cost material of high reflectivity and proven durability now generally available, although several candidates may be so identified in the near future (see section on materials). In addition, integrating a large reflective expanse into any building design—let alone one that incorporates a solar collector to start with—is enough to give even a daring designer solar nightmares. Nonetheless, these challenges have been successfully and ingeniously addressed by a handful of individuals.

One solution that has been incorporated into a comparatively large number of projects is to extend a flat tar and gravel roof, covered with white reflective stone—in front of the collec-

Continued

REFLECTIONS *continued*

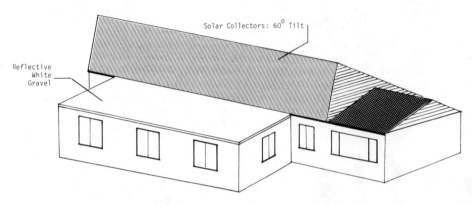

Reflective White Gravel

Solar Collectors: 60° Tilt

Figure 3: Harry Thomason's Third Solar House

tor array. Figure 3, of Harry Thomason's solar house #3, illustrates this approach. A great advantage here is that the reflective surface—white gravel—is a durable and easily understood material. A disadvantage is that the stone is highly diffuse as a reflective surface. Although as much as 80 percent of the light striking the stones is reflected, only a relatively small portion is forward-reflected onto the collector. Nonetheless, the Phoenix building of Colorado Springs, featuring two identical collector arrays that differ only in that one sits behind an expanse of white gravel, has reported an increase in heat output of 25 percent by this collector over its non-enhanced counterpart. A further advantage of a flat roof that should not be overlooked is that it provides a platform from which to work on the collector proper.

A more daring response to the design challenge posed by reflectors is the People/Space Co. solar house by Robert Shannon. Built in 1975 on speculation (it sold), the People/Space building features 400 square feet of air-cooled solar collector augmented by two 8 by 20 foot reflector arrays. The reflective panels are faced with flat aluminum sheets and held in place by

cables. Bob Shannon does not have quantified data of system enhancement due to the reflectors, but it is undoubtedly significant, and was achieved at relatively low cost, especially when compared to additional collector area. A disadvantage of sheet aluminum as a reflective material, however, is the tendency for it to lose some reflectivity over time due to weathering.

Steve Baer, Richard Crowther, Richard Falbel and others have worked with a variety of effective reflector integrations. A significant solar building that incorporates a reflective surface and a flat plate collector is the Henry Mathews house in Coos Bay, Oregon (Figure 4). It casually and successfully integrates a reflective surface into the building design. In-depth analytical work based on this building has been done by D.K. McDaniels et al at the University of Oregon at Eugene.

The Mathews house is a one-story, 37 by 89 foot building with about 1,600 square feet of heated floor space. The collector (of single-glazed, liquid-cooled non-pressurized design) is 5 by 80 feet and runs the length of the roof. The roof in front of the collector, sloping at 8° to the south, was made reflective by pressing aluminum foil into a coat of plastic roofing cement.

The University of Oregon investigators found that the increase of radiation incident on the collector due to the aluminum foil was on the order of 140 to 170 percent. *Moreover, this enhancement of incident radiation results in an effective increase of 100 percent in collector output over the heating season, due to improved collection efficiency.* A further significant finding of the Oregon team is that optimum orientation has the collector plane approximately at right angles to the reflector, with the reflector tilted between 0° and 10° above the horizontal.

Reflective Materials

There are numerous reflective materials on the market, probably none of which were developed with the goal of providing a low-cost and durable reflective surface for solar collection. A product search, however, has turned up a really interesting variety of films, sheets and boards that may have solar applications. Undoubtedly this is just a partial listing. In some instances, costs are given. If anyone knows of other good reflective materials, I would appreciate knowing about them.

Rigid Board—Charlie Wing of Cornerstones, in Maine, has uncovered a builders' material called Storm-Brace Thermo-Ply Sheathing, manufactured by Simplex Industries of Adrian, Michigan. Thermo-Ply is a 1/8 inch waterproof high-density cardboard, faced with aluminum foil on one or two sides. This material is not designed to be exposed to the weather for longer than framing-time in the construction process; however some has survived two years in Maine weather with no apparent deterioration. Its low cost ($4 to $5 for a 4 by 8 foot sheet), two-dimensional strength, relatively light weight and 95 percent reflectivity may make it a useful material in some instances.

Earle Barnhardt of New Alchemy had located a Mylar-faced fiberboard with a Tedlar protective coating, manufactured in the Boston area. It has a twenty year outdoor lifetime and retails at 40¢ per foot and I would be grateful to anyone who could identify it.

Sheets & Films—Kalwall Corporation of Manchester, N.H., can supply their SunLite material with a factory-applied Mylar coating for about $1 per square foot. A 1 or 3-mil protective coating protects the Mylar against abrasion and ultraviolet degradation. Available in 58 inch widths, this material is easily bent into curved shapes and is highly durable.

Solar Usage Now in Bascomb, Ohio, sells a self-adhesive reflective foil on a vinyl or paper backing for 35 to 55¢ a square foot. It is highly reflective but comes only 12 inches wide.

The Dorrie Process Company in Norwalk, Connecticut, manufactures the material used as trim strips for Detroit automobiles. It is an 8-mil aluminized vinyl material with a 1-mil Tedlar protective coating. Dorrie has received many inquiries about its product but is cautious about encouraging its use in the solar field, because its limited specularity is unsuited to focusing-type reflectors. Also, they are not set up to retail their product—but samples may be available upon request.

Finally the King-Seeley Thermos Company of Winchester, Massachusetts manufactures Astralon material that consists of two layers of 1-mil aluminized polyethylene bonded to both sides of a glass scrim. It is a cloth-like, highly reflective material used for space blankets. It comes in 56 inch widths and costs about 10¢ a square foot—but the minimum order is sizable. ☼

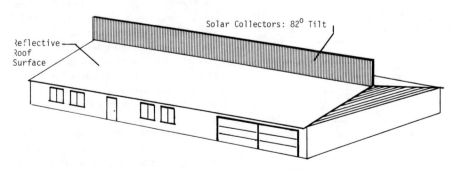

Figure 4: The Henry Mathews House in Coos Bay, Oregon

Making the most of cans for
A SOUTH-WALL SOLAR HOT AIR COLLECTOR

By Bruce Hilde and James Gangnes

The sun's energy is without a doubt the most dependable, low-cost energy source around. This construction manual will help the home owner add a solar system to his house at a very low cost, brought about by these factors:

1. The use of recycled beverage cans
2. The low cost of the suggested glazing materials
3. Use of the south wall existing on a house as the insulated support structure for the solar furnace
4. Do-it-yourself construction

The savings in energy gained from the solar furnace will pay for the cost of materials within three to four years. To be specific, the cost per 100 square feet of collector area is approximately $280. The money saved at the current utility prices in Minnesota would be $130 per heating season if you are heating electrically, $80 per season if

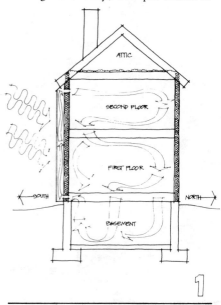

Mr. Hilde is head of the Northern Solar Power Company. Mr. Gangnes is head of the Prairie Community Design Center. Extra copies of this article can be ordered from the Northern Solar Power Company, 311 Elm St., South, Moorhead, Minn. 56560, for $2.

you use natural gas, and $60 per season if you heat your house with oil.

The principles of solar hot air heating are easy to understand. As shown in drawing 1, the solar collector is attached to the south wall of your house. The sun's rays pass through two layers of glazing material (like a thermo-pane window) and are absorbed by the beverage cans, which are painted flat black. The cans are arranged in vertical columns and attached to the south wall. There are approximately 1,500 cans per 100 square feet of collector area. Each can has jagged holes cut in the top and bottom to allow the passage of air. As the flat black paint absorbs solar radiation, the metal cans heat up, consequently heating the air inside them. (The temperature will not rise above 250°F, and will not cause a fire hazard.) The hot air rises naturally through the vertical columns on the wall, to be ducted into the house through an outlet duct located at the top. Cool air from the first floor or basement enters a lower input duct and the base of the collector again. Rock storage areas may be added to retain the heat collected during the daytime for night time use.

The system in drawing 1 is attached to the non-gabled end of the south wall of the house. The natural flow of air without a fan is accomplished by the thermosiphon principle similar to the draft caused in a chimney due to the rising hot air. In this application a series of ducts will extend across the entire top and bottom of the collector and vent through the wall into the living areas. Fans within the house may be used to move the hot air through the house.

A more efficient method of collection is the use of a fan to pull the hot air out of the collector and force it into the rooms as shown in drawing 2. This method will lower the collector operat-

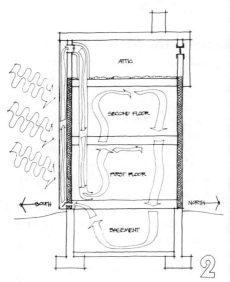

ing temperature and thus reduce the heat losses to the cold outside air.

Drawing 3 shows the solar collector during summer operation. Because of the high angle of the sun, many rays are reflected and do not enter the col-

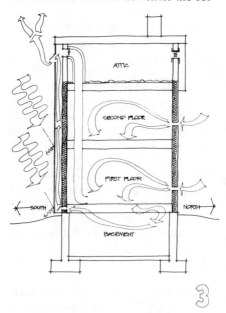

CAUTION: When circulating fan is disconnected in spring, the summer exhaust vent must be open to avoid overheating.

lector. The energy that does enter heats the air, which rises and is vented to the outside. A thermosiphon effect cools the house by drawing cool air through an open north window or basement vent, and into the base of the collector.

The construction technique that follows is only one of an unlimited number of variations and designs. Each particular house needs to have the solar collection system individually designed or adapted. The basic methods identified here should serve most houses.

In order to make construction easier and the solar heating more effective, some windows on the south wall can be covered by the collector, or only by the glazing. It is possible in other circumstances to build around and between windows at the expense of time and sacrifice in performance.

If there are any questions about construction or other applications of this solar heating system, call:

Bruce Hilde   (218) 233-2515
Northern Solar Power Company
or David Gangnes (701) 232-4001
Prairie Community Design
Center

The solar construction described below was designed for the south wall of the gabled end of a house. Floor joists running east-west and wood sheathing are preferred but not necessary and other situations can be accommodated. The suggested inlet into the collector from the basement will be a series of 1½-inch holes, drilled through the south floor joist located below the floor and above the concrete basement wall. (Fig. 4, note 1) Duct work will connect this inlet to the cold air return of your existing forced air heating system. Electrical or hot water heating systems will need additional duct work in the basement or inlet holes placed above the floor level.

A 12-inch hole cut through the south wall near the roof peak will serve as the outlet from the collector. (Fig. 4, note 2) Hot air from the collector will be ducted through the attic to the living areas by means of a combination of 12-inch flexible insulated duct tubing, the existing interior stud walls, and room vents. A blower fan and automatic controls will be located in the attic to move air from the collector into the living areas. The roof overhang will serve as a manifold to di-

rect air towards the collector outlet. A summer air vent should be constructed in front of the outlet hole in order to exhaust warm summer air and circulate cool air from the north side of the house through the rooms.

Collector construction will be divided into three phases—I: Collector base and insulation; II: Application of cans, paint, and glazing; III: Installation of fan, automatic control, and duct work.

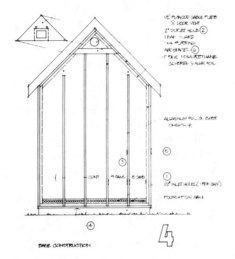

BASE CONSTRUCTION

Collector base and installation

1 Remove all siding on the south wall in order to expose the wood sheathing.
2 Make sure the wall is insulated; if not, have an insulation contractor do the needed work.
3 Staple aluminized foil[1] to the entire wall, cutting out for windows.
4 The base structure will be made by toenailing 1 by 4 inch #2 pine boards to the wall on edge. The boards will be referred to as furring. A bead of Hypalon[2] caulking should be applied under all perimeter furring to avoid air leaks into the collector. #16 smooth box nails will be used, after holes have been drilled to avoid cracking the furring boards. Nails should be alternated one side to the other, spaced every 2 feet, and the furring set on edge and nailed to the wall. (Fig. 4, note 3)
5 The first furring will be toenailed along the horizontal floor joist about 1 inch above the basement sill. (Fig. 4, note 4) Later on 1½ inch holes will be drilled into the joist to form the inlet to the collector.
6 The next furring will be placed ver-

tically starting from the east edge of the house. All vertical furring boards will have small 3/64-inch holes drilled through them for wiring the cans against the wall. The holes will lie along a line 2¼ inches from the wall edge of the furring. The first hole will be 8¼ inches from the bottom end. A 4¾-inch spacing will be laid out for the remaining length of furring.

7 The first vertical 1 by 4 at the east side should butt to the horizontal base furring. It should be mitered at the roof angle and butt to the roof overhang. #7 box nails are used to butt 1 by 4s to each other. (Fig. 4, note 5)

8 Allow a 1-inch gap for insulation, then set eight cans next to each other along the wall. The next vertical furring will start after the eighth can and extend to within 8 inches of the roof overhang. The 8 inch gap allows for air passage. (Fig. 4, note 6)

9 The third vertical furring will be at a nine can spacing from the second furring.

10 Each successive spacing will alternate eight cans, nine cans, eight cans, etc., until the west edge of the wall is reached. The last spacing should be anywhere from five to thirteen cans. Finally the roof overhang is furred.

11 At 3-foot intervals a 1 by 1-inch strip should run horizontally across all vertical furring. One inch notches will be required in the furring to flush the strips with the furring.

12 At the roof peak a summer air vent will be constructed to allow for venting of hot summer air. It will be nailed and glued as shown in Fig. 6, note 7. The ends are mitered to match the roof angle but should be located about 2 feet below the peak.

13 A layer of 1-inch Urethane[3] rigid foam insulation is now glued to the perimeter of the collector base with Dow Mastic No. II[4]. Aluminum foil should be glued over the urethane insulation to prevent deterioration due to sunlight.

14 A 12-inch hole will be cut in the peak to match the 12-inch duct collar. 1½-inch holes will be drilled through the floor joist to match each can along the bottom of the collector. The collector base is now complete. You may paint the base black if you wish.

Continued

73

HOT AIR COLLECTORS *continued*

Application of cans, paint and glazing

1   The holes in phase I, step 14 above will be covered with a light weight Aluminized Mylar[5] film to act as a one way valve. Cut the Mylar 2½ inches wide and long enough to cover all holes between each furring section. The area to be covered should be sanded to make the seal air tight and stop the cold night air from backing up. The strips of Mylar should be glued with silicone glue applied in a straight horizontal line just above the holes.

2   18 gauge expanded[6] metal lath will be used to hold the cans above the inlet holes. It should be cut to length to fit between each collector section. A sheet metal man may be necessary to bend the pieces into two 90 degree angles with a 4½-inch side and two 2½-inch flanges. The flanges rest on the in-

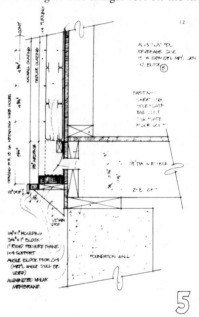

sulation and are nailed to the 1 by 2 inch nailer strip. (Fig. 5, note 8)

3   The next step is stacking the cans, but first we have to punch jagged holes in the top and bottom so as to open about 2/3 of the can's area. Special tools, can openers, or beer openers may be used.

4   All cans should be washed in trisodium phosphate[7] to clean and prepare the surface for paint.

5   The first row of cans is now placed above the expanded metal plenum. An 18 gauge wire[8] is run across the first row and through the appropriate 3/64-

inch holes as mentioned in PHASE I, step 6. The wire should be nailed to the outside of the two end furring boards.

6   Cans will be stacked on top of each other and wired until the roof overhang is reached. A minimum clearance of 8 inches should be maintained between the top can and the overhang furring now covered with tin foil and insulation. This will create a manifold for air flow to the outlet above. The cans will have a staggered appearance as they conform to the roof line.

7   A compressor and spray paint machine will be needed to spray the wall of cans with flat black lacquer[9]. Suggested paints are mentioned in the supply list. Coverage is essential but try to eliminate a heavy coating of paint. You may want to paint the cans individually. Be sure to cover the foil covering the insulation to retain its reflective surface.

8   It is now time to put on the first glazing of Tedlar[10]. Ideally, glazing should be applied on a dry day to minimize future condensation between glazing which could lessen the amount of sunlight striking the cans. A sharp razor knife will help to cut this durable plastic film. A caulking gun with Hypalon sealant will be used to put a bead on the first three vertical base structure and appropriate pieces in between. The 4-foot roll of Tedlar will be applied from the top down. Staples may be needed to hold it temporarily at the top. Press the Tedlar against the caulking to seal collector and hold the first glazing in place. Apply another bead on the third, fourth, and fifth furring and all pieces between. The third furring should have two beads of caulk with Tedlar between. Apply the second 4-foot piece of Tedlar to overlap the the first. Smooth and staple the top. As you repeat the above steps you will quickly glaze the entire wall with an air tight seal.

9   Caulk over the top of the Tedlar and furring again. 1 by 4s cut into 1-1/8 inch strips will now be nailed over all the newly caulked furring to give a 1-1/8 inch air gap. Nail with #7 box nails.

10   Apply caulking and repeat step 8 using the Kalwall glazing[11]. One-half inch wood strips should be applied over

should be used to make a pilot hole just deep enough to penetrate the ½-inch strip and Kalwall. The hole should be larger than the #7 box nail used but no longer than the nail head.

Installation of fan, automatic control, and duct work.

1   The summer air vent design will be left up to the builder; but a well insulated, weatherstripped design is essential to good performance. We suggest a design built from ½ inch plywood with 1 inch of urethane glued to it. The air vent hole will be covered with an overlapping, weatherstripped, insulated, hinged and latched cover. A spring and cable may be used to open and close from ground level. (Fig. 6, note 9)

2   The fan and duct location is variable depending on the house and the access to an interior stud wall where the heat

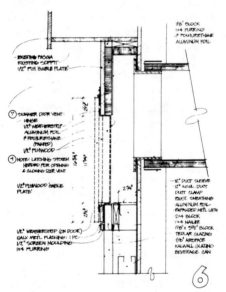

vent into the living area is formed. The use of 12 inch Thermaflex[12] insulated duct and 1½ inch Styrofoam[13] rigid insulation will serve as building materials for the duct work. Dow Mastic II will be needed to glue the fan box and ducting together.

3   The location of the fan and controller[14] will be determined by each situation. The thermostat should be mounted near the top of the collector and accessible through the summer vent cover or collector outlet hole. It should be set for 80 degrees. The fan and controller should be located near the interior stud wall. Two stud wall sections should be used to give adequate duct area.

4 In order to avoid electrician costs an extension cord may be used to connect the controller and fan to a power outlet (110 volt). See the electrical wiring diagram (Fig. 7, note 10) for wiring instructions. Load carrying wires should be #12 gauge, thermostat wire #20 gauge.

5 Cold air return ducts are most easily built by using Styrofoam between floor joists and using Thermaflex to connect with the furnace cold air return.

6 The fan speed should be adjusted to keep the collector air temperature at 100°F at solar noon.

There are many alternatives to the particular design offered in this booklet.

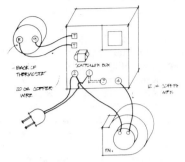

CAUTION: When circulating fan is disconnected in spring, the summer exhaust vent must be open to avoid overheating.

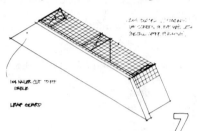

On houses without gabled south walls, an enlarged inlet and outlet vent to and from the collector and along the full length of the wall can allow air circulation without the use of fans or ducting. This is called a "passive" solar collection system.

We have suggested using interior stud (non-load bearing) walls as air ducts. If this is structurally unsound for your particular application, you may want to run the ducting from the attic through the ceiling and into a closet or chimney liner to be vented into living areas.

Other options are also available for the use of existing air distribution systems and rock storage areas, but the extensive design and construction work will be left up to the ambitious builder.

The materials used in this design are stocked in most cities and are available off the shelf. Substitutions may be used but reliability will be uncertain. We will list the items in the order they appeared in the construction steps. Likely places to find them and unit prices are also included. The amount needed for each specific application will depend on the size of the collector and will be left up to the builder. Prices are approximate.

1 Reflective aluminum foil is used for insulation and can be found in most lumber yards.

Price: $8 per 500 square feet

2 Eternaflex Hypalon Sealant is an acrylic latex sealant that dries in five days to a rubbery gasket. Other alternatives to try are Tremco Mono, a one part acrylic terpolymer, Sonolastic, a polysulfide base sealant with high temperature characteristics, or Silicone caulk. *Price: $1.95 a tube Hypalon*
$2.75 a tube Tremco
$2.85 a tube Sonolastic
$3.60 a tube Silicone

3 Urethane foam by the trade name Thurane has the highest R factor (5.8 per inch) of any insulation. It also has good heat resistance. It's available from most roofing companies. Styrofoam may be used but has lower R factor and heat resistance.

Price: 33¢ per board foot

4 Dow General Purpose Mastic No. II is a solvent-dispersed adhesive for Styrofoam and Thurane (urethane) foams. It's available in lumber yards or construction material supply houses. There are several other foam adhesives that will work and are available on the market.

Price: $1.50 a tube

5 Aluminized Mylar is a thin plastic film, micro-deposited with highly reflective 99 percent pure aluminum. The only distributor in our area is Northern Solar Power Company, 311 South Elm Street, Moorhead, MN. Phone: (218) 233-2515.

Price: 30¢ a square foot

6 18 gauge expanded metal is available at sheet metal shops or plaster suppliers.

Price: Not available

7 Tri-sodium phosphate is an excellent cleaning agent and is available at most paint stores.

Price: Modest

8 18 gauge galvanized wire is available in most hardware stores.

Price: 75¢ for 50 square feet

9 There are many flat black paints available. Some of them are: Rust-oleum, Moore Guard, Gilt Edge, etc.

Price: $4 - 5 a quart

10 Tedlar is a thin, high strength plastic film made by DuPont. It comes in 4 foot rolls, has a 20 year life, and is ultraviolet resistant. It is available through the Northern Solar Power Co.

Price: 40¢ a square foot

11 Kalwall is a fiberglass material with high solar transmittance, 20 year life, and ultraviolet resistance. This excellent glazing material is also available through the Northern Solar Power Co.

Price: 60c a square foot

12 Thermaflex III is a versatile, insulated duct system available through most heating and air-conditioning suppliers.

Price: $2 a foot

13 Styrofoam is a well known insulation material.

Price: 21¢ a board foot

14 The blower fan will depend on the size of the collector used. For collectors under 100 square feet a Brundage model #DS6 1140 RPM centrifugal blower fan will be sufficient. For collectors between 100 and 200 square feet a Brundage model #DS6 1725 RPM unit should be used. This fan is available from Adams, Inc., Fargo. A comparable model should handle 350 CFM at ¼ inch S.P. for a collector area under 100 square feet, or 600 CFM at ¼ inch S.P. for collector areas over 100 square feet, but less than 200 square feet.

Price: $80 a unit

The fan controller or switching relay is a Honeywell Tradeline Control model #RA89A are also available from Adams, Inc.

Price: $16.35 a unit

The thermostat is a 24 volt mercury switch made by Honeywell. We will use the cooling side of the thermostat to turn the fan on. The model number is #T87F-1859. Also available at Adams, Inc.

Price: $9.56 a unit

Lumber, nails, tin foil, and tools are left up to the builder.

Flat plates: liquid

SUNCEIVER/35775
manufactured by:
Halstead & Mitchell
P.O. Box 1110
Scottsboro, Ala. 35768

A fully insulated solar collector using copper tube aluminum-finned surface. Double glazing incased with an aluminum frame. **Features and options:** total absorbing fin surface is equal to 174 square feet. **Maintenance requirements:** occasional cleaning of glass. **Price:** $265.

SUNMAT
manufactured by:
Calmac Manufacturing Corp.
150 S. Van Brunt St.
Englewood, N.J. 07631
John M. Armstrong; phone: (201) 569-0420
The Sunmat is a low-cost, lightweight, flexible, freeze-tolerant, mat-type collector. The system uses a four foot wide flexible grid of rubber tubes which are cemented on site to an insulation bed and covered with plastic glazing. **Features and options:** Mat can run up to 125 feet long, which reduces problem of plumbing hookup. **Guarantee/warranty:** 5 year pro-rated warranty on tubing; 1 year warranty on all other components. **Maintenance requirements:** resurfacing of Kalwall glazing every 5-7 years. **Availability:** direct from factory. **Price:** $59/mat. $2.86/square foot.

STANDARD SOLAR COLLECTOR/SD-5
manufactured by:
Solar Development, Inc.
4180 Westroads Drive
West Palm Beach, Fla. 33407
Don Kazimir; phone: (305) 842-8935
SD-5 collectors are available in 4 x 10 foot and 2 x 10 foot sizes. They include copper absorber plates and either one or two layers of Kalwall Sun-Lite premium glazing (.025 inches). The collectors designed for 30 psf or 48 psf wind loading and can be supplied with full roof mounting hardware. Insulation is Technifoam 1 or 2 inch thickness. **Features and options:** a variety of sizes up to 4 x 14 foot are available, double or single glazed. Also, series or parallel flow patterns can be ordered depending on collector orientation (vertical or horizontal mounting). **Installation requirements:** horizontal or vertical mounting depending on application and roof pitch. Lag screws are provided to fasten brackets and pitch pans to roof rafters. **Guarantee/warranty:** limited warranty is for 5 years for hot water and space heating; one year for pool heating. **Maintenance requirements:** not given. **Manufacturer's technical services:** systems design services is provided at no cost when SDI products are used, unless the engineering is particularly extensive. **Regional applicability:** unlimited. **Price:** $347 for 4 x 10 foot single glazed.

UNIVERSAL 100 THERMOTUBE, 45 BG & 410BG
manufactured by:
Southern Lighting Manufacturing
501 Elwell Ave.
Orlando, Fla. 32803
Kevin J. Drew; phone: (305) 894-8851

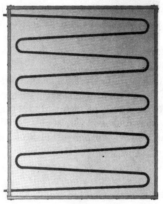

The Thermotube has copper tubing continuously soldered to an all copper plate, with inert high K insulation; housed in an insulated heavy-gauge aluminum frame with 3/16 inch tempered glass or Kalwall glazing in 4 x 5 foot or 4 x 10 foot units. **Features and options:** no soldered joints inside the collector. Single or double glazing with either tempered glass or Kalwall .040 Premium is optional. 4 x 5 foot or 4 x 10 foot units. **Installation requirements/considerations:** designed to withstand 130 mph winds. Mounting instructions given in installation manual. **Guarantee/warranty:** warranted against defects in material and workmanship for five years. Warranty is limited to repair and replacement of defective unit, except glass breakage and freeze damage. **Maintenance requirements:** drainage may be required in freezing conditions. **Manufacturer's technical services:** advise on mounting and sizing. **Regional applicability:** can be used in all regions. **Available from:** Hughes Supply, Inc throughout Florida. **Price:** $10-12 per square foot.

FLAT PLATE COLLECTORS/AP-18 and AP-6
manufactured by:
The Solaray Corp.
2414 Makiki Hgts. Dr.
Honolulu, Hawaii 96822
Lawrence Judd; phone: (808) 533-6464
Anodized aluminum panels with all copper absorber plate, header pipes and runner pipes. Nickel black chrome selective surface. Silver solder brazing all pipe connections. Single glazing frame for glass or plastic laminates using silicon for attachment. **Features and options:** two sizes available — 34¼ x 76¼ inches and 18 x 48 inches. **Guarantee/warranty:** 5 year warranty. **Maintenance requirements:** none. **Manufacturer's technical services:** on request. **Regional applicability:** southern climates due to single glazed panel. **Availability:** local supplies or direct factory shipments. **Price:** direct shipment app. $6/square foot (without glazing) depending on quantity.

LOF SUNPANEL
manufactured by:
Libbey-Owens Ford Glass Co.
1701 E. Broadway St.
Toledo, Ohio 43605
Marty Wenzler; (419) 247-4350
The LOF SunPanel is an all copper plate thermally isolated in an all welded aluminum frame available with selective or non-selective coatings, regular or low-iron glass. **Features and options:** Black Chrome or flat black paint, regular or low iron or water white glass, double or single glazing, mounting hardware. **Installation requirements/considerations:** designed to be supported from the four mounting brackets, permitting a variety of mounting arrangements without disturbing the integrity of the collector housing. Direct contact of the collector housing with dissimilar metals must be avoided to prevent galvanic corrosion. **Guarantee/warranty:** 1 year warranty. **Maintenance requirements:** none. **Manufacturer's technical service:** computer analysis, technical assistance as needed during installation. **Regional applicability:** collector is designed to withstand any environmental conditions. **Price:** on request depending on options.

FLAT PLATE COLLECTOR SHS/00L1
manufactured by:
Solar Home Systems, Inc.
12931 West Geauga Trail
Chesterland, Ohio 44026
Joseph Barbish; phone: (216) 729-9350
Thirty square foot low temperature collector for space heating and domestic hot water. **Guarantee/warranty:** limited warranty to original purchaser that collector shall be free from defects in material and workmanship for two years. **Manufacturer's technical services:** design and engineering. **Regional applicability:** North of Mason-Dixon line. **Price:** consult factory.

SOLECTOR

manufactured by:
Sunworks, Division of Enthone, Inc.
P.O. Box 1004
New Haven, Conn. 06508
Floyd C. Perry, Jr.; phone: (203) 934-6301

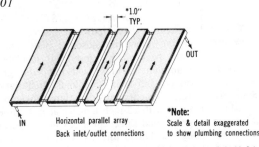

*1.0" TYP.

OUT

IN

Horizontal parallel array
Back inlet/outlet connections

*Note: Scale & detail exaggerated to show plumbing connections

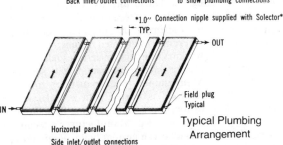

1.0" Connection nipple supplied with Solector

*1.0" TYP.

OUT

IN

Field plug Typical

Horizontal parallel
Side inlet/outlet connections

Typical Plumbing Arrangement

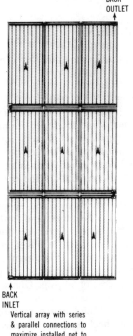

BACK OUTLET

BACK INLET

Vertical array with series & parallel connections to maximize installed net to gross ratio

Sunworks' liquid cooled, factory assembled Solector solar energy collector has been developed for residential, commercial, industrial, and institutional building systems that require high thermal efficiency, long-term performance, and minimum installed cost per Btu delivered. **Features and options:** available in draindown or internal-manifold versions with option of two sizes, single or double glazing, and choice of three types of finishes. **Installation requirements/considerations:** internal manifold Solector available with several types of plumbing connections to minimize the amount of external plumbing hardware. **Guarantee/warranty:** five year materials and workmanship guarantee (with the exception of cover glass breakage); entire guarantee statement is available upon request. **Maintenance requirements:** little, if any, servicing should be required, but in the event that it is, the glass cover may be removed or replaced independent of adjacent modules. **Manufactuter's technical services:** architectural and engineering staff, through local representatives, is available to provide consulting services to qualified architects, engineers and contractors involved in solar system design. **Regional applicability:** all regions. **Availability:** a list of all representatives is available on request from Sunworks or by calling Sweets Catalogue toll-free Buyline number: (800) 255-6880. **Price:** contact representatives.

SOLAR-PAK/SG-18-P AND DG-18-P

manufactured by:
Raypak, Inc.
31111 Agoura Road
Westlake Village, Cal. 91361
H. Byers or A. Boniface; phone: (213) 889-1500

Aluminum heat absorbing surface with copper waterways for domestic water heating, swimming pool heating, and central heating in climates that do not fall below 30°F; 20.7 square feet. **Features and options:** floated absorber, heavy duty insulation, tempered low-iron glass, replaceable desiccant. **Installation requirements/considerations:** installation and operating manuals available. **Guarantee/warranty:** one year on materials and workmanship. **Regional applicability:** representation in all states and international markets. **Availability:** from inventory. **Price:** SG-18-P: $306, DG-18-P: $360.

SERC LIQUID COLLECTOR MODULE

manufactured by:
Solar Energy Research Corp.,
701B South Main St.
Longmont, Colo. 80501
James B. Wiegand, president; (303) 772-8406

Modular panel in redwood and galvanized frame — 2 feet x 8, x 10, x 12, or x 16 feet. All copper absorber construction; 6 inches of fiberglass insulation; factory tested. **Features and options:** full wetted area collector for high efficiency; removable spray tubes; selective lacquer; silicone sealants; Lexan cover for damage resistance; light-weight: 3 lbs. per square foot. **Installation requirements/considerations;** gravity return flow to heat storage tank; collector mounting hardware adapts to roof or wall easily; no special skills to install. **Guarantee/warranty:** one year. **Maintenance requirements:** freeze-proof self-draining design allows use of pure water; clean nozzles through access to end caps; integral strainer eliminates maintenance. **Manufacturer's technical services:** custom engineered systems. See resources listing. **Regional applicability:** unlimited. **Availability:** ninety day delivery — 30 percent down on custom sizes. **Price:** $10 per square foot, F.O.B. factory.

SUNEARTH SOLAR COLLECTOR/ 3296 AND 3597A

manufactured by:
Sunearth Solar Products Corp.
RD 1, Box 337
Green Lane, Pa. 18054
H. Katz; phone: (215) 699-7892

Sunearth solar collectors come in two basic models. The first is a modular skylight mounted type (field assembly required). The second is a fully assembled box type flat plate collector for mounting on existing roofs or frames. Both collectors use water or antifreeze solutions as the heat transfer medium. **Features and options:** double glazing consisting of heat, scratch and craze resistant acrylic and heat proof Teflon film. The absorber plate is formed and coated for optimal heat transfer. Waterways are all copper with brazed connections. **Installation requirements/considerations:** model 3296 is shipped unassembled; 3597A is shipped completely assembled ready for installation. **Guarantee/warranty:** limited five year warranty on defects in material and workmanship from date of installation. **Maintenance requirements:** in case of accidental damage or vandalism glazing and collectors can be easily replaced or repaired with Sunearth Factory Parts and Service. **Manufacturer's technical services:** manual for sizing collectors in domestic hot water system. **Regional applicability:** all U.S. regions, Europe, Middle East. **Price:** contact sales department.

SUNAID

manufactured by:
Revere Copper and Brass Inc.
Solar Energy Dept.
P.O. Box 151
Rome, N.Y. 13440
William J. Heidrich; phone: (315) 338-2401

Revere Modular Collectors are prepackaged and prepiped units for retrofit applications and flat roof construction. **Features and options:** single and double glazed units are available with paint, cuprous oxide, or black chrome coating on absorber plate, units are easily hooked to plumbing by means of ½ inch internal FTP connections. **Installation requirements/considerations:** units should be securely fastened and protected against freeze-up. **Guarantee/warranty:** five years on all components except glass. **Maintenance requirements:** periodic checking to see that glass has not been obstructed or is not excessively dirty. **Manufacturer's technical services:** trained staff to provide technical advice related to design and application of the collectors. **Regional applicability:** equal for all areas of the country. **Availability:** nationwide through authorized distributors and dealers. **Price:** available upon request.

SUN ROOF

manufactured by:
Revere Copper and Brass Inc.
Solar Energy Dept.
P.O. Box 151
Rome, N.Y. 13440
William J. Heidrick; phone: (315) 338-2401

Unit was developed from the Revere laminated-panel system for use on new construction to function both as a weathertight copper roof and as an efficient solar energy collector. The owner/contractor supplys roof trusses, insulation, flashing and trim details, and supply and return piping. **Features and options:** system takes either single or double glazing; circuiting may be either grid or sinous. **Installation requirements/considerations:** freeze protection and pressure relief in excess of 50 psig. **Guarantee/warranty:** five years on all components except glass. **Maintenance requirements:** periodic checking to see that glass has not been obstructed or is not excessively dirty. **Manufacturer's technical services:** trained staff to provide technical advice related to design and application of the collectors. **Regional applicability:** equal for all areas of the country. **Availability:** nationwide through authorized distributors and dealers. **Price:** available upon request.

SOLAR COLLECTOR

manufactured by:
SolarKit of Florida, Inc.
1102 139th Ave.
Tampa, Fla. 33612
Wm. Denver Jones; phone: (813) 971-3934

Collectors have redwood frame, copper tubing, aluminum absorber plate, aluminum back plate, closed-cell insulation, aluminum mounts, and tempered glass cover. 3x5 feet; 50 pounds. **Installation requirements/considerations:** collector is made with a 2-inch-long leg at each corner, pre-drilled to accept a 3/8-inch bolt, and can be bolted directly to roof or to any sturdy frame. Because of light weight, roof reinforcement not required. **Guarantee/warranty:** warranted against defects in workmanship and materials under normal use for three years from date of purchase. **Maintenance requirements:** minimal; redwood is treated with a sealer at the factory; occasional painting or staining is recommended. **Manufacturer's technical services:** can custom build. **Regional applicability:** because of single glazing, present collectors are suitable mainly for southern half of U.S.; however, design can easily be altered to allow for double glazing; present insulation is adequate for northern climates. **Availability:** directly from factory, or Bob Vijil, Sungard-Systems, 528-B E. Brandon Blvd., Brandon, Florida 33511. **Price:** $125 to 150 per collector.

DAYSTAR '20'

manufactured by:
Daystar Corp.
90 Cambridge St.
Burlington, Mass. 01803
Paul P. Chaset; phone: (617) 272-8460

Copper flat plate liquid collector with 21 square feet of net absorber area. **Features and options:** straight-through internal manifolding eliminates reverse return and separate headers. Unistrut mounts for simple roof or stand mounting. **Installation requirements/considerations:** available in left-or right-hand supply and return. **Guarantee/warranty:** one year, parts and labor on defects in manufacture. **Maintenance requirements:** minimal. **Manufacturer's technical services:** full engineering and computer sizing. **Regional applicability:** all regions. **Availability:** send for name of closest dealer. **Price:** write for quote.

FLAT PLATE COLLECTOR

manufactured by:
O.E.M. Products, Inc.
Solarmatic Div.
220 W. Brandon Bvld.
Brandon, Fla. 33511

Absorber plate is aluminum with imbedded copper tubes. Case is extruded, anodized aluminum. **Features and options:** single or double glazing; serpentine or manifold grid pattern. **Price:** consult manufacturer.

MARK M FLAT PLATE COLLECTOR

manufactured by:
Mark M Manufacturing
R.D.#2, Box 250
Rexford, N.Y. 12148
Mark Urbaetis; phone: (518) 371-9596
Offer a variety of units using aluminum or copper absorber plates with flat black or selective coatings; housed in steel pans and glazed with Kalwall or glass. Polycarbonate available upon request. **Features:** pool heater using polycarbonate also available. **Installation requirements/considerations:** all collectors can be mounted using standard techniques of building construction. **Guarantee/warranty:** five years limited warranty on flat plates using copper manifolds. **Maintenance requirements:** limited to corrosion and Ph requirements. **Manufacturer's technical services:** design and checks and pH requirements. **Manual applicability:** entire northeast, especially eastern N.Y. and western Vermont and Massachusetts. **Price:** from $200 up.

COLUMBIA REDI-MOUNT COLLECTOR/77-3376 and 76-3496

manufactured by:
Columbia Chase Solar Energy Div.
55 High St.
Holbrook, Mass. 02343
Walter H. Barrett; phone: (617) 767-0513
This collector is designed for high performance and ease of installation. It is composed of a one-piece molded fiberglass frame, copper collector plate, and shatterproof glazing. There is a full perimeter mounting flange built into the frame. **Features and options:** light weight—45 pounds. Options include selective coating, double glazing, rails filled with foam. **Installation requirements/considerations:** the collector should face true south at a 45 degree angle for domestic hot water heating. For home heating the angle should be latitude plus 10. They can be installed on a roof or on the ground. **Guarantee/warranty:** one year. **Maintenance requirements:** minimal, periodic inspection. **Manufacturer's technical services:** all drawings and instructions included. Assistance given in estimating project equipment requirements. **Regional applicability:** shipping worldwide. **Availability:** see distributors section. **Price:** $306.

SOLARGIZER SOLAR COLLECTOR

manufactured by:
Solargizer Corp.
220 Mulberry St.
Stillwater, Minn. 55082
William Olson; phone: (612) 439-5734
Copper solar collectors, 48 inches by 96 inches, designed for long life; fiberglass case and cover; aluminum frame. **Guarantee/warranty:** One year on collector, longer on specific components. **Regional applicability:** all areas. **Price:** consult factory.

PAC OPEN FLOW COLLECTOR

manufactured by:
PAC, a division of
People/Space Co.
49 Garden Street
Boston, Mass. 02114
Robert Shannon, phone: (617) 742-8652

An open flow collector with a single cast body of polyester resin, a gel-coat of conductive plastic, single or double Kalwall cover plates. No metal parts, corrosion free, self-draining. **Options:** fire retardant or non-fire retardant resins, color of sides and back, one or two cover plates. **Feature:** can use pure water for efficient, trouble-free collection. **Installation requirements /considerations:** lightweight unit requires minimal structure easily plumbed with CPVC pipe where non-potable water system is used. **Guarantee/warranty:** one year. **Maintenance requirements:** replace cover plates after 12 to 15 years for higher efficiency. **Regional applicability:** any region of USA. **Price:** $216 per collector; $6.75/square foot

SOL-RAY SOLAR COLLECTOR

manufactured by:
Unit Electric Conrol,Inc./Sol-Ray Division Division
130 Atlantic Dr.
Maitland, Fla. 32751
Maurice S. Stewart, phone:(305)831-1900
All copper collector plate and tubing, high-strength aluminum housing, high temperature polyurethane foam insulation, tempered glass, for domestic hot water, swimming pool and space heating. **Installation requirements/considerations:** should be done by a plumber or someone familiar with solar water heater installation. **Guarantee/warranty:** three years from date of installation on materials and workmanship, excluding paint and glass. **Maintenance:** none. **Manufacturers technical services:** consulting. **Regional applicability:** no regional limitations — should be used in conjunction with a heat exchanger or drain down system in areas that have extended periods of freezing weather. **Price:** $250 (discounts available on volume purchases).

MEDIUM TEMPERATURE SOLAR COLLECTOR

manufactured by:
Chamberlan Manufacturing Corp.
R & D Div., P.O. Box 2545
East 4th and Esther Streets
Waterloo, Iowa 50705
William H. Sims; phone: (319) 232-6541

Solar collectors adaptable to a wide range of systems. The units have been independently tested in compliance with ASHRAE 93-77. Complete printed material is available. **Features and options:** two sizes, 3 by 7 or 3 by 8 feet, fully assembled modules. The absorber plate is steel construction with approximately 95 percent wetted surface. Black paint and black chrome coatings; one or two glazings.

Installation requirements/considerations: installation mounting points are provided on the collector. Normal installation would require a flat surface spaced about 7 feet apart to coincide with the ends of the collector. **Guarantee/warranty:** one year on material and workmanship. **Maintenace requirements:** normal routine fluid maintenance program. **Manufacturer's technical services:** engineering staff and computer analysis capabilities. **Regional applicability:** national. **Availability:** manufacturer's representatives. **Price:** jobs individually quoted.

SUNWAVE 420 SOLAR COLLECTOR

manufactured by:
Acorn Structures, Inc.
P.O. Box 250
Concord, Mass. 01742
Stephen V. Santoro; phone (617) 369-4111
Designed for simple, cost-effective use in low temperature systems. Collectors are 4 by 20 feet, reducing the number of required connections, flashing, and seals to minimize installation costs. Each collector has a net area of 70 square feet. Absorber plates are aluminum with copper tubes attached mechanically. **Features and options:** self-draining, light weight, fewer connections minimize leaks, can stagnate without damage, internal manifolding. **Guarantee/warranty:** two years. **Maintenance requirements:** washing of glazing material as required by environmental conditions. **Manufacturer's technical services:** available for a fee on customer's installations. **Regional applicability:** anywhere the sun shines. **Price:** $10 per square foot.

SOLAR HEAT EXCHANGER

manufactured by:
Fafco Inc.
138 Jefferson Drive
Menlo Park, Cal. 94025
Larry Hix; phone: (415) 321-6311
Unglazed solar collector for heating swimming pools, hot water, aquaculture, hydroponics, and other low temperature requirements. **Features and options:** panels are modular and can be clamped together to achieve any desired collector area, making it possible to heat any size pool. Inside the panel's plastic extrusion are hundreds of parallel channels that allow the water to pass under virtually every square foot of panel area exposed to the sun. **Installation requirements/considerations:** compatible with existing pool equipment and plumbing. **Guarantee/warranty:** five year limited warranty when defective parts are returned to the factory. **Maintenance requirements:** panels should be drained in the winter in areas where the temperature drops below 25°F. **Manufacturer's technical services:** factory trained dealer-installers. **Price:** consult manufacturer.

SUNSPOT SOLAR COLLECTOR

manufactured by:
El Camino Solar Systems
5330 Debbie Lane
Santa Barbara, Cal. 93111
Allen K. Cooper; phone: (805) 964-8676
Flat plate collector, aluminum absorber fin, copper waterways, Tedlar-coated acrylic cover. **Features and options:** design to resemble an active low profile skylight. Counter flashing moldings to become available in mid-1977. **Installation requirements/considerations:** low weight (100 pounds) simplifies handling by two men on pitched roof. **Guarantee/warranty:** five year limited guarantee. **Maintenance requirements:** none. **Manufacturer's technical services:** consultation for special installation problems. **Regional applicability:** unlimited. **Availability:** currently from manufacturer only. National distribution expected in 1977. **Price:** $296—FOB Santa Barbara.

COPPER SOLAR COLLECTOR

manufactured by:
Ametek, Inc., Power Systems Group
One Spring Ave.
Hatfield, Pa. 19440
Frank W. Gilleland; phone: (215) 822-2971
The Ametek copper flat plate solar collector is liquid-cooled. Selective surface can produce high temperature (200°F plus) water for heating and cooling applications. **Features and options:** selective coating. **Guarantee/warranty:** limited warranty. **Maintenance requirements:** none known

BURTON SOLAR COLLECTOR

manufactured by:
Burton Industries, Inc.
243 Wyandanch Ave.
North Babylon, N.Y. 11704
Burton Z. Chertok; phone: (516) 643-6660

Liquid flat plate collector incorporating modular absorber plate of tongue and groove aluminum extrusions lined with copper (see Sunstrip absorber plate). Absorber plate is mechanically strong and resistant to thermal stresses and has good fin efficiency and liquid side heat transfer. Glazing is two covers of Kalwall. **Features and options:** Optional cover plate materials. **Installation requirements/considerations:** standard for liquid flat plate collectors. **Guarantee/warranty:** one year. **Maintenance requirements:** routine inspection. **Manufacturer's technical services:** applications and technical consulting. **Regional applicability:** no limitations. **Availability:** from manufacturer. **Price:** $10.25 per square foot basic.

SOLAR COLLECTOR
PANEL/SC-200

manufactured by:
Solar Innovations
412 Longfellow Blvd.
Lakeland, Fla. 33801
Ron Yachabach; phone: (813) 688-8373
3¼ by 22¾ by 96¼ inch flat plate liquid panel with copper Roll-Bond absorber plate and Kalwall glazing. **Features and options:** a universal mounting bracket is available to make installation on any roof possible. **Guarantee/warranty:** five year guarantee, void if used with corrosive liquids. **Price:** consult factory.

FLAT PLATE COLLECTOR

manufactured by:
Payne, Inc.
1910 Forest Drive
Annapolis, Md. 21401
Fred W. Hawker; phone: (301) 268-6150
Flat plate trickle collector, vacuum-cast from ultra-high-strength lightweight concrete. Single or double glazing of Tedlar/Teflon film. Designed for low cost and long life; 2 feet wide x 7 feet high. Weighs approximately 55 pounds. **Installation requirements/considerations:** manifolds are external, so only single supply and return pipes are required. Weight is approximately 4 lbs per square foot. **Guarantee/warranty:** two years against defects. **Maintenance requirements:** none. **Regional applicability:** will install in Washington, D.C. and Baltimore area. Will ship to anywhere in the continental U.S. **Price:** $70.

SOLARGENICS COLLECTOR/
SERIES 76-77

manufactured by:
Solargenics, Inc.
9713 Lurline Ave.
Chatsworth, Cal. 91311
Rowen Collins; phone: (213) 998-0806
The Series 76 flat plate collector (selective or non-selective) makes use of structural aluminum and copper fluid channels to achieve low installed weight and long service life. The collector can be provided in any length up to 25 feet providing maximum installed collector in any area. **Features and options:** each collector has an integral control/instrumentation sensor and an automatic air bleed device. Double glazing is available as an option. Clear, bronze or other anodize finishes optional. **Installation requirements/considerations:** collector is a structural unit—the 10 foot unit spans 10 feet, the 15 foot unit spans 13 feet, the 20 foot unit spans 16 feet. Mounting hardware is available. Installed weight is 5.2 pounds per square foot. **Guarantee/warranty:** five year limited warranty. **Maintenance requirements:** none. **Manufacturer's technical services:** complete installation and service available on residential projects. Full on-site support on industrial projects. **Regional applicability:** worldwide. **Availability:** distributors and representatives. **Price:** contact factory for nearest distributor.

SERIES 200 SOLAR COLLECTORS

manufactured by:
Energy Converters, Inc.
2501 N. Orchard Knob Ave.
Chattanooga, Tenn. 37406
David Burrows; phone: (615) 624-2608
Collectors consist of low-iron glass cover plates, aluminum or copper tube-in-sheet, commercial flat black coating or proprietary selective surface, and durable, coated galvanized steel pan, with 4 inches of high temperature, low binder fiberglass insulation. Size is 3 x 8 foot x 5 inch. **Features and options:** drain-back operation; single or double glazing. **Guarantee/warranty:** 5 years. **Maintenance requirements:** dependent on system application. **Manufacturer's technical services:** available on request. **Regional applicability:** all climates. **Availability:** through distributors or direct from factory. **Price:** $284-$396.

SEP FLAT-PLATE COLLECTOR, CU30

manufactured by:
Solar Energy Products, Inc.
1208 N.W. 8th Ave.
Gainesville, Fla. 32601
The CU 30 collector emphasizes easy installation, simplified servicing, and high performance. Design exceeds the HUD, NASA, and NBS standards and is approved by ERDA for federally-funded projects. Size: 29.65 square feet. Typical applications include open and closed water, space, pool, and process heating systems. **Features and options:** structural integrity—wind loads to 139. Thermal performance stability to 300°F. Design life—thirty years; working pressures to 150 psi; water-white glass. Choice of color. **Installation requirements/considerations:** standard orientation, tilt, non-shading requirements; may be mounted vertical or horizontally. **Guarantee/warranty:** five years. **Maintenance requirements:** automatic operation; no routine maintenance. **Manufacturer's technical services:** technical assistance. **Regional applicability:** regional limitations. **Price** $325.

SOLAR PANEL/CUS30 and CUP30

manufactured by:
Gulf Thermal Corp.
P.O. Box 13124
Airgate Branch
Sarasota, Fla. 33578
Dudley Slocum; phone: (813) 355-9783

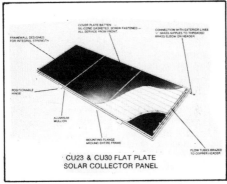

CU23 & CU30 FLAT PLATE
SOLAR COLLECTOR PANEL

Panels are constructed for rugged, dependable performance, long life, and minimum maintenance (access to all components through top of panel with simple hand tools). Panels are designed to withstand no-flow temperatures up to 300 degrees F. **Features and options:** components for flat or pitched roof mounts, fixed angle or adjustable available from collector manufacturer. **Guarantee/warranty:** guaranteed against all defects in construction (except glass and copper damage due to aggressive water) for a period of five years. Gulf Thermal will replace any defective parts free of charge. **Maintenance requirements:** none. **Availability:** Solar Energy Products, Inc., Gainesville, Fla. and Solar Service, Inc., Hendersonville, N.C. **Price:** consult manufacturer.

FLAT PLATE COLLECTOR

manufactured by:
O.E.M. Products, Inc.
Solarmatic Div.
220 W. Brandon Blvd.
Brandon, Fla. 33511

Absorber plate is aluminum with embedded copper tubes. Case is extruded, anodized aluminum. **Features and options:** single or double glazing; serpentine or manifold grid pattern. **Price:** consult manufacturer.

FLAT PLATE SOLAR COLLECTORS

manufactured by:
PPG Industries Inc.
One Gateway Center
Pittsburgh, Pa. 15222
Phone: (412) 434-3555

PPG's flat plate solar collectors may be specifically designed to meet a particular need. Extensive computer programs are designed to assist the engineer in determining the most cost effective units. **Features and options:** number and type of glazings are open; either aluminum or copper absorber plates; selective or non-selective surface; three configurations and three pipe options. Total possible variations over 250. **Guarantee/warranty:** two year limited warranty. **Maintenance requirements:** none. **Maufacturer's technical services:** computer-backed sizing and installation at this time. **Regional applicability:** continental U.S. **Availability:** direct from Ametek. **Price:** quotes on request.

DELTA T/MODEL 15

manufactured by:
Delta T Company
2522 West Holly St.
Phoenix, Ariz. 85009
James L. Hoyer; phone (602) 272-6551

Copper tubes mechanically bonded to aluminum fins designed for high operating pressures: factory leak tested at 200 psig: 17.138 square feet per absorber. **Features and options:** copper tubes and copper headers to eliminate internal corrosion problems; aluminum fins completely surround copper tubes; al copper tube connections made with silver solder. **Installation requirements/considerations:** connections to copper heads may be made effectively with soft copper tubing. Absorbers weigh approximately 1.35 pounds per square foot; design permits field connections through side, top, or rear of collector casing; either face of absorber may be positioned to face sun. **Guarantee/warranty:** Twenty months on materials and workmanship. **Availability:** avilable for shipment. **Price:**

quantity	unit price	app. price per square foot
25 to 99	$94.79	$5.50
100 to 199	$86.68	$5.03
200 to 299	$77.71	$4.51
300 to 399	$74.17	$4.30
400 or more	$73.72	$4.28
24 or less	$99.10	$5.75

design assistance. **Regional applicability:** worldwide. **Availability:** distributors. **Price:** $160 to $370 each: $8.89 to $20.56 per square foot.

SUNSTREAM SOLAR COLLECTOR/ 50

manufactured by:
Sunstream, a Div. of Grumman Houston Corp.
P.O. Box 365
Bethpage, New York 11714

A 25 square foot net area liquid cooled collector with an arched acrylic cover plate. **Features and options:** aluminum or aluminum with copper tube absorber plates. **Price:** consult factory.

ALL SUNPOWER/3784

manufactured by:
All Sunpower, Inc.
10400 S.W. 187th St.
Miami, Fla. 33157
E.M. Kramer; phone: (305) 233-2224

Closed-loop flat plate collector for use with non-toxic, non-freezing, non-corroding heat transfer fluid. Aluminum absorber and tubing. **Installation requirements/con-**

LENNOX MODULE/LSC18

manufactured by:
Lennox Industries Inc.
P.O. Box 250
200 South 12th Ave.
Marshalltown, Iowa
Phone: (515) 754-4011

The Lennox LSC18 solar collector module is a glass-covered, selectively-coated copper tube-steel plate absorber, adaptable to any type and size of solar system. **Features and options:** available with single (LSC18-1S) or double (LSC18-1) tempered low iron glass covers. The glass is etched to enhance solar transmission and the absorber is

siderations: standard for flat plate collectors. **Guarantee/warranty:** ten years. **Manufacturer's technical services:** available. **Regional applicability:** nationwide. **Price:** consult manufacturer.

REYNOLDS SOLAR COLLECTOR

manufactured by:
Reynolds Metals Co.
Torrance Extrusion Plant
2315 Dominguez St.
Torrance, Cal. 90508

4 by 8-foot flat plate collector with integrally finned tube aluminum absorber plate. Flat black non-selective paint; two layers of film glazing. **Features and options:** continuous serpentine extruded absorber plate. **Installation requirements/considerations:** may be installed in parallel or series. All connections of the aluminum collectors to dissimilar metal piping must be done by use of non-metallic fittings to minimize the possibility of galvanic corrosion. **Guarantee/warranty:** one year. **Manufacturer's technical services:** available on request. **Regional applicability:** all regions. **Price:** $184 to $264 each FOB Torrance, Cal. (Price based on quantity).

coated with a highly selective black chrome. **Installation requirements/considerations:** two-position mounting brackets on each corner of the enclosure permit installation in separate supports or frames constructed of wood or metal. Collectors can be installed individually or in multiple banks end-to-end and/or side-by-side, and assembled in parallel, series, or series-parallel combinations. **Guarantee/warranty:** one year after date of delivery; or one year after date of installation if installed by Honeywell. **Availability:** through a large network of Lennox distributors. **Price:** consult Lennox.

SOLAR COLLECTOR PANELS/200 G

manufactured by:
Alten Corporation
2594 Leghorn St.
Mountain View, Cal. 94043
Klaus Heinmann; phone: (415) 969-6474
150° to 200°F solar collector with extruded aluminum-finned plate, copper tube imbedded. **Features and options:** single or double glazed. **Installation requirements/considerations:** simple. **Guarantee/warranty:** two year full part and labor. Ten years against leakage. **Regional applicability:** western U.S. **Price:** $430 per unit; $15.36 per square foot.

THE NORTHRUP FLAT PLATE SOLAR COLLECTOR/NSC-FPIT

manufactured by:
Northrup, Inc.
302 Nichols Dr.
Hutchins, Tex. 75141
E.C. Ricker; phone: (214) 225-4291
Each flat plate has 22 square feet net collecting area. Copper piping carries fluid heated by black aluminum absorber. Single glazing of Tedlar, with no-glazing option. **Features and options:** available in hot water and space heating systems, with built-in freeze protection. Framing and mounting apparatus available as an option. **Installation requirements/considerations:** systems come complete except for copper tie-in tubing; requires floor or similar drain and 120 volt hook-ups. **Guarantee/warranty:** eighteen months from shipment or twelve months from installation, whichever occurs first. Limited warranty. **Maintenance requirements:** negligible. **Manufacturer's technical services:** inquire regarding specific jobs. **Regional applicability:** works best in moderate freeze or non-freezing environments. **Availability:** contact factory. **Price:** consult manufacturer.

COLE SOLAR COLLECTOR/410AT or 410A

manufactured by:
Cole Solar Systems
440A East St. Elmo Rd.
Austin, Texas 78745
Warren Cole; phone: (512)444-2565
Copper tube collector with spot welded aluminum fins; flat black surface. Outer glazing of ⅛ inch acrylic; inner glazing of FEP Teflon. 4×10 feet or 4×8 feet. **Features and options:** 3½ inches of mineral fiber insulation standard; 410AT is double glazed; 410A is single glazed with acrylic only. **Installation requirements/considerations:** mounts with four bolts; adjustable mounting brackets available. **Guarantee/warranty:** five years on parts and labor if returned to factory. **Maintenance requirements:** virtually none. **Regional applicability:** nationwide. **Availability:** 3 weeks delivery. **Price:** consult manufacturer.

MARK V HYDRONIC COLLECTOR

manufactured by:
Solar Corporation of America
100 Main St.
Warrenton, Va. 22186
Walter Sutton; phone: (703) 347-7900
3 feet x 6.5 feet x 5 inches collector panel with copper tubed aluminum fin absorber plate enclosed in a galvanized steel container and insulated to an R-15 value. The outer glazing is low-iron tempered glass. Collector incorporates system supply and discharge manifolding. **Features and options:** internal manifold ¾ inch OD - 2½ inches OD copper tube; selective absorber coatings available, t=.91, glass seals and manifold hose connections are high temperature rubber. **Installation requirements/considerations:** collector is compatible with both wood and steel supports. Install collectors with 1 inch drop per 20 foot horizontal run to take advantage of gravity drain. **Guarantee/warranty:** one year warranty from date of delivery to the buyer against defects in materials and workmanship and to perform satisfactorily. **Maintenance requirements:** change system filter to prevent dirt and sludge buildup in collector. **Manufacturer's technical services:** system design and costing; installation and custom collector design. **Regional applicability:** continental United States. **Price:** 1-10 units; $10.95 per square foot; 11-50 units; $9.08 per square foot; 50+ units: factory pricing.

MARK III HYDRONIC COLLECTOR

manufactured by:
Solar Corporation of America
100 Main St.
Warrenton, Va. 22186
Walter Sutton; phone: (703) 347-7900
4 foot x 8 foot x 5 inches collector panel with copper-tubed aluminum fin absorber plate enclosed in a galvanized steel container and insulated to an R-15 value. The outer glazing is low-iron tempered glass. Collector incorporates system supply and discharge manifolding. **Features and options:** internal manifold ¾ inch OD - 2½ inch OD copper tube; selective absorber coatings available, t=.89, glass seals and manifold hose connections are high temperature rubber. **Installation requirements/considerations:** collector is compatible with both wood and steel supports. Install collectors with 1 inch drop per 20 foot horizontal run to take advantage of gravity drain. **Guarantee/warranty:** one year warranty from date of delivery to the buyer against defects in materials and workmanship, and to perform satisfactorily. **Maintenance requirements:** change system filter to prevent dirt and sludge build up in collector. **Manufacturer's technical services:** system design and costing; installation and custom collector design. **Regional applicability:** continental United States. **Price:** consult factory.

Flat plate: air

SDI SOLAR AIR HEATER
manufactured by:
Solar Development, Inc.
4180 Westroads Drive
West Palm Beach, Fla. 33407
Don Kazimir; (305) 842-8935

The absorber plate for this air collector is a vitreous screen that causes the necessary turbulence for high performance. The panels are normally mounted in series parallel groups on a vertical wall facing south—with or without rock storage. **Features and options:** the air panels are particularly cost effective when mounted vertically on the south wall of a "day shift only" industrial building. Installation is inexpensive and mounting and ducting is simple. **Installation requirements/considerations:** mount two panels in series with each two-panel group in parallel. Air flow should be approximately 200-300 cfm per panel. **Guarantee/warranty:** three year limited warranty. **Maintenance requirements:** minimal. **Manufacturer's technical services:** engineering services available for system sizing and layout. **Regional applicability:** unlimited. **Price:** $210 for 29.5 square feet of net area.

SOLECTOR
manufactured by:
Sunworks, Division of Enthone, Inc.
P.O. Box 1004
New Haven, Conn. 06508
Floyd C. Perry, Jr.; phone: (203) 934-6301

Sunworks' air cooled, factory assembled solar energy collector has been developed for residential, commercial, industrial, and institutional building systems that require high thermal efficiency, long term performance, and minimum installed cost per Btu delivered. **Features and options:** available with two types of ducting, also available with single or double glazing, and choice of three finishes. **Installation requirements/considerations:** available with side-only duct connectors or side/back connections to minimize or eliminate need for external ducting. **Guarantee/warranty:** five year materials and workmanship guarantee (with the exception of cover glass breakage); entire guarantee statement is available upon request. **Maintenance requirements:** little if any servicing should be required, but in the event that it is, the glass cover may be removed or replaced without disturbing the collector unit installation or flashing; module may also be removed independent of adjacent module. **Manufacturer's technical services:** architectural and engineering staff, through local representatives, is available to provide consulting services to qualified architects, engineers, and contractors involved in solar system design. **Regional applicability:** all regions. **Availability:** a list of representatives is available on request from Sunworks or by calling Sweets Catalog toll-free Buyline number: (800) 255-6880. **Price:** contact representatives.

FLAT PLATE COLLECTOR SHS-00Ai
manufactured by:
Solar Home Systems, Inc.
12931 West Geauga Trail
Chesterland, Ohio 44026
Joseph Barbish; phone: (216) 729-9350
30.02 square foot air collector for space heating. **Guarantee/warranty:** limited warranty to original purchaser that the collector shall be free from defects in materials and workmanship for two years. **Manufacturer's technical services:** system design and engineering. **Regional applicability:** North of Mason-Dixon line. **Price:** consult factory.

SUNPOWER SELECTOR PANEL
manufactured by:
Sunpower Industries, Inc.
10837 S.E. 200th
Kent, Wash. 98031
Al Abramson; phone: (206) 854-0670
The Sunpower Solar Energy Selector Panel screens out ultra-violet rays and the long, low-frequency infra-red rays, selecting only the high intensity energy from the "white" light spectrum. This energy is transported in the passing air stream to a rock storage bin. **Features:** outer glazing is acrylic greenhouse fiberglass (4 oz.); DuPont Tedlar under-glazing (1 mil.). **Guarantee/warranty:** one year limited on materials and workmanship. **Installation requirements/considerations:** installed with typical construction materials and methods. Complete installation instructions included. **Price:** consult factory/see listing under complete system.

AIR COLLECTOR
manufactured by:
American Solarize Inc.
19 Vandeventer Ave.
Princeton, N.J. 08540
Joe Beudis; phone: (609) 924-5645
Air collector with aluminum honeycomb absorber panel to cut reflectivity and increase absorbing area. Sizes are 3x7 feet x 7 inches, or 4x8 feet x 7 inches. **Installation requirements/considerations:** HVAC contractor can install collector and related duct work; all necessary parts supplied. **Guarantee/warranty:** ten years. **Maintenance requirements:** none. **Manufacturer's technical services:** design and sizing services, field supervision when necessary. **Regional applicability:** well suited for humid zones, no temperature limitations. **Availability:** direct f.o.b. from factory. **Price:** $150 to $275.

SOLAR-KAL AIRHEATER
manufactured by:
Kalwall Corp.
1111 Candia Road
Manchester, N.H. 03103
Scott Keller; phone: (603) 668-8186

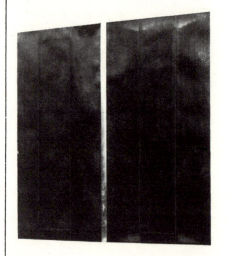

For use as an active (using blowers) or passive (thermosiphon) solar air collector. Designed to be installed on south facing walls or roofs. **Features and options:** constructed from non-corrosive materials; two basic ducting options; clamp-tight installation system. **Installation requirements/considerations:** insulation is field applied as and if required. Easily installed as the weather wall of a building as well as the solar collector. **Guarantee/warranty:** none implied or expressed. **Maintenance requirements:** At five year intervals the cover sheet may be refinished with Kalwall Weatherable Surface to provide surface erosion protection. **Manufacturer's technical services:** sales and technical staff are available to discuss customers requirements at the above phone number. **Regional applicability:** nationwide. **Price:** $4.31 to $5.81 per square foot.

SOL-AIRE PANELS

manufactured by:
McArthurs, Inc.
P.O. Box 236
Forest City, N.C. 28043
W.H. McArthur; phone: (704) 245-7223
Flat plate air collectors either 36 x 96 or 36 x 120 inches in size. Alcan .019-inch black baked enamel embossed absorber plate. **Features and options:** single or double tempered glass or Kalwall glazing. **Installation requirements/considerations:** mounting cleats and angles furnished. **Price:** consult manufacturer.

SOLAR HEAT COLLECTOR

manufactured by:
Custom Solar Heating Systems
P.O. Box 375
Albany, N.Y. 12201
Donald Porter; phone: (518) 438-7358
Flat plate air collector designed for mounting on the roof, ground-level area, or south wall of a building. **Features and options:** is directional for maximum heat attainment, can be ducted directly into hot air systems. **Guarantee/warranty:** one year on electrical components and five years on collector. **Maintenance requirements:** painting of collector if needed. **Regional applicability:** 200 mile radius of Albany, N.Y. for installation. **Price:** $495 for collector, installation and designing extra.

SILICONE RUBBER PRODUCTS

manufactured by:
Chase-Walton Elastomers, Inc.
27-S Apsley St.
Hudson, Mass. 01749
Steven W. Page, Jr.; phone: (617) 485-5600
Design and fabrication of custom elastomeric components including silicone rubber fabric-reinforced hose, gaskets, tubing, sealants, and die-cut parts. **Features and options:** many applications in solar systems; long-term thermal stability and ozone or weather resistance. **Manufacturer's technical services:** upon request. **Price:** consult factory.

SUN*TRAC SOLAR COLLECTORS/ 232, 296, 228 AND 260

manufactured by:
Future Systems, Inc. 12500 West Cedar Dr.
Lakewood, Colo. 80228
William V. Thomson; phone: (303) 989-0431
Vertical-vane, fixed flat plate air collector, glazed with two panes of 3/16 inch tempered glass. **Features and options:** extruded absorber plate surface for maximum absorption of solar energy. Modular units from 32 square feet to 160 square feet are available. Aluminum reflectors are optional. **Installation requirements/considerations:** wood supports at 4 foot intervals provide for structural mounting. Air ducts, 4 by 20 inches, can be provided with oval adapters to 9 inch circular duct. **Guarantee/warranty:** five years on material and workmanship except for glazing and reflector shields. **Maintenance requirements:** none. **Manufacturer's technical services:** engineering services; network of distributors, dealers. **Regional Applicability:** U.S. and Canada. **Availability:** contact Future Systems, Inc. **Price:** manufacturer's suggested retail prices: #232—4 feet wide, $600; #296—12 feet wide, $1,236; #228—16 feet wide, $1,596; #260—20 feet wide, $1,950.

THE NORTHRUP SUNDUCT™ SYSTEM/SDC-8 & SDC-12

manufactured by:
Northrup, Inc.
302 Nichols Dr.
Hutchins, Tex. 75141
E.C. Ricker; phone (214) 225-4291
Eight and 12-foot long air collectors, each 33 inches wide, heat air for space heating, make-up, drying, or process requirements. Solad woth necessary plenums and U.L. listed air handler. **Installation requirements/considerations:** ventilating contractor will fabricrate ductwork, set up air handling unit, and mount collectors on wall or roof. **Guarantee/warranty:** Eighteen months from shipment or 12 months from installation, whichever occurs first. Limited warranty. **Maintenance requirements:** negligible. **Manufacturer's technical services:** inquire regarding specific jobs. **Regional applicability:** worldwide. **Availbility:** consult factory. **Price:** consult manufacturer.

FLAT PLATE COLLECTORS/SERIES 2000

manufactured by:
Solaron Corporation
300 Galleria Tower
720 S. Colorado Blvd.
Denver, Colo. 80222
Ray T. Williamson
A flat-plate, double glazed, air heating collector with a porcelain enameled steel absorber plate. For space heating, service and domestic water heating, process heat such as agricultural product drying. **Guarantee/warranty:** a 10 year limited warranty covering collector performance and a 1 year limited warranty on equipment supplied by Solaron. **Maintenance requirements:** none. **Regional applicability:** all climates and geographic areas. **Price:** see distributor/dealer.

SERIES IV SOLAR COLLECTOR

manufactured by:
Contemporary Systems, Inc.
68 Charlonne St.
Jaffrey, N.H. 03452
The Series IV collector is a lightweight, structural unit suitable for roof or wall installation with any warm air solar space heating system. It has removable double glazing and optically blackened absorber plate mounted in an extruded aluminum chassis. **Features and options:** special double glazing system (patent pending) provides positive air seal without heat distortion of glazing materials; low cost per square foot Btu output. **Installation requirements/considerations:** shipped without insulation to allow site-specific application to be made as an integral, structural component of roof or wall that reduces costs and provides improved appearance in new construction; also suitable for retrofit applications; available in any length required (by 2 feet wide). **Guarantee/warranty:** limited one year warranty. **Maintenance requirements:** bi-annual visual inspection and cleaning if necessary. **Manufacturer's technical services:** computerized project analysis and component integration. **Regional applicability:** designed specifically for temperate Northeast latitudes but suitable for use in all temperate regions. **Availability:** currently available. **Price:** $5.85 per square foot active area, or $10.18 per linear foot (nominal 2 feet wide).

ADJUSTABLE MOUNTING BRACKET/MK-103

manufactured by:
Solar Innovations
412 Longfellow Blvd.
Lakeland, Fla. 33801
Ron Yachabach; phone: (813) 688-8373

Adjustable aluminum mounting bracket for solar water heating collector panels, for easy adjustment of collector angle. **Features and options:** adapts to any style collector panel, on any style roof. **Price:** consult factory.

DIFFERENTIAL THERMOSTATS
for solar systems

By Edward S. Peltzman

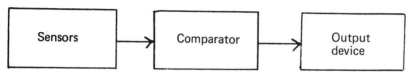

Sensors → Comparator → Output device

Fig. 1

The objective of any solar energy system, whether that system is intended for space heating or cooling, or for domestic hot water, is to collect as much energy from the sun as possible, and to transfer this energy to storage for retention and use as required. All other factors being equal, the ability of a solar energy system to perform effectively is dependent on the operation of the system's controls.

The device that regulates the flow of fluid to the collectors and back to the storage tank is called a differential thermostat controller. It activates the pump when there is heat available at the collectors, thus causing fluid to circulate to them, and it shuts the pump off when no sunlight is available. It can also perform other functions, such as keeping the fluid from freezing in the collectors, or determining if the storage temperature is hot enough for space heat. A differential thermostat controller generally consists of three parts: the sensors, a differential comparator, and an output device (Figure 1).

Sensors

The function of the sensors is to monitor the temperature of the fluid in the collector and in the storage tank. Temperature-sensing elements commonly used in the industry are listed in Table 1, along with their most important characteristics. Only the sensors provided with a particular controller should be used; interchanging sensors from other manufacturers or other units may cause your controller to malfunction.

A variety of sensor housings have been developed for use in collectors and storage tanks or beds (see Figure 2). The cost and the ease of installation of a housing are

E.S. Peltzman is vice president for engineering, of Rho Sigma, Inc.

often inversely proportional. Basic sensors are inexpensive, but they tend to be difficult to mount and weatherproof in an installation. More-expensive units come in weatherproof housings, and many are compatible with standard electrical and plumbing fittings, making installation that much easier. Sensors may be connected to the differential comparator by means of either a terminal strip or wire nuts.

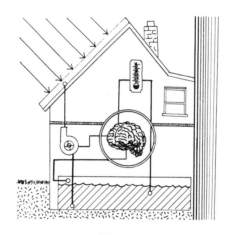

Fig. 2

Differential comparators

An input circuit—most commonly a balanced bridge—connects the sensors to the comparator. Figure 3 shows a typical circuit. This design employs a basic differential comparator circuit hooked up with the sensors which are in a common series bridge. The common point between the two temperatures sensors is connected to one input of the comparator (−). This represents the temperature differential between collector and storage. A fixed voltage, Vref, is connected to the other input (+) to determine the magnitude of the difference. As the signal from the sensor changes with

temperature, the differential voltage becomes greater than the reference level, and the output is switched on.

The circuitry in the differential thermostat compares the inputs from the sensors to each other and/or to fixed temperatures. In other words, when the temperature in the collector is only slightly hotter than the fluid in the storage tank, the controller will shut off the drive circuit, stopping the pump that circulates the working fluid to the collector, since there is little or no heat to be collected. Or, when the collector temperature nears the freezing point of the working fluid, the controller may activate a valve that allows the fluid to drain from the collector before it can freeze.

Many controls, then, not only perform the basic comparison function, but can also provide additional control functions, such as those listed in Table II. Each function may control its own output, or it may combine with other functions to control a single output. To provide these additional functions, the bridge circuit is modified to measure absolute temperature as well as temperature differentials. The bridge outputs are routed to additional comparators to provide these functions. A typical circuit of this type is shown in Figure 4.

A hysteresis (variable bias) is introduced by a resistor to each comparator circuit to prevent oscillation of the system. In effect, this creates two sets of on/off points to activate the circuits. Thus, it takes one set of temperatures to turn on a function and a second to turn it off. The difference between the turn on/turn off level may be as much as 20°F or as little as 3°F, depending upon the application. Without this difference, very small temperature variations could cause the system to excessively cycle on and off.

Output devices

The comparators control the driver circuits (outputs), which may be either relays or a solid state device. When matching a control to a load, such as a pump motor, a valve, or a blower, the electrical ratings and isolation

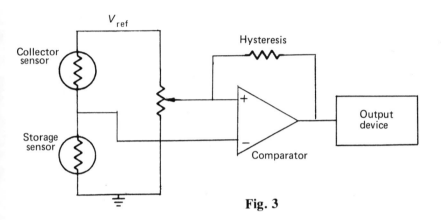

Fig. 3

References

Solar Energy Engineering & Product Catalog, Rho Sigma, Inc., 15150 Raymer St., Van Nuys, Calif. 91450.

The Omega Temperature Measurement Handbook, Omega Engineering Inc., Box 4047, Stamford, Conn. 06907 (1975)

AMF Potter & Brumfield General Catalog: Electromechanical, Dry Reed and Mercury-Wetted, Solid State, Hybrid and Time Delay Relays, Potter & Brumfield, 200 Richland Creek Dr., Princeton, Ind. 47671.

of the output circuits must be considered, to ensure compatibility and safety.

The ratings of a differential thermostat controller shoud be given in terms of maximum current, and also in motor horsepower. If the controller is not rated in horsepower, then the starting current and efficiency of fractional-horsepower motors (for pumps, blowers, etc.) will have to be matched to the output circuit. Table III shows approximate horsepower ratings for various relay contacts. If the load is more than the rating allows, then an additional booster relay will be required.

If a solid state output device is used it must be rated in operating voltage and in horsepower of the pump or valve. A 120-v ac solid state device will not work in a 240-V circuit. A problem with some solid state devices is electrical isolation. The solid state output is sometimes common to one of the input power lines and sensor circuits. Thus, if it is wired incorrectly, the solid state controller does not have sufficient isolation between the output voltage and the input line and sensors, or if it is wired incorrectly, then the sensors will be placed at a 120-V ac potential above ground instead of 24 V or less. This would constitute a serious shock hazard. Thus, when selecting a differential thermostat, carefully check to see that isolation of the output circuit from the sensor circuits and input power, is specified on the device.

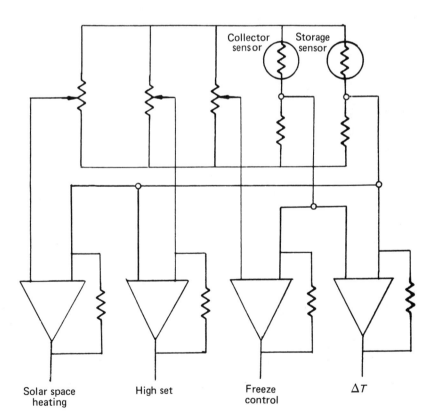

Fig. 4

Elements of this design and application are covered by U.S. Patent NO. 3,986,489, issued to Rho Sigma, Inc.

Controls

DIFFERENTIAL THERMOSTAT WITH FREEZE PROTECTION/RS 104

manufactured by:
Rho Sigma, Inc.
15150 Raymer St.
Van Nuys, Cal. 91405
Robert Schlesinger; phone: (213) 342-4376

Two independent outputs provide differential pump control and independent anti-freeze protection of collectors. **Features and options:** second output can function based on signal from storage rather than collector sensor. This mode allows the RS 104 to switch space heating systems to back-up systems when the temperature of the storage tank is too low. **Installation requirements/considerations:** standard electrical procedures. **Guarantee/warranty:** one year warranty against defects in parts and materials. **Maintenance requirements:** none. **Manufacturer's technical services:** provide design assistance from a control point of view on multi-pump/valve space heating system. **Regional applicability:** national.
Price: trade price $93.54 — wholesale and OEM prices available on application.

FIXFLO AND VARIFLO CONTROLLERS/H-1503, H-1505, H-1510, H-1512

manufactured by:
Hawthorne Industries, Inc.
1501 S. Dixie Highway
West Palm Beach, Fla. 33401
Ray Lewis; phone: (305) 659-5400
On/off type differential temperature controllers designed for solar energy applications. Models H-1505 and H-1512 are variable flow rate controllers that regulate not only flow when present, but also the amount of flow through the collector array based upon temperature differential. **Features and options:** recirculating freeze protection, upper temperature limit, dual outlet control (H-1505 and H-1512 only). Activator control available for fixed-point

temperature sensing and draindown freeze protection (H-1504, H-1511). **Installation requirements/considerations:** plug-in power conncetions, foam tape mounting, designed for quick, simple installation. No professional trades required. Variable flow control only applicable to specified pumps and blowers. **Guarantee/warranty:** one year repair or replacement warranty. Out-of-warranty repair available at minimal cost. **Maintenance requirements:** none. **Manufacturer's technical services:** control systems design for special applications, generally utilizing stock controls. **Regional applicability:** no limitations. **Availability:** wholesale factory-direct. Shipments from stock. For retail consult factory for distributors. **Price:** consult factory.

DIFFERENTIAL THERMOSTAT/RS 106

manufactured by:
Rho Sigma, Inc.
15150 Raymer St.
Van Nuys, Cal. 91405
Robert Schlesinger; phone: (213) 342-4376

Applicable to space heating systems. Procured by NASA in support of ERDA's solar demonstration program. **Features and options:** adequate space in high voltage compartment to house a booster relay to handle horsepower-rated fans and pumps. **Installation requirements/considerations:** compatible with standard electrical hardware. **Guarantee/warranty:** one year warranty against defects in parts and materials. **Maintenance requirements:** none. **Manufacturer's technical services:** recommendations on various ways to interface the RS 106 with large capacity, horsepower rated fans and pumps. **Regional applicability:** national.
Price: trade prie $81.82 — wholesale and OEM price availble on application.

DIFFERENTIAL THERMOSTAT/RS#240

manufactured by:
Rho Sigma, Inc.
15150 Raymer St.
Van Nuys, Cal. 91405
Robert Schlesinger; phone: (213) 342-4376

Designed for the wide variety of systems currently used to heat swimming pools and spas, this unit will control solenoid valves and auxiliary pumps. **Features and options:** direct control of an auxiliary pump to circulate water from the pool to the collector. Customer adjustable thermostatic control sets upper limit to pool temperature. **Installation requirements/considerations:** wired to load side of pump timer. **Guarantee/warranty:** one year warranty against defects in parts and materials. **Maintenance requirements:** none. **Manufacturer's technical services:** provides information on vacuum breaker/air vent, check valves and solenoid valves. **Regional applicability:** national. **Price:** trade price $76—wholesale and OEM prices available.

PROPORTIONAL CONTROL WITH DUAL OUTPUT/RS 500P SERIES

manufactured by:
Rho Sigma, Inc.
15150 Raymer St.
Van Nuys, Cal. 91405
Robert Schlesinger; phone: (213) 342-4376

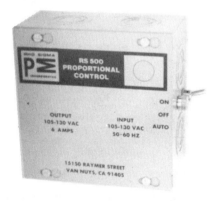

The RS 500P series increases solar energy collection efficiency by modulating the speed of fans and pumps in solar energy systems. The increased efficiency is particularly pronounced and valuable during fall, winter, and spring months or other times when the available solar energy is low. **Features and options:** two independent outputs; high-temperature and low-temperature detection circuits. Switch meets local code requirements. Light provides visual indication of pump speed. **Installation requirements/considerations:** standard electrical procedures or plug-in. **Guarantee/warranty:** one year warranty against defects in parts and materials. **Maintenance requirements:** none. **Manufacturer's technical services:** will fit the RS 500P to customer's particular requirements for dual-pump, drain down, or other system configurations. **Regional applicability:** national.
Price: trade price $74.56 — wholesale and OEM prices available on application.

CONTROLS/02A

manufactured by:
del sol Control Corp.
11914 U.S.#1
Juno, Florida 33408
Rodney E. Boyd; phone: (305) 626-6116

A differential control sensitive to the resistance offered by two sensors. Solar applications include: pumps, fans, electric valves and transformers used in: hot water space, and swimming pool heating. **Features and options:** a control/pump system completely wired, only requiring insertion in a ½ inch water line. No electrician, just plug it in the wall. Ideal for potable hot water heaters. **Guarantee/warranty:** one year guarantee. **Maintenance requirements:** none. **Manufacturer's technical services:** application assistance. **Regional applicability:** nationwide. **Availability:** In Florida: Huges Supply Inc. **Price:** $39.95 each at 100 and up.

SOLID STATE DIFFERENTIAL CONTROLLER/SPC-2000 SERIES

manufactured by:
Solarics
P.O. Box 15183
Plantation, Fla. 33318
Ronald Stein

The SPC-2000 series are totally solid state differential controllers for use in hot water and space heating systems. Two 6 amp controlled outputs (15 amps available) with anti-freeze control are standard features. Two matched sensors are included with each unit. **Features and options:** power consumption less than 3 Watts; optional lightning protection circuit, manual override and/or AC outlets. **Installation requirements/considerations:** mounts to wall; professional installation not required with optional external AC outlets; internal terminal strip provided for professional installation. **Guarantee/warranty:** 1 year replacement or repair warranty. **Maintenance requirements:** none. **Manufacturer's technical services:** custom engineering services available for special control requirements. **Regional applicability:** no restrictions. **Availability:** factory direct. **Price:** $34.50; add $5.50 for two manual override circuits; add $5.50 for two external AC outlets; add $2.50 for lightning protection.

SOLAR HEATING AND COOLING CONTROLS

manufactured by:
Solar Control Corp.
5595 Arapahoe Rd.
Boulder, Colo. 80302
Thomas B. Kent; phone: (303) 449-9180

Home heating and cooling solid state logic and switching controls for air or liquid systems. **Features and options:** One or two motor systems or a one to eight motor system. Controllers provide status outputs to allow monitoring of operational modes and temperatures of the collector and storage outputs. **Installation requirements/considerations:** 8 by 7 by 4 inches; 3 pounds. **Guarantee/warranty:** one year on material and workmanship. **Price:** consult manufacturer.

DELTA-T/DTT-70 thru ATT-3413

manufactured by:
Heliotrope General
3733 Kenora Drive
Spring Valley, Cal. 92077
Sam Dawson; phone: (714) 460-3930

The Delta-T differential thermostat is an automatic motor control that turns on and off the circulator pump/blower in a solar heating or solar hot water system when the collector temperature exceeds the storage temperature by present differentials. **Installation requirements/considerations:** enclosure is standard 4-11/16 x 4-11/16 inches. **Guarantee/warranty:** 1 year. **Maintenance requirements:** none. **Manufacturer's technical services:** yes. **Regional applicability:** any region.

Availability: factory direct or through dealers. **Price:** $31.50 to $69.50.

SOLAR CONTROLS/AD-101 AND AD-102

manufactured by:
AeroDesign Co.
P.O. Box 246
Alburtis, Pa. 18011
R.G. Flower; phone: (215) 967-5420
The AD-101 and AD-102 are both compact and capable of functioning as either differential controls or temperature limit switches. More complex control functions are available to the user simply by interconnecting any number of AD-101's or AD-102's. **Features and options:** AD-101 is housed in metal enclosure, intended for hand-wiring using standard electrical parts. AD-102 is housed in black phenolic enclosure, intended for shelf-mounting, and uses standard plug and socket connections. **Installation requirements/considerations:** complete installation guidelines are provided. **Guarantee/warranty:** five year limited warranty against defects in material and workmanship. **Maintenance requirements:** none. **Manufacturer's technical services:** consultation is provided on special problems of temperature monitoring and control. **Regional applicability:** no restrictions. **Availability:** from factory. **Price:** AD-101 $40 in single quantity. AD-102 $50 in single quantity. Quantity discounts available.

SOLAR AIR MOVER

manufactured by:
Solar Control Corp.
5595 Arapahoe Road
Boulder, Colo. 80302
Thomas B. Kent; phone: (303) 449-9180
The Solar Air Mover (SAM) provides total system air flow and operation mode control in one package. Eliminates complicated ductwork, dampers, and field wiring. **Features and options:** storage "chimney effect" losses are prevented by automatically closing the storage inlet duct. Collector inlet duct is also automatically closed during non-collection modes including the power off condition. Heat dissipated by the blower motor is used to heat the house directly. Available with or without solid state logic and switching controls; compatible with other off-the-shelf control systems. **Guarantee/warranty:** one year on materials and workmanship. **Price:** consult manufacturer.

TEMPERATURE DIFFERENTIAL SWITCH

manufactured by:
West Wind Co.
Box 1465
Farmington, N.M. 87401
Geoffrey Gerhard; phone: (505) 325-4949
Switch has two probes; when the temperature of one is greater than that of the other a power switch is activated. Maximum energy extraction from solar collectors is thus obtained. Can also be used to operate movable insulation panels, curtains, etc. **Features and options:** integrated circuit used for temperature sensing, heavy-duty relay for power switching. AC or DC models. Heat-cool switch available that allows unit to operate in either mode without reversing probes. **Guarantee/warranty:** two years on all parts and labor, after which units will be repaired for the cost of parts plus shipping. **Maintenance requirements:** none. **Manufacturer's tecnical services:** engineering, consulting for applications. **Regional applicability:** all regions. **Price:** $35 (includes shipping in U.S.)

DIFFERENTIAL
TEMPERATURE THERMOSTAT

PATENT PENDING

. . . for solar heating and
solar hot water system control

DELTA-T ™

DESCRIPTION

The DELTA-T™ Differential Temperature Thermostat is an automatic motor control which turns on and off the circulation pump/blower in a solar heating or solar hot water system when the collector temperature exceeds the storage temperature by preset differentials.

Also, in reverse, the DELTA-T™ may be utilized in solar nocturnal cooling systems to control summer night cold pick-up. By reversing the sensor locations (high temperature sensor at storage and low temperature sensor at cool exterior location), the pump or blower will go on when the exterior is colder than storage.

Enclosure is standard 4-11/16" by 4-11/16" UL Listed electrical outlet box (11.9mm x 11.9mm).

The DTT-90, 290 and 690 Series are designed to be hooked-up without the need to hire an electrician as they are equipped with a grounded power supply cord and a receptacle outlet in the box cover. The 690 series has two receptacles, one for the pump and one for the freeze protection valves.

A feature of the DTT-100 series is an AUTOMATIC OFF condition when the collector temperature is below 80°F. This feature prevents the pump turning on at night when a large amount of heat has been extracted from storage. — MODE 1

A circuit to turn the pump on during freezing temperatures is incorporated into the DTT-200 series. This feature prevents damage to the collector from freezing by circulating the warmer storage water. The pump turns on at 36°F and off when the collector is heated up to 37.5°F. — MODE 2

The DTT-400 series is designed with a high limit control of the storage temperature. When the storage tank is 160°F or higher, the pump will not run. By keeping the water temperature below 160°F, the fear of scalding is eliminated and a precautionary mixing/tempering valve is not needed. — MODE 4 (Other temperatures are available upon special order.)

For the Freeze-Fail-Safe™ system the DTT-690 series is available which incorporates a second 120V receptacle output to hold electric valves in position to allow water flow through system. (The motor control is also a receptacle outlet and the box is equipped with a 120V line cord input.) Upon power failure or freezing conditions, 42°F, the valve circuit turns off allowing the valves to return to their normal position which causes the collector to drain. The sensor for the Freeze-Fail-Safe™ system is a low voltage, hermetically sealed bi-metalic element designed for military/space applications under U.S. Government Spec. MIL-S-24236. — MODE 6

To add versatility and more accurate control of functions the DTT-70 and DTT-790 series offer an externally adjustable Turn-On Differential with a fixed Turn-Off of 1.5°F. Another adjustable model is the DTT-3410 series which incorporates an external adjustment for the High Limit, MODE 4, off temperature.

All models without receptacles have a bypass switch which allows the pump to be turned OFF or ON, bypassing the thermostat.

Construction is all solid state electronics except the UL Listed 10 Amp, 1/3 HP, 120V relay which is designed for 10 million mechanical cycles. 220V is also available.

All models are optionally available with a Normally Closed relay function by specifying "NC" after the part number.

For applications requiring higher than 1/3 HP, or 24V AC output, compatible relays and transformers are shown on the reverse side.

An indicator light on the front plate of the DELTA-T shows when the temperature differentials are in an ON condition.

(over)

HELIOTROPE GENERAL, 3733 kenora drive, spring valley, ca. 92077
(714) 460-3930

INSTALLATION

Installation is easily accomplished by running 115V to the box and from the box to the motor. Thermistor Ends for the sensor leads are furnished with the unit. The customer provides the interconnecting 18/2-lead wire which connects the low voltage DC sensor leads to the Thermistor Ends for which connectors are furnished. Hardware to attach Thermistor Ends to pipes, flat surfaces or curved surfaces is also furnished.

SPECIFICATIONS, PRICING

DIFFERENTIAL TEMPERATURE THERMOSTATS Output Control ⅓ HP @ 120 VAC (for higher HP use with R-120)

MODEL	ON differential @ 50°C °F	°C	OFF differential @ 50°C °F	°C	Electrical Hookup Thru box knockouts	Line Cord/ Cover Recp.	Electrical Input 120VAC 50/60 Hz	12 VAC 12 VDC	MODE 1 OFF Below 80°F	MODE 2 ON Below 36°F	MODE 4 OFF Above 160°F	MODE 6 Valve Recp. OFF Below 42°F	UL Listed	PRICE Includes Thermistor Ends
DTT-80	9.0	5.0	3.0	1.7	x		x						Yes	$38.00
DTT-81	4.5	2.5	1.5	0.8	x		x						Yes	38.00
DTT-82	15.0	8.3	5.0	2.8	x		x						Yes	38.00
DTT-83	15.0	8.3	1.5	0.8	x		x						Yes	38.00
DTT-90	9.0	5.0	3.0	1.7		x	x						pending	41.50
DTT-91	4.5	2.5	1.5	0.8		x	x						pending	41.50
DTT-92	15.0	8.3	5.0	2.8		x	x						pending	41.50
DTT-93	15.0	8.3	1.5	0.8		x	x						pending	41.50
DTT-100	9.0	5.0	3.0	1.7	x		x		x				Yes	43.00
DTT-101	4.5	2.5	1.5	0.8	x		x		x				Yes	43.00
DTT-102	15.0	8.3	5.0	2.8	x		x		x				Yes	43.00
DTT-103	15.0	8.3	1.5	0.8	x		x		x				Yes	43.00
DTT-200	9.0	5.0	3.0	1.7	x		x			x			Yes	43.00
DTT-201	4.5	2.5	1.5	0.8	x		x			x			Yes	43.00
DTT-202	15.0	8.3	5.0	2.8	x		x			x			Yes	43.00
DTT-203	15.0	8.3	1.5	0.8	x		x			x			Yes	43.00
DTT-290	9.0	5.0	3.0	1.7		x	x			x			pending	46.50
DTT-291	4.5	2.5	1.5	0.8		x	x			x			pending	46.50
DTT-292	15.0	8.3	5.0	2.8		x	x			x			pending	46.50
DTT-293	15.0	8.3	1.5	0.8		x	x			x			pending	46.50
DTT-400	9.0	5.0	3.0	1.7	x		x				x		Yes	43.00
DTT-401	4.5	2.5	1.5	0.8	x		x				x		Yes	43.00
DTT-402	15.0	8.3	5.0	2.8	x		x				x		Yes	43.00
DTT-403	15.0	8.3	1.5	0.8	x		x				x		Yes	43.00
DTT-500	9.0	5.0	3.0	1.7	x			x					No	35.00
DTT-501	4.5	2.5	1.5	0.8	x			x					No	35.00
DTT-502	15.0	8.3	5.0	2.8	x			x					No	35.00
DTT-503	15.0	8.3	1.5	0.8	x			x					No	35.00
DTT-690	9.0	5.0	3.0	1.7		x	x				x	x	pending	68.00
DTT-691	4.5	2.5	1.5	0.8		x	x				x	x	pending	68.00
DTT-692	15.0	8.3	5.0	2.8		x	x				x	x	pending	68.00
DTT-693	15.0	8.3	1.5	0.8		x	x				x	x	pending	68.00
DTT-300	Cust. Spec.		Cust. Spec.		Customer Specify		Customer Specify		Cust. Spec., One Only				No	53.00

80, 100, 200, 400, and 500 Series.

90 and 290 Series.

690 Series.

For normally closed relay contact function specify "NC" after part number. Add $5.00 to price.
For 240 VAC Input with 240 VAC Double Pole Output Control specify "240V" after part number, add $7.50 to price.

DIFFERENTIAL TEMPERATURE THERMOSTATS ADJUSTABLE, Output Control: ⅓ HP @ 120 VAC

MODEL	ON differential @ 50°C °F	°C	OFF differential @ 50°C °F	°C	Electrical Hookup Thru box knockouts	Line Cord/ Cover Recp.	Electrical Input 120VAC 50/60 Hz	12 VAC 12 VDC	MODE 1 OFF Below 80°F	MODE 2 ON Below 36°F	MODE 4 OFF Above 160°F	MODE 6 Valve Recp. OFF	UL Listed	PRICE Includes Thermistor Ends
DTT-70	Adjustable		1.5	0.8	x		x						Yes	$48.00
DTT-71	Adjustable		1.5	0.8	x		x		x				Yes	53.00
DTT-72	Adjustable		1.5	0.8	x		x			x			Yes	53.00
DTT-74	Adjustable		1.5	0.8	x		x				x		Yes	53.00
DTT-790	Adjustable		1.5	0.8		x	x						pending	51.50
DTT-791	Adjustable		1.5	0.8		x	x		x				pending	56.50
DTT-792	Adjustable		1.5	0.8		x	x			x			pending	56.50
DTT-794	Adjustable		1.5	0.8		x	x				x		pending	56.50
DTT-3410	9.0	5.0	3.0	1.7	x		x				Adjustable		Yes	53.00
DTT-3411	4.5	2.5	1.5	0.8	x		x				Adjustable		Yes	53.00
DTT-3412	15.0	8.3	5.0	2.8	x		x				Adjustable		Yes	53.00
DTT-3413	15.0	8.3	1.5	0.8	x		x				Adjustable		Yes	53.00

70 and 3410 Series.

790 Series.

For normally closed relay contact function specify "NC" after part number. Add $5.00 to price.
For 240 VAC Input with 240 VAC Double Pole Output Control specify "240V" after part number, add $7.50 to price.

THERMISTOR END SENSOR (included with DTT-series)

TES-1 $4.00

10" long, teflon wire, 3000 ohm @ 25°C. Includes wire nuts for electrical hookup and silicone fiberglass tape for surface mounting. (For replacement use with DTT-series or with Electronic Thermometer ET-0/350.)

BP-25 $6.50

3000 ohm @ 25°C, Thermistor Sensor with 1/4" MPT.

BP-50 $7.00

3000 ohm @ 25°C, Thermistor Sensor with 1/2" MPT.

RBP-50 $7.00

Same as BP-50 except leads extend from reverse end. For direct immersion into tanks. Utilize customer furnished 1/2" coupling and any length of pipe. Wire connections are enclosed inside pipe.

FREEZE SENSOR (included with DTT-690 series)

FS-1 $20.00

"ON" above 42°F ± 5°F. "OFF" below 52°F rising temperature. Hermetically sealed non-adjustable bimetalic type. Includes strap hardware for attachment to pipe. (May also be used in conjunction with DTT-400 and HELIO-MATIC series to provide additional feature of freeze recirculation protection.)

WEATHERPROOF CASE

2T 511 GA $14.30

INTERMATIC, NEMA type II, hasp for lock, hinged door with rubber seal. (For use as enclosure when the DTT-series is to be mounted outside.)

TRANSFORMERS

T-120/24, knockout mounted	$4.50
T-120/24P, plug mounted	$5.00

120V AC Primary, 24V AC Secondary. 20 VA. UL listed. (Use when DTT signal output needs to be 24V AC.)

T-240/12, knockout mounted	$7.50

240V AC Primary, 12V AC Secondary, 20 VA.

RELAYS with ENCLOSURES

R-120 $14.30   R-24 $14.30   R-12 $18.20

AC Coil Voltage indicated in part number. All Are DPDT, 25 Amp, 1 HP. UL listed. Two piece enclosure has four 1/2" conduit knockouts.

FREEZE-FAIL-SAFEtm VALVE PACKAGE

VP-1 $72.50

Use in conjunction with DTT-690 series. Five connections (to and from collectors, to and from storage tank, and drain). Collector will automatically drain upon removal of electrical signal.

VACUUM RELIEF VALVE

VRV-36A $6.95

Valve required on all draining type systems. 3/4" MPT. Mount at highest location above collector. (Allows atmospheric air to enter collector so that collector will drain.)

AIR VENTING VALVES, Float Type

AV-426, 1/4" MPT, 150 Max PSI $11.90
AV-67, 1/8" MPT, 35 Max PSI $ 7.60

Valve required on all draining type systems. Mount at highest point above collector. (Allows trapped air to be released from system when pump is started.)

STOREXtm TANK with double-wall heat-exchanger.

For closed loop systems where solar heated antifreeze solutions heat water in tank through integral heat exchanger.
All tanks have auxiliary 4.5 KW heater.

65 Gallon	10 sq. ft. Exchanger	TC-65E-DW	$228
80 Gallon	15 sq. ft. Exchanger	TC-80E-DW	$263
120 Gallon	20 sq. ft. Exchanger	TC-120E-DW	$344

Separate data sheet available upon request.

EXPANSION TANK

ET-442 $20.50

Tank required on all closed loop systems. 1/2"MPT, 8" dia. x 13" high for systems up to 25 gallons of Heat Exchange Fluid capacity, diaphragm type captive air. (When Heat Exchange Fluid becomes heated the expansion is absorbed by the Expansion Tank.)

TEMPERING VALVE

TV-526 $11.90

Externally adjustable from 120-160°F, 3/4" sweat fittings, completely non-ferrous construction, thermostatic element replaceable without removing valve. (For use with domestic hot water tank. Cold water is mixed with hot to regulate exit water temperature at valve setting, preventing scalding water temperatures at the faucet. Lengthens out the delivery of hot water from tank by mixing cold at the tank, not the faucet. Necessary for all solar systems.)

TEMPERATURE AND PRESSURE RELIEF VALVE

T&P-100XL $5.12

3/4"MPT, 150 lbs., 210°F.

GRUNDFOS CIRCULATION PUMPS

CAST IRON, Closed loop systems only, including two valves.

UPS-20-42F $60.20

1/20HP @ 115V AC, 5-position variable flow control by external adjustment plus two-speed motor control.

UP-26-64F $68.60 Same as above except 1/12 HP and without two speed.

STAINLESS STEEL, Open loop, potable water systems, including two valves.

UM-25-18SU $75.15

1/35 HP @ 115V, AC, Single speed without flow control.

UP-25-42SF $91.00 Same as above except 1/20 HP.

UP-25-64F $107.70 Same as above except 1/12 HP.

LINE CORD with pump, add $3.00 to pump price. Specify "with line cord" after any pump part number and the pump will be pre-wired with a line cord to simply plug into the DTT-90, 290, 690 and 790 series.

ELECTRONIC THERMOMETER

ET-0/350 $74.50

Temperature range 0° to 350° displayed on a four inch wide dial. Dial shows temperature in both Fahrenheit and Celcius. Thermistor sensors may be located up to 1,000 feet away from thermometer. Customer provides interconnecting two-conductor wires (18 to 24 ga. recommended) which are attached to the electronic thermometer terminal strip and to the Thermistor End Sensors. Twelve position rotary switch allows temperature sensing at up to 11 different locations plus an "OFF" position. Response time is immediate upon changing switch location. Thermometer comes with three Thermistor End Sensors. For sensing at additional locations, purchase one TES-1, listed above, for each additional location. 115V AC input with three-wire line cord. Dimensions: 11" wide x 4" high x 8 1/4" deep. Separate data sheet available upon request.

Ask about our two location strip chart recorder and a hand-held battery powered digital.

SOLAR HEATED SWIMMING POOL CONTROL SYSTEM

HELIO-MATIC I tm

Assembly of Delta-T, 12V transformer, Weatherproof Case, and 12V Electric Valves with complete instructions. Both 120 and 240 Voltage inputs available. Separate data sheet available upon request.

INSOLATION METER

IM-5 $57.50
Hand-held meter, indicates solar intensity.

A message about our product reliability:

All Heliotrope products, whether manufactured by us or for us, are of the highest quality available consistent with sensible cost considerations.

For instance: Every Delta-T unit is electrically and heat aged for over 100 hours in order to precipitate premature electronic component part failures. / Another burn-in is true for the Thermistor End Sensors to stabilize their characteristics and assure consistent differentials. / Inside the Delta-T you will note that every unit is identified with an inspection label which shows its own test reading number. We can relate this number to within one-tenth of a degree fahrenheit. / And field rejections are failure-analyzed and the inspection report for each one is individually reported to the president of the company. / Underwriters Laboratories Listing has been obtained on many products assuring the customer of safety and code compliance.

All of us at Heliotrope General assure you that our products have been precision engineered for dependable performance and accurate operation, every time, all the time.

Respectfully,

Sam Dawson

Sam Dawson, President

ORDER FORM to HELIOTROPE GENERAL, 3733 Kenora Drive, Spring Valley, CA. 92077 • (714) 460-3930

Date_____

Name _____

Company Name _____

Address_____

City_____

State _____ Zip _____ Phone_____

QUANTITY	DESCRIPTION AND PART NUMBER	PRICE	TOTAL
	California Residents add 6% Sales Tax		
	Add $3.50 Shipping Charge for each DTT ordered		
	Ship: C.O.D. ☐ or Check Enclosed ☐		

SOLID STATE DIFFERENTIAL TEMPERATURE CONTROLLER/SPC-1000 SERIES

manufactured by:
Solarics
P.O. Box 15183
Plantation, Fla. 33318
Ronald Stein

The SPC-1000 series are totally solid state differential controllers with one 6 amp controlled output (up to 15 amps available). Two matched sensors are included with each unit. **Features and options:** power consumption less than 3 Watts; optional lightning protection, manual override and/or external AC outlet. **Installation requirements/considerations:** mounts to wall; professional installation not required with optional external outlet; internal terminal strip provided for professional installation. **Guarantee/warranty:** 1 year replacement or repair warranty. **Maintenance requirements:** none. **Manufacturer's technical services:** custom engineering services available for special control requirements. **Regional applicability:** no restrictions. **Availability:** factory direct. **Price:** $29.50; add $3.00 for manual override; add $4.00 for external AC outlet; add $2.45 for lightning protection.

DIFFERENTIAL THERMOSTAT/E10

manufactured by:
Energy Converters, Inc.
2501 N. Orchard Knob Ave.
Chattanooga, Tenn. 37406
David Burrows; phone: (615) 624-2608

All solid state differential thermostat that can be used to turn on pumps or other devices with the aid of two appropriately placed thermistors (supplied). Present on-off temperature settings set to minimize unnecessary cycling of solar equipment. Other settings can be specified. **Features and options:** optional on-off temperature differential settings available. **Installation requirements/considerations:** handles up to 8 amps continuously or 10 amps intermittently and operates on standard 120 volt AC household electricity. **Manufacturer's technical services:** available upon request. **Availability:** through distributors or direct from factory. **Price:** $75.

THERMO-MATE™ DELUXE CONTROLLER, DC-761

manufactured by:
Solar Energy Research Corp.
701B South Main St.
Longmont, Colo. 80501
James B. Wiegand, president; (303) 772-8406

A completely self-contained, four probe, solar controller and data screen access system. Integral house thermostat, manual overrides, mode indicator lights, digital temperature readouts, probe calibrations,

and adjustable differentials, bottom tank to collector and top tank to house. Use with air or liquid systems. **Features and options:** standard features include liquid crystal display, digital, adjustable, two-stage house temperature thermostat with easy set-back feature and field selection of standby heat cut-in temperature. **Installation requirements/considerations:** calibrate probes with lead wires as used in field before covering probes. Provide 120V AC outlet. **Guarantee/warranty:** one year factory; defective units replaced. **Maintenance requirements:** none. **Manufacturer's technical services:** unit may be ordered with two extra sensors for temperature monitoring or with strip chart outputs at extra charge. **Regional applicability:** unlimited. **Price:** $500.

SOLAR CONTROL/SC100

manufactured by:
Natural Power Inc.
New Boston, N.H. 03070
Rick Katzenberg; phone: (603) 487-2426

Senses temperatures at selected points in system and responds to demands, considering availability of energy supply, by activating pumps, blowers, valves, dampers, shutters, and heaters. Provides total automatic control. **Features and options:** meter and L.E.D. display-show current operating mode and give information on conditions at various locations within the system. Functional with either liquid or air system. **Price:** $400.

SUN*TRAC 100

maufactured by:
Energy Applications, Inc.
830 Margie Drive
Titusville, Fla. 32780
Napoleon Salvail; phone: (305) 269-4893

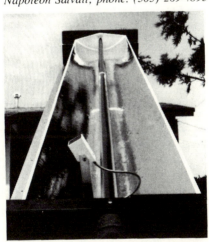

A complete orientation control system for solar tracking devices. Weighs 3½ ounces. Entire subminiature tracking control circuitry housed in less than 2 cubic inches of space. Painted aluminum shell, stable glass filter elements, and potted circuitry make the unit rugged and reliable for outdoor use. **Features and options:** tracking accuracy

under clear sky conditions is 15 to 20 minutes of arc with an optional accuracy of 7 arc minutes built into the unit. Circuit design prevents both directional relays from closing at the same time. Includes reed relay output for simple interface with any reversible drive motor. Optional interface module/power supply for directional control of a permanent split capacitor gear motor. Optional Dayton gear motor for tracking drive assemblies. **Installation requirements/considerations:** two tapped holes are provided on either side of the unit for mounting; the unit is mounted on the device for tracking the sun at the proper angle, unshaded by the device during the day. **Guarantee/warranty:** one year on parts and labor. **Maintenance requirements:** none. **Maufacturer's technical services:** product technical consultant. **Regional applicability:** areas with considerable constant haze can make some adjustment but may find it unsuitable if haze is heavy. **Availability:** from manufacturer. **Price:** 1 to 3: $59.95 each. 4 to 19: $56.95. 20 to 39: $53.95. Over: $48. Optional interface module/power supply (S/T-100-1)—$59.95. Optional Dayton drive motor—$27.

AUTOMATIC FLOW CONTROL VALVE/UP-050 (½ in.) through UP-200 (2 in)

manufactured by:
Aqueduct Component Group
1537 Pontius Ave.
Los Angeles, Cal. 90025
Herman Wertheimer; phone: 213/479-3911

The product will hold the design flow to within plus or minus 4 percent no matter what the system pressure variation, and works on the pressure compensation principal. The system is completely and automatically balanced the moment it is installed on-line. **Features and options:** union connection for easy access; components may be changed later to reflect design flow changes; available with pressure taps, screwed, or with solder cup ends. **Installation requirements/considerations:** pipe wrench or solder joints; may be installed on line in same pipe diameter as line. **Guarantee/warranty:** one year as to material and workmanship replaced free FOB the factory. **Maintenance requirements:** if system is contaminated, components may have to be cleaned once every year. **Manufacturer's technical services:** full engineering staff at factory with reps in every state. **Regional applicability:** may be used in any region. **Availability:** factory reps. **Price:**

½ inch	$23 each
⅝ inch	$25 each
¾ inch	$29 each
1 inch	$36 each
1½	$42.50 each
1½ inch	$53 each
2 inch	$72 each

SOLAR COMMANDER/SD-10

manufactured by:
Robertshaw Controls Co.
100 W. Victoria
Long Beach, Cal. 90805
Preston Welch; phone: (213) 638-6111

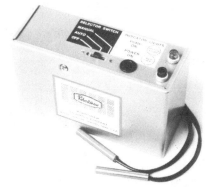

Differential controller provides sensitive temperature response and solid state switching to effectively operate a circulating pump in a liquid filled solar heat storage system. Solid state thermistor sensors are used to signal temperature differential to the control circuit from the solar collector and the storage tank with accuracy from extended distances up to several thousand feet if necessary. **Features and options:** fixed pump cut-in differential or adjustable cut-in differential. **Installation requirements/considerations:** equipped with access terminal strip for the low voltage sensor circuit and line lead conduit adapter access. **Guarantee/warranty:** 1 year from date of installation. **Maintenance requirements:** none. **Regional applicability:** national. **Availability:** all Uni-Line Dealers and Wholesalers. **Price:** consult manufacturer.

THERMO-MATE™ STANDARD CONTROLLER, SC-762

manufactured by:
Solar Energy Research Corp.
701B South Main St.
Longmont, Colo. 80501
James B. Wiegand, president; (303) 772-8406

A four-sensor solar space heating controller containing two differential thermostat circuits, both with adjustable differential settings. Use with water or air solar systems. House-top differential circuit in series with two-stage house temperature thermostat for most efficient use of stored heat at reduced house temperatures. **Features and options:** solid state relay optional; adjustable differential 2° to 15°F in both circuits standard. **Installation requirements/considerations:** wire four probes and relay: 120V power, 10 Watts. **Guarantee/warranty:** one year on parts and labor—return unit to SERC for repair or replacement. **Maintenance requirements:** none. **Regional applicability:** unlimited. **Price:** $150.

DIFFERENTIAL THERMOSTAT/RS 500S SERIES

manufactured by:
Rho Sigma, Inc.
15150 Raymer St.
Van Nuys, Cal. 91405
Robert Schlesinger; phone: (213) 342-4376

Two independent solid state outputs designed to conrol fans and pumps having other than permanent capacitor or shaded pole motors. Designed for small capacity fans and pumps. **Features and options:** two independent outputs. High temperature and low-temperature detection circuits. Switch meets local code requirements. Light provides visual indication of pump speed. **Installation requirements/considerations:** standard electrical procedures or plug-in. **Guarantee/warranty:** considerations: standard electrical proce- and materials. **Maintenance requirements:** none. **Manufacturer's technical services:** will fit the RS 500S to customers particular requirements for dual-pump, drain down, or other system configurations. **Regional applicability:** national. **Price:** trade price $65 — wholesale and OEM prices available on application.

LCU-110 DIGITAL LOGIC CONTROL UNIT

manufactured by:
Contemporary Systems, Inc.
68 Charlonne St.
Jaffrey, N.H. 03452
John Christopher; phone: (603) 532-7972

A logic control unit that samples collector and storage temperatures, collector/storage differential, and thermostat demand, and chooses the operating mode; heating from collectors, heating from storage, storing heat, heating from back-up system or standby. **Features and options:** includes a de-icing cycle control for occasions when a severe ice storm glazes the collectors (this is a manual mode). **Installation requirements/considerations:** the installation is compatible with current trade skills and standards. **Guarantee/warranty:** limited one year warranty. **Maintenance requirements:** none. **Manufacturer's technical services:** logic system integration with Series IV hardware. **Regional applicability:** unlimited. **Availability:** currently available. **Price:** LCU-110: $276. Sensor probes and house thermostat: $81.

UNIVERSAL SWITCHING UNIT-A

manufactured by:
Contemporary Systems, Inc.
68 Charlonne St.
Jaffrey, N.H. 03452
John Christopher; phone: (603) 532-7972
The USU-A is an integrated blower/switching system that incorporates air valves and blower into one component, reducing specialized on-site assembly labor and maxi-

mizing air transport efficiency in one pre-engineered package. **Features and options:** an automatic return feature should electrical power fail during operation; the air circulating device is of top quality, and while first cost is high, operating costs over design life are low, yielding overall savings. **Installation requirements/considerations:** connects to conventional insulated ducting that interconnects collectors, storage and supply/return trunk lines. **Guarantee/warranty:** limited one year warranty. **Maintenance requirements:** normal lubrication requirements for blower and damper motors. **Manufacturer's technical services:** computerized project analysis, system integration, site supervision and follow up analysis. **Regional applicability:** unlimited. **Availability:** currently available. **Price:** 800-1600 CFM 1.6 square feet duct area: $1,341. 1600-2800 CFM 2.8 square feet duct area: $1,665. 2800-4000 CFM 4.0 square feet duct area: $2,407.

SIMONS SOLAR DIFFERENTIAL THERMOSTATS

manufactured by:
Simons Solar Environmental Systems
24 Carlisle Pike
Mechanicsburg, Pa. 17055
C. John Urling; phone: (717) 697-2778
All solid state differential thermostat using optical couplings and triacs rather than mechanical relays. Each unit is supplied with encapsulated matched sensors supplied with 6 feet of Teflon insulated wire. **Features and options:** wired for plug in, elapsed time meter, private labeling, 6 amp or 20 amp control, freeze preventer. **Installation requirements/considerations:** standard mounting for 8 inch x 6 inch x 4 inch electrical box. **Maintenance requirements:** none. **Manufacturer's technical services:** custom designs and modifications. **Regional applicability:** no limitation. **Price:** consult factory.

AIR HANDLING UNIT/AU-40-50

manufactured by:
Solaron Corporation
300 Galleria Tower
Denver, Colo. 80222
Ray T. Williamson
A combination package of motor-blower, water heating coil, 2 dampers, and a controller enclosed in a sheet steel case. **Features and options:** AU-40 handles 300-1400 CFM and AU-50 handles 1400-2000 CFM. **Installation requirements/considerations:** should be installed in a heated area. **Guarantee/warranty:** 1 year limited warranty on materials and workmanship. **Maintenance requirements:** adjust fan belts and lubricate motor as recommended by the manufacturer. **Regional applicability:** applicable to all climatic conditions and geographic areas. **Price:** consult manufacturer.

If you're considering Controls —
Consider these, from Natural Power

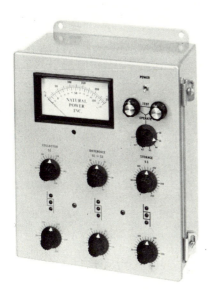

S10/SOLAR CONTROL

- Modular Assembly — each control tailored to YOUR needs.
- Differential, over-temp., freeze, auxiliary and other controls as required.
- Up to 10 independent inputs and 5 outputs are standard, more available.
- Fully user adjustable.
- Continuous performance and status monitoring.
- Self-test and system exercise capability.

S25/DIFFERENTIAL THERMOSTAT

- Ideal for DHW, pools and simpler space heating systems.
- Two temp. inputs, 10 Amp relay output.
- Independently adjustable ON and OFF temp. settings.
- System exercise and self-test capability.
- Continuous performance and status monitoring.

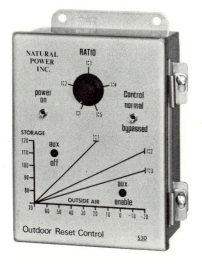

S30/OUTDOOR RESET CONTROL

- Monitors outdoor and storage temperatures.
- Lowers auxiliary heat cut-in point as outdoor temperature rises.
- Increases any system's efficiency.
- Includes ratio control, bypass control & status indicators.

FAFCO SOLAR CONTROL

manufactured by:
Fafco Incorporated ·
235 Constitution Dr.
Menlo Park, Cal. 94025
Phone: (415) 321-3650
High accuracy differential controls for all solar applications. Simple differential or setpoint scale with heat-or-cool logic. Output options are one valve, two valve, or 3 horsepower motor control relay. **Installation requirements/considerations:** can be installed indoors or outdoors. **Guarantee/warranty:** one year. **Maintenance requirements/considerations:** none scheduled. **Regional applicability:** all regions. **Availability:** name of local distributor on request. **Price:** set by distributor.

DIFFERENTIAL THERMOSTATS

manufactured by:
Solar Control Corp.
5595 Arapahoe Road
Boulder, Colo. 80302
Thomas B. Kent; phone: (303) 449-9180
Modular differential thermostats to control single and multiple modes of operation: allow for the design of custom systems to control swimming pool and domestic hot water installations. They can also be used where heat pumps are used to provide auxiliary heating and air conditioning. **Features and options:** logic priorities can be integrated with temperature comparators to create predetermined conditions for operation. Multiple hook-ups allow as many priorities to be met as required for the operation of a given output. **Guarantee/warranty:** one year on materials and workmanship. **Price:** consult manufacturer.

SOLAR TEMPERATURE CONTROL/ R7406A

manufactured by:
Honeywell, Inc.
Residential Div. Customer Service
1885 Douglas Drive North
Minneapolis, Minn. 55422
Phone: (612) 542-7500
The R7406 solar temperature control provides automatic control of circulating pumps, valves, dampers, motors, and other accessories used in solar energy systems. A solid state differential amplifier with thermistor sensors operates a 3-pole load switching relay. **Features and options:** plug-in resistors permit changing on and off differential and adapting control for single-function temperature control. **Installation requirements/considerations:** mounts in any position on a standard 4 by 4 inch junction box. **Price:** consult customer service.

Pumps

ELECTRONICALLY CONTROLLED CIRCULATING PUMP SYSTEM/PA-201

manufactured by:
Solar Innovations
412 Longfellow Blvd.
Lakeland, Fla. 33801
Ron Yachabach; phone: (813) 688-8373

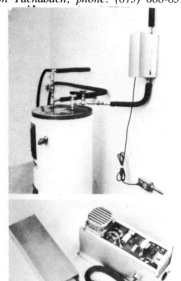

Integral pump, valve, and electronic control assembly housed in aluminum box. Designed to suit solar water heating systems. Solid state controls. **Features and options:** electric valve prevents night heat loss. **Guarantee/warranty:** five year guarantee. **Price:** consult factory.

PUMP, HOT WATER CIRCULATOR/ UMS 20-28

Grundfos Pumps Corp.
2555 Clovis Ave.
Clovis, Cal. 93612
Mrs. Evelyn Graham, sales assistant; phone: (209) 299-9741
Hot water circulation pump with 1/25 HP energy saving split-capacitor motor. Maximum head of 9 feet and maximum flow of 23 GPM. All internal water touched parts of high grade stainless steel. Self venting and water lubricated. **Features and options:** two-speed, variable head control, automatic thermal reset protection, aluminum oxide bearings, water lubricated, noiseless and sealless. UL recognized. **Installation requirements/considerations:** for closed systems only; not fresh water. Can be mounted with vertical or horizontal shaft. Available with ¾ or 1 inch unions or ¾, 1, 1¼, or 1½ inch flanges. Weight: 9 lbs. **Guarantee/warranty:** Eigh-

teen months from date of purchase. Any pump that is defective, providing it was handled and installed properly, will be replaced with a new pump. **Maintenance requirements:** none—including no lubrication. Can be used in systems with 230°F water temperature. **Manufacturer's technical services:** solar applications engineer available for technical assistance plus four regional sales offices across United States (Clovis, Chicago, Pittsburgh, and New Jersey). **Price:** $100 list.

PUMP, HOT WATER CIRCULATOR/ UP 25-42 SF

manufactured by:
Grundfos Pumps Corp.
2555 Clovis Ave.
Clovis, Cal. 93612
Mrs. Evelyn Graham, sales assistant; phone: (209) 299-9741
Hot water circulation pump with 1/20 HP energy saving permanent split-capacitor motor. Maximum head of 14 feet and maximum flow of 23 GPM. All water-touched parts of high grade stainless steel. Self-venting and water-lubricated. **Features and options:** No variable head control, single speed, automatic thermal reset protection, aluminum oxide bearings, water lubricated, noiseless and sealless. **Installation requirements/considerations:** for open potable, fresh water systems. Can be mounted with vertical or horizontal shaft. Available with ¾ or 1 inch union connections or ¾, 1, 1¼, or 1½ inch flanges. Weight: 7 lbs. **Guarantee/warranty:** Eighteen months from date of purchase. Any pump that is defective, providing it was handled and installed properly, will be replaced with a new pump. **Maintenance requirements:** none—including no lubrication. **Manufacturer's technical services:** solar applications engineer available for technical assistance plus four regional sales offices across the United States (Clovis, Chicago, Pittsburgh, and New Jersey). **Price:** $171 list.

PUMP, HOT WATER CIRCULATOR/ UPS 20-42 F

manufactured by:
Grundfos Pumps Corp.
2555 Clovis Ave.
Clovis, Cal. 93612
Mrs. Evelyn Graham, sales assistant; phone: (209) 299-9741
Hot water circulation pump with 1/20 HP energy saving permanent split-capacitor motor. Maximum head of 14 feet and maximum flow of 23 GPM. All internal water-touched parts of high grade stainless steel. Self-venting and water-lubricated. **Features and options:** variable head con-

Continued

trol, two speed, automatic thermal reset protection, aluminum oxide bearings, water lubricated, noiseless, and sealless. Can be used in systems with 230°F water temperature. **Installation requirements/considerations:** for closed systems only; not fresh water. Can be mounted with vertical or horizontal shaft. Available with ¾ or 1 inch union connections or ¾, 1, 1¼, or 1½ inch flanges. Weight: 10 lbs. **Guarantee/warranty:** Eighteen months from date of purchase. Any pump that is defective, providing it was handled and installed properly, will be replaced with a new pump. **Maintenance requirements:** none—including no lubrication. **Manufacturer's technical services:** solar applications engineer available for technical assistance plus four regional sales offices across United States (Clovis, Chicago, Pittsburgh, and New Jersey). **Price:** $100 list.

PUMP, HOT WATER CIRCULATOR/ UP 26-64 F

manufactured by:
Grundfos Pumps Corp.
2555 Clovis Ave.
Clovis, Cal. 93612
Mrs. Evelyn Graham, sales assistant; phone: (209) 299-9741
Hot water circulation pump with 1/12 HP energy-saving permanent split-capacitor motor. Maximum head of 20 feet and maximum flow of 32 GPM. All internal water-touched parts of high grade stainless steel. Self-venting and water-lubricated. **Features and options:** Variable head control, single speed, automatic thermal reset protection, aluminum oxide bearings, water lubricated, noiseless, and sealless. **Installation requirements/considerations:** for closed systems only; not fresh water. Can be mounted with vertical or horizontal shaft. Available with ¾ or 1 inch union connections or ¾, 1, 1¼, or 1½ inch flanges. Weight: 12 lbs. **Maintenance requirements:** none—including no lubrication. **Manufacturer's technical services:** solar applications engineer available for technical assistance plus four regional offices across United States (Clovis, Chicago. Pittsburgh, and New Jersey). **Price:** $112 list.

PUMP, SUBMERSIBLE/SP 2-10

manufactured by:
Grundfos Pumps Corp.
2555 Clovis Ave.
Clovis, Cal. 93612
Mrs. Evelyn Graham, sales assistant; phone: (209) 299-9741
Stainless steel submersibles provide an opportunity to install the pump in the fluid being pumped for pressures up to 1000 psi and flow of 700 GPM. Primarily designed for overcoming high pressure losses quickly in draindown systems. **Features and options:** pump can be adjusted by adding or deleting stages to give an exact pressure and flow. Priming screw to prevent burn-up and built-in check valve are standard.

Installation requirements/considerations: ca be mounted in storage tank or piping, in horizontal or vertical position. Minimum noise and vibration. **Guarantee/warranty:** Eighteen months from date of purchase. Any pump that is defective, providing it was handled and installed properly, will be replaced with a new pump. **Maintenance requirements:** none—including no lubrication. **Manufacturer's technical services:** solar applications engineer available for technical assistance plus four regional sales offices across United States (Clovis, Chicago, Pittsburgh, and New Jersey). **Price:** $437 list.

PUMP, MULTI-STAGE/ CP 2,3,8; CR 30

manufactured by:
Grundfos Pumps Corp.
2555 Clovis Ave.
Clovis, Cal. 93612
Mrs. Evelyn Graham, sales assistant; phone: (209) 299-9741
Both horizontal and vertical multi-stage pumps for systems requiring up to 1,000 psi or 200 GPM. Primarily designed for recharge draindown-type systems. **Features and options:** all stainless steel or bronze chambers available. **Installation requirements/considerations:** available in self-priming or non-self-priming; can be vertically or horizontally mounted. **Guarantee/warranty:** eighteen months from date of purchase. Any pump that is defective, providing it was handled and installed properly, will be replaced with a new pump. **Maintenance requirements:** none—including no lubrication **Manufacturer's technical services:** solar applications engineer available for technical assistance plus four regional sales offices across the United States (Clovis, Chicago, Pittsburg and New Jersey). **Price:** varies.

This space reserved!

It's time we stopped filling up every square inch of space in our cities with new construction and start preserving space. Space to walk. Room to be.

Protect the human and environmental quality of your community by joining the National Trust for Historic Preservation. Historic preservation means more than saving old houses. Write: National Trust for Historic Preservation, Department 0606, 740 Jackson Place, NW, Washington, DC 20006.

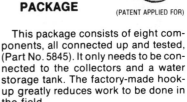

CONCENTRATING COLLECTORS

By John I. Yellott, P.E.

Mr. Yellott is visiting professor in architecture at Arizona State University in Tempe, Arizona.

Figure 1: *Shuman and Boys' solar steam generator used at Meadi, Egypt, 1913-1916. Photo courtesy of the Smithsonian Institution, Astrophysical Observatory.*

The first attempts to make technological use of solar energy involved the generation of high temperatures, first for purely scientific experiments like those conducted by Lavoisier before the French Revolution. One hundred years later there was another awakening of interest in the use of solar energy at relatively high temperatures, this time for the operation of pumps. Mouchot in France, Ericsson in the United States, and a few other experimenters around the world began using paraboloidal concentrators or parabolic troughs, again with the generation of steam as their objective.

We are not quite sure who invented the flat-plate collector, but Tellier of France began experimenting with flat-plate collectors in about 1885, for the purpose of generating ammonia vapor to run small vertical reciprocating engines. There is a question as to whether he actually built the devices he described in his writings, but he was probably the first to attempt a low pressure flat-plate approach to the collection of solar energy.

Truncated cones, lined with mirrors, were used with varying degrees of success by A. G. Eneas, in Arizona and California, and, after these tests, a number of flat-plate collectors, generally running at very low pressure, were tested in the deserts of Arizona and California.

Interest in solar energy reached a very low point between World Wars I and II, with only Dr. Charles Abbot, then director of the Smithsonian Institution, continuing to work, with a few other enthusiasts in various parts of the world, including Russia.

Collectors may be divided into a number of types, depending upon the service to which they are going to be put. First, the pressure at which they will operate is important, and, second, they may be divided into flat plates, generally fixed in one orientation, and concentrating collectors, which must follow the sun to attain highest effectiveness but can do a good job with less frequent adjustments if lower degrees of concentration are acceptable.

The flat-plate collector of the tubular variety, now used all over the world, seems to have originated almost simultaneously in a number of locations. They were in use in Australia, Japan, California, Arizona, and Florida before World War I, and, after the establishment of the nation of Israel, they proliferated there in a manner unmatched

Figure 2: *Parabolic trough in polar mounting used by Dr. C.G. Abbot on Mt. Wilson in 1924. Photo courtesy of the Smithsonian Institution.*

elsewhere in the world.

Solar radiation arrives at the earth with most of the energy in the form of direct rays from the sun, and, generally, a much lower proportion comes in the form of diffuse or sky radiation. The direct component can be concentrated by using either reflection or refraction. The first large concentrators, used by Lavoisier, had ingenious refractors, consisting of convex sheets of glass, sealed at their edges, and filled with white wine, to create a very large double-convex lens. It is doubtful whether Lavoisier recognized that a substantial proportion of the infrared radiation would be absorbed by the fluid within his lens, but it operated with sufficient effectiveness to allow him to ignite diamonds, and to perform other experiments that required very high temperatures. The same general idea, using a multiplicity of double-convex lenses, was used much more recently at California Institute of Technology, but, again, the objective was to produce a relatively small focal zone with very high rates of irradiation.

Since large lenses become unwieldy, heavy, and difficult to manipulate, attention was turned quite early in the development of concentrating collectors to the use of reflecting surfaces.

Polished metals and mirrors were used, with silver or mercury applied to front or back surfaces, depending upon the demands of the installation, and some very large installations were completed and operated nearly a century ago.

The concept of the parabolic trough concentrator is also extremely old. John Ericsson used one in New York in 1883 to generate steam and thus run a small steam engine. He used an equatorial mounting, with the axis of rotation adjustable so that it pointed towards the North Star, but the inclination of the reflecting surface could be adjusted to compensate for changes in the declination. He and those of his era were well aware of the astronomical nature of the earth's motion around the sun, so there was no mystery about the mechanism needed to track the sun in its apparent motion around the earth.

It became obvious early in the search for effective concentrators that horizontal paraboloidal troughs were effective if they possessed the capability of partial rotation, to compensate for the fact that the solar declination varies throughout the year from +23.47 degrees on June 21 to -23.47 degrees on December 21. The most spectacular example of this technology was the gigan-

tic pumping engine, Fig. 1, erected by Frank Shuman, from Philadelphia, U.S.A., and Dr. C. V. R. Boys, from London, England. The objective of their work was to generate steam that could in turn operate an engine, and this would pump, they hoped, large quantities of water for irrigation purposes. The Shuman and Boys troughs had a total area of 13,269 square feet for solar collection, and the concentration ratio was 4.5 to 1. Each collector was 205 feet long, and they used six of these monsters. Analysis of their test reports indicates that the collection efficiency was approximately 40 percent, which, considering the fact that there was little opportunity to insulate the heat absorber, was a very respectable performance

Small Concentrators

The attempt to make bigger and bigger concentrators appears to have subsided with the ending of the Shuman-Boys experiments in Egypt in 1916, and much smaller sun-following parabolic troughs, Fig. 2, were used by the late Dr. Charles Abbot to provide heat for cooking in his observatory on the top of Mount Wilson, late in the 1920s. Dr. Abbot continued his interest in the use of parabolic trough concentrators, and, in 1936, he produced a unit that operated a 1/2 horsepower steam engine for demonstration at the International Power Conference. This used three relatively small-diameter parabolic troughs, about 6 feet long, arranged so they could follow the sun with an altazimuth mounting. Later, he turned his attention back to the polar mounting, which required only one degree of rotation, and his last large steam generator, tested in the late 1950s at the University of Arizona, was fabricated in his own basement, when he was well past ninety years of age.

With the sudden increase in interest in solar energy, a number of new en-

Continued

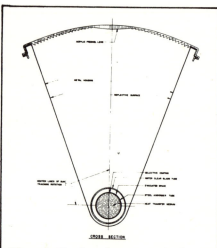

Figure 3: *Northrup linear Fresnel solar collector. Photo courtesy of Northrup, Inc., Hutchins, Tex.*

trants have come into this particular field. Some are using horizontal mountings, with the axis of their concentrators running in an east-west direction, with only enough rotation capability to compensate for the changing declination of the sun. The major problem encountered in these devices is getting proper insulation for the heat absorbing pipe that runs along the focal line. Dr. Abbot was one of the pioneers in the concept of using a glass tube to surround his steel absorber pipe, with the space between the steel and the glass tubes evacuated to minimize or, he hoped, to eliminate heat loss by convection. Other manufacturers are producing sun-following parabolic troughs mounted in the polar manner, with reflection accomplished either by using chemically polished aluminum, or aluminized foil.

A whole family of fixed concentrators has now come into the field, generally using parabolic surfaces that can accept solar radiation over a reasonably wide angle, and, by multiple reflections, direct it to an absorbing tube at the bottom. One of the first of these was patented by Alwin Newton, now of York, Pennsylvania, who added the interesting feature of making his reflecting surfaces adjustable, to compensate again for solar declination.

Refracting Concentrators

The principle of refraction as a means of concentrating the sun's direct rays has been investigated frequently. The flat Fresnel lenses that are available from science supply companies can be used to produce small hot spots, but these have had only limited usefulness. Recently, a new approach to the subject of refractive concentration has been introduced by Northrup, Incorporated, of Dallas, Texas (Fig. 3). Using a carefully calculated series of steps in a linear version of the Fresnel lens, this system can successfully concentrate a 12-inch-wide band of direct solar radiation onto a steel tube of no more than 2-inch diameter. The illustration shows the second generation of Northrup collectors, in which the steel absorber tube is coated with a selective surface to minimize radiative heat loss, and is surrounded by a high-transmittance glass tube. The space between the two tubes is evacuated, just as it was in the case of Dr. Abbot's earlier concentrators, but the Northrup device

Figure 4: *Martin-Marietta concentrating segmental heliostat, viewed from the sun-facing side. Photo courtesy of Martin-Marietta Corp., Denver, Colo.*

has the advantage that rotation of the refracting element around the absorber tube is all that is necessary to keep the sun's rays concentrated on their target. The earlier Northrup collectors used rotating joints, but, with an advanced development of the lens, it is now possible to keep the absorber tube fixed, while the concentrator rotates about it to follow the sun.

Applications of Concentrating Collectors

The most obvious application today of concentrating collectors is in the field of absorption refrigeration, where temperatures of 200° to 230°F are needed to make conventional lithium-bromide-water machines operate at their full capacity. These temperatures are difficult if not impossible to attain for more than very brief periods by means of flat-plate collectors, and so the largest of the absorption installations currently being made under the ERDA demonstration program will use the linear Fresnel lens concentrator.

Another obvious target for more advanced versions of concentrating collectors is the operation of engines on either the Rankine cycle, for vapors, or the Sterling-Brayton-Ericsson cycle. The latter are variations of the gas turbine concept, and, if concentrators can be developed that can produce temperatures in the range of 600°F consistently, very respectable thermal efficiencies can be obtained.

Setting our sights slowly lower, Rankine cycle engines or turbines can be operated with vapors at temperatures in the range of 300° to 400°F. The cycle efficiency will be low in such cases, but the major objective is to make such devices take over the job of water pumping in arid areas, where energy is currently being derived from natural gas-driven pumping engines. History is again repeating itself, with water pumping the most important application.

Second to that is a means of producing refrigeration with a higher coefficient of performance that can be obtained with the absorption system, and with, it is hoped, a lower total cost. The absorption refrigeration system has a coefficient of performance substantially below 1.0, and at first sight, a Rankine cycle compression system looks much more attractive. However, when one considers the fact that the cycle efficiency of the engine is not likely to exceed 10 percent, and the coefficient of performance of the compression system will probably not be any greater than 4.0, the overall use of the collected solar energy is very little better for the Rankine cycle than for the absorption device. The problem at this point is to come up with a relatively inexpensive and foolproof system for concentrating at a moderate degree, sufficient to give, in most climatic regions, temperatures in the range of 200° to 230°F, thus enabling standard commercial absorption refrigeration machines to operate satisfactorily.

For obtaining very high temperatures, a multiplicity of slightly curved concentrating mirrors is as old as Archimedes, but only now are such devices being brought to a practical scale. Fig. 4 shows the heliostat developed by Martin-Marietta Corporation, of Denver, in which apparently flat glass mirrors are held in a planar configuration, on the altazimuth mount. Actually, the mirrors are not flat but, instead, they are slightly curved by being supported at their edges, as shown in Fig. 5, and pulled in slightly at their centers. Over the very long focal distances envisioned in this system, the image becomes circular, and approximately twice the size that would be produced by a perfect parabola. By concentrating the images of a large number of these individual concentrators, very high solar irradiation can be accomplished. Although the Martin-Marietta system is designed primarily for use with very large power generating systems, there is no fundamental reason why it cannot be used for a smaller system, and this would seem to be one feasible answer to the problem of obtaining sufficient energy to run relatively small engines.

In conclusion, it should not be overlooked that concentrators, as we know them today, make little if any use of the diffuse radiation from the sun, and, in humid locations or on cloudy days, from half to 100 percent of the solar irradiation is diffuse. A great challenge remains to be answered, and that is the development of a concentrator that can harness the diffuse as well as the direct rays of the sun. ☼

Figure 5: *Rear of Martin-Marietta concentrating heliostat, showing method of mounting individual mirrors. Photo courtesy of Martin-Marietta Corp.*

Concentrating Collectors

PARABOLIC TROUGH CONCENTRATOR

manufactured by:
Hexcel
11711 Dublin Blvd.
Dublin, Cal. 94566
George P. Branch; phone: (415) 828-4200

Parabolic trough concentrator in module lengths of 20 feet with apertures up to 9 feet. Tracking features include shaded photo-transistor sensor, accuracy to plus or minus ½°, full inversion of trough for storage. Troughs are aluminum honeycomb with aluminum skins; reflective film is aluminized modified acrylic. **Features and options:** all metal frame. Working fluids can be air, water, or oil. Control options include day/night operation sensor, over-temperature protection, adjustable cloud and/or insolation operational limits, and no-flow protection. **Price:** consult factory.

FES DELTA CONCENTRATING COLLECTOR

manufactured by:
Falbel Energy Systems Corp.
472 Westover Rd.
Stamford, Conn. 06902
Gerald Falbel; phone: (203) 357-0626
Aluminum or copper 4 by 8-foot by 6-inch concentrating collector for hot water and swimming pool heating systems. Achieves a net of 2.3:1 concentration for both direct and diffuse solar radiation. No tracking is required. Send for catalog. **Price:** consult manufacturer.

SLATS™

manufactured by:
Sheldahl, Advanced Products Div.
Northfield, Minn. 55057
A. J. Wendt; phone: (507) 645-5633
A line-focus solar collector with 1 by 20-foot mirror segments for focusing the sun's

rays. Ten parallel segments are mounted between end supports in a planar configuration. Pivoting on a long axis, the segments focus on a single fixed receiver and rotate face down to protect the reflective surface when not in use or in severe weather. Gang-driven by a single motor, the unit automatically sun tracks and has safety features to stop storm damage, frost buildup, and provide protection against overheating of the receiver. **Features and options:** operating temperatures to 600°F. **Price:** consult manufacturer.

SOLAR TUBE SYSTEMS

manufactured by:
Heilmann Electric
127 Mountainview Road
Warren, N.J. 07060
phone: (201) 757-4507
Solar Tubes are each 6 feet long and 4 inches in diameter; tube is made of Lexan. Absorber surface is aluminum with 5/8 inch copper tubing bonded to it and backed with fiberglass insulation. **Features and options:** lightweight and breakage resistant. Tubes may be rotated during and after installation to maximize solar collection. **Installation requirements/considerations:** one man can install an entire system using ordinary hand tools. **Guarantee/warranty:** five year guarantee on material and workmanship. **Availability:** prompt service and quick delivery. **Price:** minimun order 54 linear feet. 54-108: $4.25 per linear foot. 109-270: $3.50 per linear foot. 271-650: $3.25 per linear foot. Over 650: $2.95 per linear foot.

NORTHRUP CONCENTRATING COLLECTOR/NSC-P

manufactured by:
Northrup, Inc.
302 Nichols Dr.
Hutchins, Texas 75141
E.C. Ricker; phone: (214) 225-4291
Radiation is concentrated onto a black chrome-coated copper tube and produces high temperatures at high efficiency. Collectors are ganged together and track the sun. Manifolding, framing, and support apparatus sold with collectors and trackers. **Installation requirements/considerations:** virtually all necessary parts included in ''erector-set'' like kit. **Guarantee/warranty:** Eighteen months from shipment or twelve months from installation, whichever occurs first. Limited warranty. **Maintenance requirements:** negligible. **Manufacturer's technical services:** inquire regarding specific jobs. **Regional applicability:** worldwide. **Availability:** consult factory. **Price:** consult manufacturer.

SOLARLAB

MND Inc.
P.O. Box 15534
Atlanta, Ga. 30333
A complete teaching lab for the use of solar energy. Aluminum and copper flat plate collectors. Fiberglass and plastic glazing; a 40 to 1 concentrator; an air heater and a photovoltaic panel, completely instrumented with insulated test bed and adjustable mount. **Installation requirements/considerations:** water supply and 110 V 5 amp line. **Guarantee/warranty:** none. **Maintenance requirements:** none. **Manufacturer's technical services:** advice and components. **Regional applicability:** no limitation. **Price:** varies with options desired.

SUNTREK-7 SOLAR ENERGY LABORATORY

Alpha Solarco
Suite 2230, 1014 Vine St.
Cincinnati, Ohio 45202
M. Uroshevich; phone: (513) 621-1243
A parabolic concentrating tracking collector for high temperature solar energy collection to be used primarily for research. demonstration, and data aquisition. **Features and options:** steel pipe receiver/absorber with black chrome selective coating mounted in Corning evacuated glass tube. Optional models offer vacuum pump attached permitting experiments with variable vacuum. No-flow boiler mode of operation can produce temperatures up to 1000° F. Many options available. **Price:** consult factory.

KTA CORP. SOLAR COLLECTORS

KTA Corporation
12300 Washington Ave.
Rockville, Md. 20852
Ted Knapp, phone: (301) 468-2066
KTA's solar collectors employ a tubular design that exhibits efficiencies of 50-60 percent under U.S. National Bureau of Standards test criteria (NBSIR 74-635). It consists of a parallel array of double glass cylinders. The outer tube is half-silvered with a mirror finish that concentrates sunlight on a copper conduit that carries the heat transfer fluid. **Features:** because of the parallel array of the reflectorized tubes, the elements can be factory-rotated to a fixed angle that absorbs the optimum amount of solar insolation. Collectors are lightweight and low in cost. **Installation requirements/considerations:** panels are easily installed, can be integrated into most plumbing systems. Pitch of roof not a problem due to rotation of tubes. **Guarantee/warranty:** KTA warrants collectors to be free of manufacturing defects for a period of thirteen months following date of shipment. **Maintenance requirements:** it is recommended that field repair of KTA solar collectors not be attempted, but that they be returned to the factory for warranty or post-warranty repair. **Price:** ranges from $149 to $412.

SUNPUMP/SOLAR COLLECTOR MODULE SCM-100

manufactured by:
Entropy Limited
5735 Arapahoe Ave.
Boulder, Colo. 80303
Henry L. Valentine; phone: (303) 443-5103

A non-tracking focusing collector that heats water to a vapor state within a heat pipe. The vapor can be transported to a condensor (heat exchanger) where the energy of the latent heat of vaporization (540 calories per gram) is transferred. **Features and options:** requires small diameter tubing for water supply and steam manifold. Water volume is less than 1/10 that of a comparable flat plate collector. **Installation requirements/considerations:** collectors can be mounted on horizontal, vertical, or sloping surfaces—either on or remote from a building. **Guarantee/warranty:** one year free from defects in material and workmanship. Design life twenty years minimum with periodic maintenance. **Maintenance requirements:** periodic inspection and cleaning of outer glazing. **Regional applicability:** anywhere. **Price:** consult factory.

CONCENTRATING COLLECTOR/ SKI-MK-4 LINEAR PARABOLIC

manufactured by:
Solar Kinetics, Inc.
147 Parkhouse Street
Dallas, Tex. 75207

A medium temperature (450°F) collector—tracking in one plane, with a concentration ratio of 40:1. The receiver tube is glass encapsulated. All aluminum unit construction allows high wind stress and long life. **Guarantee/warranty:** one year. **Maintenance requirements:** minimal. **Manufacturer's technical services:** available on request. **Regional applicability:** all. **Price:** consult factory.

SOLAR EXPANDER

manufactured by:
Dutcher Industries, Inc.
Solar Energy Div.
7617 Convoy Court
San Diego, Cal. 92111
E.S. Cox; phone: (714) 279-7570

A solar-powered steam engine useable for any shafy power requirements. Develops 5 l.c. at 1,100 rpm using 14 pounds per hour of steam. **Features and options:** is L-head poppet valve engine with push-rod operation from crankshaft driven polydyne cam. Has hydraulic valve lifter, self-relieving valves, and 10:1 expansion ratio. **Installation requirements/considerations:** must have water supply from solar collector at 350°F and up inlet, supply of cooling water, and concrete mounting pad. **Guarantee/warranty:** ninety days on parts and labor. **Maintenance requirements:** crank-

case drain and fill every six months. **Manufacturer's technical services:** installation and repair both in field and factory. **Regional applicability:** generally in the southwestern U.S. but useable in other countries of similar insolation rates. **Availability:** factory only. **Price:** consult manufacturer.

SUN-TRACKING RESIDENTIAL SOLAR HOT WATER SYSTEM/1000

manufactured by:
Energy Applications, Inc.
830 Margie Drive
Titusville, Fla. 32780
Napoleon P. Salvail; phone: (305) 269-4893

A residential, north-south oriented, daily sun-tracking solar collector system that focuses the sun with rotating parabolic reflectors on Pyrex-jacketed stationary absorbers. The 36 square foot unit comes in kit form for homeowner installation. **Features and options:** fully automatic operation; lightweight—2.5 to 3 pounds per square foot; structure of aluminum, plastics and Pyrex to reduce corrosion and deterioration in humid environments. Add-on tracking modules (12 square feet each) are optional as is support structure for ground and flat roof installations. **Installation requirements/considerations:** collector faces south with a fixed inclination equal to the latitude of the location. **Guarantee/warranty:** one year on parts and labor. **Maintenance requirements:** none. **Manufacturer's technical services:** product technical consultant. **Regional applicability:** not advisable for areas with frequent ice storms at present time. **Availability:** from manufacturer. **Price:** $950 (optional add-one tracking modules of 12 square feet $150 each).

ACUREX CONCENTRATING COLLECTOR

manufactured by:
Acurex Aerotherm
485 Clyde Ave.
Mountain View, Cal. 94042
Uwe Schmalenbach; phone: (415) 964-3200

Concentrating parabolic trough collector designed to heat liquids or gases to temperatures between 140° and 500°F. Modules are 10 feet long with 6-foot apertures; eight modules can be coupled together forming a line of collectors driven by a single tracking system. **Features and options:** receiver tube is 1.25-inch stainless steel coated either with black paint or with black chrome over nickle plate. The tube is enclosed in a Pyrex jacket. Aluminum lighting sheet mounted on a painted structure is the reflective surface. Possible working fluids include water, organic liquid, or air. Applications include industrial water and air heating, shaft power for irrigation pumping or electrical generation, steam generation, and space cooling. **Price:** consult manufacturer.

CONCENTRATING AND SUN-TRACKING COLLECTOR/2000

manufactured by:
Energy Applications, Inc.
830 Margie Drive
Titusville, Fla. 32780
Napoleon P. Salvail; phone: (305) 269-4893

For commercial and industrial process heat and solar air conditioning. The model 2000 is a north-south oriented, daily sun-tracking collector that focuses the sun with rotating parabolic reflectors on a sealed, glass-jacketed stationary absorber. The 110 square foot unit operates at 200°F plus, with efficiencies greater than 50 percent. **Features and options:** fully automatic operation with maximum control flexibility and convenient system status monitoring with ground interface panel. Collector weight of 2.5 to 3 pounds per square foot; extensive use of Pyrex, aluminum, and plastics to reduce materials corrosion and deterioration in humid environments. A ground control interface panel for one to ten arrays and a support structure for flat roofs are optional. **Installation requirements/considerations:** on-site installation requires two semi-skilled workers approximately four hours per 110 square foot array; collector components are light enough for two workers to handle: reflector modules weigh 25 to 30 pounds each, lower bearing support—25 feet long—weight 40 pounds. **Guarantee/warranty:** one year on parts and labor. **Maintenance:** none. **Maufacturer's technical services:** product technical consultant. **Regional applicability:** no restrictions. **Availability:** from manufacturer. **Price:** $15 per square foot (110 foot array). $14 per square foot for two to five arrays. Optional ground control interfacr panel, $150. Optional support structure for flat roof, $350.

VACUUM TUBE COLLECTOR/TC-100

General Electric Company
P.O. Box 13601
Philadelphia, Pa. 10101
William F. Moore; phone: (215) 962-2112

The TC-100 collector is an evacuated tube array within a steel box that uses GE fluorescent lamp tubes fabricated into a thermos bottle unit. The tubes lie in a tray that serves as a concentrator. Thermal energy is removed from the glass tubes by an independent copper fluid system. **Features and options:** high-temperature up to 300°F, high efficiency at high temperature, high Btus pr dollar. **Installation requirements/considerations:** simple four corner mounting with two hydraulic connection points. **Guarantee/warranty:** one year for materials and workmanship except glass. **Maintenance requirements:** none. **Regional applicability:** anywhere in the United States. **Availability:** delivery late 1977. **Price:** consult manufacturer.

5 big reasons why Solar Age keeps you one step ahead...

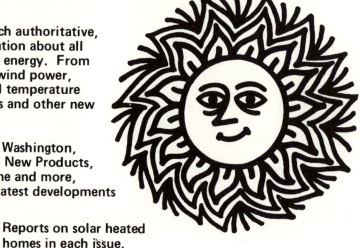

1 Nowhere else will you get such authoritative, accurate and current information about all aspects of solar and alternate energy. From solar heating and cooling to wind power, bioconversion, ocean thermal temperature conversion, photovoltaic cells and other new energy sources.

2 Regular columns like: From Washington, Local Report, Business Beat, New Products, Required Reading, Solar Scene and more, will keep you abreast of the latest developments in solar applications.

3 Reports on solar heated homes in each issue.

4 Interviews with outstanding solar people in future issues

5 Articles coming up on: Sun Rights, Examining Collector Covers, Rock versus Water Storage, Utility Rate Structures, Energy Plantations, Heat Pumps, Design Check List for Solar Heating and Cooling System, plus many others.

HEAT PIPES

*Simple, sophisticated gadgets
that should find a use in solar energy*

By John Kusianovich

First, what is a heat pipe? It's simply an evacuated container, usually a tube, inside which a liquid is sealed. When heat is applied to the outside of one end, the liquid boils; the vapor travels to the other end and condenses back into a liquid, giving up its heat. This end of the heat pipe then gives off the heat to its surroundings. The liquid travels back to the hot end by gravity or by capillary action through a wick. It's a continuous, automatic, passive process that requires no outside source of power, other than heat. The heat pipe does not create heat; it merely transports it. However, if you remember your high school science, because the heat pipe uses the latent heat of vaporization of a liquid it transfers heat much more effectively than normal fluid flow and at a nearly constant temperature. Thus, it is a hundred times more effective than a solid copper rod of the same dimensions.

Water is the most common heat pipe working fluid in the temperature range normally needed to heat houses with solar energy (100° to 300°F). For lower temperatures alcohol can be used, and at high temperatures—liquid metals.

Because the air pressure has been evacuated from the tube, the liquid will boil at a temperature lower than it would at atmospheric pressure. Another nice feature derives from the fact that very little of the internal volume of the heat pipe is actually occupied by the liquid. This allows freezing to take place with no damage.

What are the drawbacks? Cost. A heat pipe must be a perfectly sealed container or the liquid will escape. Any impurities, such as trapped air or a dirty wick, will seriously impair the heat pipe's performance.

Mr. Kusianovich, who lives in a house heated by solar concentrators in Albuquerque (see Solar Age. January 1977) is a former heat pipe manufacturer.

What are the possible uses of a heat pipe in solar energy? Primarily it can be used to carry heat from a collector to storage. For example, one end of the heat pipe can be soldered to the absorber plate and the other end immersed in a storage tank. Heat will be continuously transferred as long as the collector is hotter than the storage. Theoretically, a heat pipe will work against gravity; however, at the current state of the art they work much better if the lower end is heated. This implies a thermosiphoning solar collector. The heat pipe will have little back-flow at night and can carry more heat, but will cost more.

The increasing use of concentrating collectors will probably increase the demand for heat pipes. There, high heat carrying capacity can be a distinct advantage.

If heat has to be transferred long distances, though, a standard pump and pipe arrangement is still required, because a heat pipe more than a few feet long is difficult to obtain. The wick used in common heat pipes can lift the liquid only a few inches against gravity. Research is being conducted at the University of New Mexico on a heat pipe using a 'liquid pump' wick that may be able to lift the liquid many feet. This would allow a heat pipe to connect a roof-top collector with a basement storage tank. Another idea that would do the same thing has been patented by Steve Baer of Zomeworks. This is the use of a semi-permeable membrane as a 'wick,' the same method a tree uses to lift its sap. Height is no problem, but the cost and low flow through commercially available membranes is.

Some of the companies involved in heat pipes include Dynatherm on the east coast and TRW and Hughes Aircraft in L.A.

There are not a lot of heat pipes being used in solar energy yet, but they are being used in energy conservation, for instance to recover heat from the ventilation air being exhausted from large buildings such as hospitals. Heat pipes are arranged with one end in the intake duct and one end in the exhaust duct. This creates a heat exchanger that allows air from the two ducts to transfer heat without mixing. A leading company in this field is Q-Dot Corp. of Dallas.

Most heat pipes are made under laboratory conditions using vacuum pumps and exotic welding techniques. But a crude heat pipe can be made as follows:

Take a copper tube, seal one end by soldering on a cap. Make a wick by rolling up a piece of 100-mesh copper screen into a tube that will fit tightly inside the copper tube. Heat both tube and wick in an oven until slightly oxidized. This will allow the working fluid to 'wet' the surfaces. Insert the wick into the tube, then solder a Hoke-type vacuum valve to the open end. Fill the tube about 1/5 full of water. All components, including the water, should be very clean. With the valve open and at the upper end, heat the lower end until the water inside begins boiling vigorously. After it has been boiling for a few moments, close the valve and remove from the heat. Now try out your heat pipe by immersing one end in boiling water and feel how long it takes for the other end to get hot. Then compare this to a copper rod of the same dimensions. If the heat pipe does not get hot much faster than the rod you did something wrong. CAUTION, NEVER HEAT THIS UNIT TO MORE THAN 250°F!

If you use ½-inch pipe this heat pipe should be able to carry 100 watts of heat easily.

Higher power commercially made heat pipes have more exotic wicks, such as 'thread arteries.' This means that the inside of the pipe is finely threaded, and mesh tubes are attached along the inside walls. There are many more sophisticated variations on the heat pipe theme, beyond the scope of this little article. Heat pipes have been around for more than thirty years, and are still not widely used. Solar energy may be the field that will find a place for this simple device. ☼

HEAT PUMPS

The Carnot Cycle can be a friend to solar systems and to energy conservation. Here's what heat pumps are, what they do, how to choose one.

By Herman G. Barkmann

What is a heat pump, and how does it work? First off, the term 'heat pump' is merely an explanatory name for standard vapor-compression equipment, the heart of our refrigerators and air conditioners. In such a system energy in the form of heat is moved or pumped from one place to another. The vapor-compression cycle has the unique ability to pump energy from a heat source at one temperature to a heat sink at a higher temperature, contrary to what we have understood as the laws of nature, that require heat energy to flow from a high temperature to a lower temperature by conduction, radiation, or convection.

This capability is based on the functions of the three primary heat pump components: the evaporator, the compressor, and the condenser, and on the ability of a liquid to absorb considerable energy upon change of phase to a vapor form, and of vapor to give off that same energy in the process of changing back to a liquid. The liquid refrigerant expands and evaporates in a coil or heat exchanger, and in so doing absorbs energy from the medium surrounding the heat exchanger, thereby cooling that medium (food-stuff, air, or water). The expanded vapor is pulled into the compressor that is pumping the refrigerant through the system. The refrigerant is compressed to a high-density gas at a temperature much higher than that at which it entered the compressor. This hot gas is then circulated through a condenser or heat exchanger that acts, in effect, to reverse the evaporator. The condenser converts the hot gas into a liquid by rejecting the heat in it to a cooling medium, air or water, surrounding the condenser. This medium is lower in temperature than the hot gas, yet considerably warmer than the medium sur-

rounding the evaporator. The refrigerant is then ready to be evaporated by being pumped through an expansion valve into the evaporator. Thus the cycle starts again.

This cycle, known as the Carnot Cycle, allows a refrigerator to absorb heat from water colder than 32°F (0°C), and to reject that heat to room air at 70°F (21°C); or it allows an air conditioner to absorb heat from a room at 80°F (27°C) and reject that heat to outside air at 100°F (38°C).

A refrigerant is any liquid that will evaporate and recondense. Water is

one, but the most commonly used refrigerants are the fluorocarbons known as Freons. One important property that determines choice of refrigerant for a particular duty is its latent heat of vaporization, which governs the ability of a liquid to absorb heat upon change to vapor phase. Other important properties are the pressure-temperature relationships at different phases of the refrigerant, determining in turn the possible operating temperatures and pressure of a system.

Any refrigeration system is a heat pump. The term, however, is common-

Shook's run, in Colorado, a solar-assisted water-to-air pump system...

...supplying 80 percent of the heat and hot water to twelve apartments...

Mr. Barkman is a consulting engineer based in New Mexico.

ly used to describe rather specialized equipment capable of pumping energy in either direction. The energy required to pump the refrigerant through its different changes of state, as well as to operate fans or pumps at the heat absorption and rejection points on the cycle, is normally supplied to electric motors running the compressor and fan/pump motors. The compressor absorbs the major part of this energy.

The effectiveness of the system is measured by the amount of energy absorbed, transferred, and then rejected, compared with the energy supplied to the compressor and fans; in other words, the ratio of energy gained to energy invested. For refrigeration and air conditioning equipment this has been called the E.E.R., Energy Efficiency Ratio, and is in units of Btus per Watt. For heat pump equipment, however, the term C.O.P., or Coefficient of Performance, is used and is ex-

energy is to be used for space heating purposes, the heat pump needs only from a fourth to a half the energy that would have been used for normal electrical resistance heating. If a heat pump has a C.O.P. of 2.5, and assuming a power plant efficiency of 33 percent, one Btu of energy at the power plant will produce .825 Btus of heating energy at the heat pump condenser. This compares favorably with the use of resource energy by a gas furnace with an overall efficiency of .6 to .7.

Because heat pumps use low temperature sources of heat, they allow the use of low temperature materials, non-insulated piping, and low technology.

They also lend themselves readily to retrofits, as it is relatively easy to add one to an existing system. Residential air conditioning systems can readily be converted to heat pumps, as can many large commercial chilled water systems.

pressor, an outdoor air-cooled condenser, with either an expansion valve or capillary tube used to control expansion of the refrigerant into the evaporator coil. A reversing, or four way valve was added to allow the outdoor condensing coil to function as the evaporator, and the indoor evaporator coil to act as the condenser. Thus the equipment could cool the interior of a residence in summer by absorbing indoor heat and rejecting it to the outside air, and in the winter, by reversing the refrigerant flow, the same equipment could heat the residence. These first-generation heat pumps were beset with maintenance problems, and gained a poor reputation.

A new generation of equipment now available has been designed to withstand the heating cycle duty and is proving much more satisfactory from a service standpoint. There is much to be said for the argument that this heat pump is truly a solar heating system, using the earth's atmosphere as a collector.

A second but less common air-to-air heat pump reverses the air flow rather than the refrigerant circuit, thus avoiding the complication and losses of a reversing valve.

The air-to-water is a rather uncommon type of heat pump. In this system outside air is used for summer heat sink and winter heat source, but water is cooled and heated for distribution either to a multizone interior environmental system, or for domestic or industrial purposes.

...uses two storage tanks to give the system temperature flexibilty.

pressed as a ratio of Watts per Watt, or Btus per Btu. The theoretical C.O.P. for the heat pump cycle may be 20 or greater. Due to friction losses, heat losses, and other causes for inefficiency, in actual practice heat pumps operate with a C.O.P. normally between 2 and 4. This means that for an expenditure of one Watt in input energy, four Watts may be transferred by the refrigeration effect and used for heating. (It should be noted that for a given heat pump cycle the heat rejection C.O.P. is always greater than the cooling C.O.P. because friction losses in equipment, compressor, fans, etc., add to the heating effect and detract from the cooling effect.) It can be seen, then, that if electrical

Heat source to heat sink

There are many kinds of heat pumps, with names that normally denote first the heat source, then the heat sink. Probably the best known type is the air-to-air heat pump, where the source of energy is outside air and the sink or medium to be heated is the inside circulating air. This type, now available nearly everywhere, is manufactured by nearly all manufacturers of residential air conditioning equipment. The units originally developed from residential central system air conditioners consisting of an indoor evaporator coil, a com-

A third type, the water-to-air heat pump, is gaining popularity for installation in multizone environmental control systems. One adaptation, also called closed-loop heat pump and electrohydronic system, is designed with a water-refrigerant heat exchanger that acts as condenser during the cooling mode and evaporator during the heating mode. Several heat pumps are tied to a closed water loop that circulates between the pumps, some of which may be in interior zones, some in exterior zones. This system has the advantage that zones requiring cooling may reject heat into the water system, for use in zones requiring heat. The air side of this heat pump either absorbs heat from the occupied area on the

Continued

HEAT PUMPS

continued

cooling side, or rejects heat into it on the heating cycle.

A water-air heat pump that could be designated earth-air uses well water either as a heat sink or a source. Other types of earth-air heat pumps have been developed to use heat exchangers buried underground.

Water-to-water heat pumps have been used rather extensively in large commercial buildings. The condenser side of conventional water chillers can provide hot water for areas that need it.

One last type of heat pump that really belongs in the first class has just recently been put on the market. It is a heat-only air-to-air heat pump. This system has a higher C.O.P. and capacity due to the use of components optimized for the heating cycle, and it includes no reversing valve.

There are some drawbacks

The most common heat pump, the residential air-to-air system, has one great disadvantage: as the outside air temperature decreases and the building heat load, accordingly, increases, both the capacity and the C.O.P. of the heat pump goes down (Figure 1). The out-

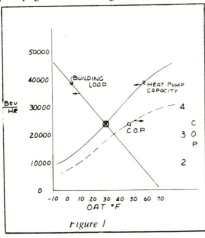

Figure 1

side air temperature at which the heating capacity of the heat pump just equals the heating demand of the house is called the balance point of the installation (point X). In the past, these heat pumps have been sized to meet the cooling load, allowing the balance point to fall where it may. Recently, however, heat pumps are being sized so that the balance point falls at the temperature of greatest heating load frequency. When the temperature of the

outside air drops below this balance point, resistance heat must be added to make up the deficit in capacity. This will, of course, decrease the overall C.O.P. One other detraction is the fact that when the outside air drops to 40°F or below, frost forming on the outside air coil degrades the operation of the system. Defrosting is accomplished by operating the system in reverse for a short period of time.

When the energy use of a heat pump is analyzed, taking into account the use of resistance heating and defrosting, an overall seasonal performance factor is determined. This is apt to be between 1½ and 2 rather than the values of 3 to 4 that can be achieved for short terms.

The heat pump, because of its complexity, tends to be more expensive in almost all forms than a more conventional system. A residential heat pump would probably cost more than three times as much as a gas furnace, and a third more than an air conditioning system. Until the recent introduction of the heat-only heat pump, it was difficult to justify the residential heat pump system economically unless air conditioning was a necessity. Costs and economic justification of the heat-only pump are yet to be determined.

Solar assists

It appears now, with curtailment of fossil fuel supplies, that electrical energy will provide an ideal auxiliary energy source for solar heating systems. Because of the relative ease of transporting electricity, this is particularly true for solar-heated buildings in areas remote from natural gas lines. Studies performed at the University of Pennsylvania have shown that a solar-augmented heat pump makes a minimum use of resource energy. Other studies at Ohio State University indicate that the heat pump is the ideal auxiliary energy system for a solar heating system.

One primary reason for joining the two techniques is that both work very well at low temperatures. Heat pumps work with C.O.P.s of up to 4 at temperatures near room temperature. Solar collecting systems in simplest form work exceptionally well when working fluid temperatures can be maintained at a level no greater than 100°F. The simplest low-cost collector operating at temperatures from 80° to 100°F may supply considerably more Btus per

dollar than many more sophisticated collectors. It has even been proposed by some brave souls that with the advent of improved heat pumps, unglazed collectors may be used. This would of course allow great nighttime heat rejection possibilities.

A simple residential air-to-air solar-augmented heat pump system might use outside air temperatures down to 50°F to operate the heat pump as a conventional system, while a solar collector system is collecting and storing heat. When outside air temperatures are below 50°F, heat could be supplied directly from storage for some time. Then, when storage temperatures drop to some predetermined point, for instance 80°F, the solar storage would provide a source for the heat pump. This heat pump would operate with a C.O.P. between 2.5 and 4, and by using direct solar space heating a seasonal performance factor of well over 3 could be obtained. Proper design procedures of course must be followed to maintain proper storage temperatures. It should be pointed out that with current energy and solar system costs, it is virtually impossible to compare such a system economically, even over a twenty-year period, with a conventional heat pump.

Choosing the system

Once owner and design engineer have decided to use a heat pump as the energy source for a heating system, the process of proper equipment sizing follows. The most commonly used heat pump for residential applications, well suited to solar assist, is the air-to-air heat pump. And the common way of classifying air-to-air heat pumps is by their cooling capacity, measured in tons—you have heard of a two-ton, or three-ton, or five-ton heat pump. A ton of air conditioning capacity is equal to 12,000 Btus per hour of total heat capacity. The heating capacity of a heat pump is generally close to the cooling capacity, at a specified outdoor air temperature—usually 45°F. Thus, the heat pumps above would be capable of supplying some 20,000, 30,000, or 50,000 Btus per hour of heating when the outside air is 45°F.

One recommended method for sizing a heat pump is to choose a unit that attains a balance point—the point

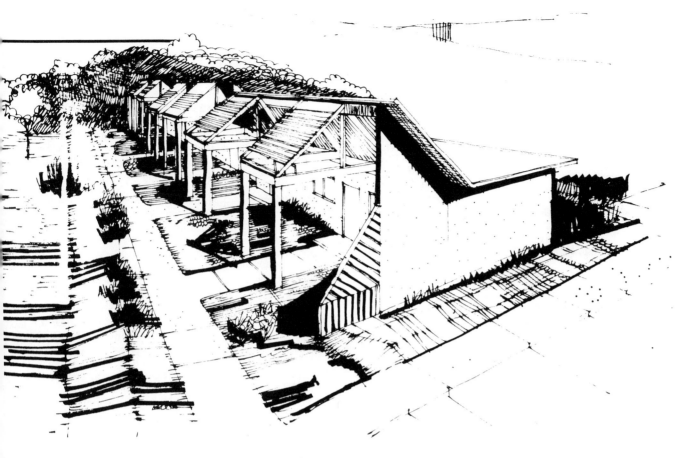

at which the heat pump curve crosses the building load curve—at the outside air temperature where the largest annual heating load is expected. This may be determined from weather data, with each temperature given as a function of hourly frequency, over as many years as possible. A histogram may then be constructed by determining the heat load or temperature difference, and multiplying it by the number of hours at that temperature, usually in five degree bins. The maximum value may then be determined, and equipment chosen with a balance point not to exceed that point.

You might choose air-to-air heat pumps with balance points as much as 10°F below that temperature, to decrease the amount of resistance heat required. This choice of higher capacity heat pump is becoming more and more acceptable where the primary requirement is for heating.

Other considerations enter the choice of a heat pump: proper air flow for the space, C.O.P. at operating temperatures, size, and cost. It is often valuable to plot load versus capacity and (integrated) C.O.P. for several heat pumps because the shape of the curve for capacity and C.O.P. versus temperature may vary among manufacturers.

Once the curves of capacity and load versus temperature are plotted for the unit chosen, the required amount of auxiliary heat may be determined (see Figure 2 where the auxiliary heat required is indicated by the vertical dis-

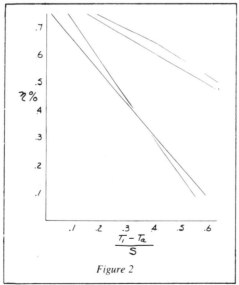

Figure 2

tance between the capacity curve and load curve). For most air-to-air heat pumps this auxiliary heat is supplied by resistance heaters in the air streams and may be provided in one or more stages.

Water-to-air heat pumps will have their performance data presented as a table that gives cooling or heating capacity as a function of the water supply temperature to the condenser/evaporator coil. When choosing a water-to-air heat pump the balance point is not a consideration. A heat pump that has a heat capacity at a minimum water temperature that is at least equal to the heat load at design temperature should be chosen, just as standard heating equipment is chosen. In this type of heat pump, the auxiliary heat is added through a boiler, or through another energy supply like a solar collector, to the circulating water system, in order to maintain the water temperature within the prescribed limits. Complete instructions for choosing this type of heat pump are included in design manuals supplied by the manufacturers. Most heat pump manufacturers, in fact, notably General Electric, have excellent instructions for equipment selection, and such information should be carefully examined.

When air-to-air heat pumps are intended for use in a solar system, the choice of heat pump may depend on the design of the outdoor unit. The newer round outdoor units are most convenient if one attempts to duct air from either the outside or a storage bin, while older rectangular models are easily adaptable. ☼

Heat pumps

WESTINGHOUSE HEAT PUMPS
manufactured by:
Westinghouse Electric Corp.
Commercial-Industrial Air Conditioning Div.
P.O. Box 2510
Staunton, Va. 24401
W.. Fennessey; phone: (703) 886-0711

Large centrifugal compressor-driven chilled-water heat pumps and centrifugal and reciprocating compressor-driven Templifier™ heat pumps used commercialy and industrially to recover heat from otherwise waste fluid sources. **Features and options:** all units are factory-prepackaged and refrigerant-charged to simplify installation requirements. **Installation requirements/considerations:** prior detailed engineering evaluation to ascertain the existence of a valid, substantial heat source is required. **Guarantee/warranty:** standard Westinghouse product warranty. **Maintenance requirements:** routine chilled-water unit maintenance is required, normally provided by the customer direct or through contract with Westinghouse field service. **Manufacturer's technical services:** engineering assistance in the selection, evaluation, application, and maintenance of equipment. **Regional applicability:** national. **Availability:** product is marketed through direct sales offices and independent sales representatives. **Price:** consult manufacturer.

SELF-CONTAINED HEAT PUMP SYSTEMS
manufactured by:
Bard Manufacturing Company
P.O. Box 607
Bryan, Ohio 43506
C.E. Hoffman; phone: (419) 443-2993
Year-round comfort control units housed in a single compact package. Desired temperature is maintained automatically by setting wall thermostat. Finished galvannealed steel cabinets house compressor-condensor, evaporator, and controls that heat and cool living spaces. **Features and options:** large

finned-coil surfaces disperse or absorb heat as system requires; compressor is equipped with off-cycle crankcase heat and is protected with internal overload and anti-slug device. Electric heat strips are optional. **Installation requirements/considerations:** units fit almost anywhere with no structural changes necessary. **Guarantee/warranty:** five years on motor compressor; one year on materials and workmanship. **Maintenance requirements:** cleaning of coils, water drains, motor lubrication. **Availability:** numerous models available from Bard dealers. **Price:** consult manufacturer.

WATER -TO-AIR HEAT PUMPS
manufactured by:
Koldwave Div. Heat Exchangers, Inc.
8100 N. Monticello
Skokie, Ill. 60076
Lloyd L. Ludkey; phone: (312) 267-8282
Water-to-air heat pumps in vertical and horizontal series with capacities ranging from 9,500 to 96,000 Btus per hour. Models are self-contained, requiring only water and electrical connection. **Features and options:** refrigeration system safety controls; Scrader valves; thermostats for remote mounting; sound absorbent insulation and reilient mounts. **Installation requirements/considerations:** units require electrical and water connections to be made. All connections, i.e., wiring boxes and plumbing connections, are external of the units. **Guarantee/warranty:** manufacturer will furnish replacements for any defective parts for one year following date of installation. Replacements for defective components in the sealed refrigeration system will also be furnished during the ensuing second through fifth year. Labor allowances furnished for repair to the refrigeration system during first ninety days. **Maintenance requirements:** standard air-conditioning equipment maintenance. **Manufacturer's technical services:** application engineering assistane from engineering department. **Regional applicability:** nationwide. **Availability:** nationwide network of distributors. **Price:** consult manufacturer.

Heat pipes

HEAT PIPES
manufactured by:
Gould Laboratories
540 East 105th St.
Cleveland, Ohio 44108
E.L.Thellmann; phone: (216) 851-5500
Heat pipe with porous metal wicking material metallurgically bonded to a wrought backing. Send for catalogue. **Price:** consult manufacturer.

SUN BUCKET
manufactured by:
Mann-Russell Electronics, Inc.
1401 Thorne Rd.
Tacoma, Wash. 98421
George Russell; phone: (206) 383-1591
A net 25-square foot solar collector for schools or colleges for instruction or research, or for individual home or plant use. Proprietary pick-ups compose twenty modified copper vacuumized heat pipes each 8 centimeters in diameter and 1.5 meters long and containing vacuum pressure guages. Each is nickel-plated and surfaced with black chrome. **Features and options:** double plastic glazing; rated a 7.500 Btus per hour peak output for equatorial sea level. Tracking system powered by silicon solar cells. **Price:** consult manufacturers.

Heat Pipe Manufacturers

ACME MANUFACTURING
7500 State Rd.
Philadelphia, PA.
(215) 338-2850

BRY-AIR, INC.
Rt. 37W
Sunbury, Ohio 43074
(614) 965-2974

ISOTHERMICS, INC.
Dept. PM Box 86
Augusta, N.J. 07822
(201) 383-5000

E.B. KAISER, INC.
2114 Chestnut
Glenview, IL.
(312) 724-4500

NOREN PRODUCTS, INC.
3511 Haven Ave.
Menlo Park, CA. 94025
(415) 365-0632

SIGMA RESEARCH, INC.
2952 George Washington Way
Richland, Wa. 99352
(509) 946-0663

GET MOVING, AMERICA!

It feels good to feel fit . . .
physically, mentally, emotionally.
So learn a skill you can play for life.

Sponsored by
American Alliance for Health, Physical Education and Recreation
1201-16th Street, N.W., Washington, D.C. 20036

MOBILE AND MODULAR HOUSES,
a fine place for solar collectors

By J. Douglas Balcomb

Manufactured housing—mobile homes and modular houses—presents a market opportunity for solar energy use quite different from that for site-built housing. We have concluded that there are significant opportunities for reducing the cost of solar energy systems in the manufactured housing market—though there are also some characteristics of that market that require care in assessment, among them identification of the potential customer willing to consider life-cycle costing in the purchase of his solar-heated manufactured house. We have concluded that the greatest opportunity for sales of solar energy systems in manufactured housing lies in the higher-priced units, particularly the 24-foot-wide units frequently manufactured in two 12-foot-wide modules and assembled on-site. Systems that use air-heating collectors or passive solar heating concepts may be buildable at a lower cost and thus be more advantageous than systems with liquid-heating collectors.

A manufactured house is one assembled in major modules or as a whole unit in a factory, transported over the highways to the desired location, and erected on the site. In a mobile home the unit is built in such a way that it can be moved from the site by adding a set of wheels and a draw bar at the front. The major axial support structure for the house remains in place after the house is located on the site; the wheels may be taken off and the unit remains, technically, a mobile home. Generally speaking, mobile homes are purchased at prices of about $15 per square foot, by individuals who plan, in general, to live in them only a few years. Even though they may not be moved during

The LASL solar mobile/modular home project under the New Mexican sun.

their lifetimes (only 1 percent of double-wides are ever moved), there is comfort in the knowledge that they could be moved.

Modular houses, by contrast, are usually more permanent structures. They are usually built in widths of 24 feet, or more, out of modules that are generally 12 feet wide. Modular houses may sell in a price range from $16 to $20 per square foot delivered. Frequently, they are better insulated and better built than mobile homes. Modular houses may be set down on a stem wall or on a slab, and frequently conform to the Uniform Building Code. They can thus qualify for longer-term financing, twenty-five or thirty years at lower interest rates.

Prefabricated houses, and panelized construction are other approaches that can be used to reduce costs by building major components, like wall sections, in the factory, to be assembled on the site.

A brief history of the market in the mobile home industry will give some insight into the nature of that market today. From 1947 until about 1957 the market was dominated by 8-foot-wide mobile homes, and total sales approximated 80,000 units a year. During the next eight years, from 1957 to

1965, 10-foot-wide units dominated the market, selling approximately 100,000 units a year. In 1961 double-wide units and 12-foot-wide units were introduced. By 1967 12-foot-wide units had captured roughly 85 percent of the market and approximately 250,000 units were sold. The five years from 1967 to 1972 were incredible growth years for the mobile home industry. The total sales increased to 575,000 units a year. Fourteen-foot units were introduced in 1968 and by 1972 accounted for 16 percent of the total market. The sale of double-wides had also grown to capture 17 percent of the market by 1972.

Recession, 1973 through 1975, had a drastic effect on the mobile home market. During this period single-family housing construction dropped from 1,309,000 units a year to 829,000 units. The mobile home industry was even harder hit. It dropped from 44 percent of single-family construction to 25 percent; sales fell to 213,000 units in 1975. By 1976 it appeared that the mobile home market may have stabilized. The growth rate was around 30 percent, with the total number of shipments in the first two quarters, at a seasonally-adjusted annual rate, running about 250,000 units.

Dr. Balcomb is head of Q Group, the Solar Energy Division at Los Alamos Laboratories.

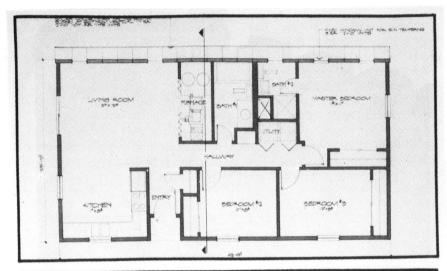

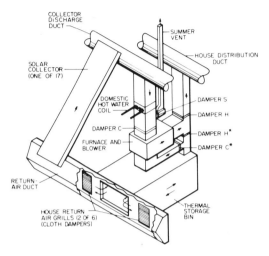

Floor plan of the manufactured house, top, above. The heat storage for the solar system takes up the bottom section of the furnace room, to a height of 30 inches. It consists of 1,536 pint glass jars, above, filled with water, sealed, and stacked in layers through which air passes.

The composition of a mobile home market has changed very significantly during this period. Fourteen-foot-wide units now dominate the market, making up about half of the total. Double-wides and 12-foot-wides account for about one-quarter of the market each. In two states where 14-foot-wide units cannot be shipped, California and Florida, the number of double-wides is roughly twice the number of 12-foot-wide units. The vast majority of mobile homes are sold in the southern half of the United States, though there are significant sales in a few northern states.

More stringent standards, fortunately, have been imposed by the U.S. Department of Housing and Urban Development. Mobile homes are now more fire and wind resistant and generally much better insulated and constructed. This leads to higher resale values. Many units are also built to be more in conformance with normal architectural styles. Multiple-module units are frequently hardly to be distinguished from site-built houses. These do cost more than the single-wides but still are far less expensive than site-built houses.

The entire mobile home industry is extremely competitive. A few hundred dollars difference in price can be crucial in the sale of a mobile home. The average wholesale price of a 70-foot-long by 14-foot-wide unit is approximately $9,000 FOB factory. This is somewhat less than $10 per square foot. Double-wides, by comparison, are aimed at a different market, generally a buyer looking for better quality in a house where he plans to live longer, and they cost appreciably more.

Generally speaking, people think of mobile homes in mobile home parks. While this is true of roughly one-half of all mobile homes, there is a general trend toward location of double-wide mobile homes on standard lots in mobile home subdivisions, or simply as individual buildings on all kinds of sites.

Manufactured housing presents an opportunity for lowering the installation costs of a solar energy system significantly. For the same reasons that manufactured housing costs less than site-built housing, the costs of the solar
Continued

MODULAR HOUSES *continued*

energy systems in manufactured housing can be reduced. Let us look at these reasons:

* Centralized buying without middlemen. Manufacturers buy in massive quantity direct from the producers.

* Assembly-line construction. Parts are in the right place at the right time; everything fits, there is no lost time in trimming and fixing. Parts are moved out quickly—the total construction time ranges from a day and a half for a mobile home assembly line to a few weeks at the most on a modular home construction lot.

* Lower wage costs. Workers for housing manufacturers generally are paid less than workers on site-built housing. Combined with efficiency of operation, this holds total labor costs to a small fraction of those for site-built housing.

All of these advantages apply to the solar energy system in a manufactured house. One of the realities of solar heating has been a realization by all concerned that a predominant fraction of the cost of system installation is in components other than the collector—piping, insulation, controls, storage, etc. Every installation is a special case—so special time must be taken to fit it together so it works correctly, and to train a new group of people practically every time. Plumbing costs, especially, are high, which leads to high costs for systems that use liquid-heating collectors.

The three major types of solar heating systems that might be considered for manufactured housing are: 1) systems using liquid-heating collectors, water tanks for thermal storage, and either forced air, radiant heating, or convector heating for distribution of heat to the building; 2) systems using air-heating collectors, a rock bed or other form of thermal energy storage, and forced-air distribution to the building; 3) passive systems that make use of natural means to achieve energy flow.

The system that uses liquid-heating collectors faces a problem: collectors are usually relatively expensive, costing $8 to $12 per square foot of collector area. Many types of liquid-heating collector are on the market; most approaches have been tried by several dif-

ferent manufacturers, and it is unlikely that these costs will be reduced greatly even in a mass-production situation. An advantage to the liquid-heating collector system is that the thermal storage is an insulated tank of water. Although this may be relatively large, from 500 to 1,000 gallons, it will take up less space in the mobile home than thermal storage for the other two types.

Systems that use air-heating collectors are particularly attractive for manufactured housing. The collectors can be built for from $4 to $6 per square foot. The system requires a great deal of attention to design and system layout in order to minimize the amounts of ducting and insulation used, but this engineering and layout, once done, can be used repetitively and easily in the assembly line.

Thermal storage for an air-heating collector system presents a special problem for manufactured housing. The normal form of thermal storage for these systems is a pebble bed or rock bed. A rock bed can be located in a basement or underneath a house and is an unusually economical form of solar storage—but where can a rock bed be located in a house to be transported over a road? Besides the question of weight, a rock bed requires roughly three times the space of a water tank in order to achieve equivalent thermal storage. One obvious solution is to locate the thermal storage under the mobile home by installing the rock bed in a pit, moving the unit into place after the rock bed is finished, and connecting up the ducting appropriately. This solution has a number of disadvantages. One of the drawing points of manufactured housing is the minimum of site work and the speed with which the housing can be installed. The installation of a rock bed decreases this advantage. Another problem is that rock is not available at all sites, particularly in many locations throughout the southeast part of the U.S., where the mobile home market is very large indeed. An alternative solution is to use water or a phase-change material in small containers to minimize space and weight. This has been done in the Los Alamos Scientific Laboratory's (LASL's) mobile/modular home to be described later.

Passive solar heating systems are not nearly as well developed or as standardized as the two types of active solar systems. Passive systems operate on the principle that solar energy incident through the glass in a home can be stored directly by absorption into the floor, the back wall, in a wall immediately adjacent to the window on the south side, or in the ceiling. Incorporation of passive solar heating into manufactured housing, therefore, presents an unusual design challenge because passively solar-heated solar buildings have frequently used massive construction such as masonry for thermal storage. Massive building construction does not seem attractive if the house must be portable. An attractive alternative is to use water in containers—drums, bottles, or bags—to be filled after location, for thermal storage.

The market in mobile homes must be carefully assessed to judge the potential for solar energy use. I do not believe there is a very significant potential among the majority of buyers of the 12-foot-wide and 14-foot-wide mobile home units. It would appear that the potential market would be much larger among buyers of double-wide units, who are usually looking for a longer-term investment in their housing and thus are willing to consider life-cycle costing in their approach. Many are older couples looking for a house for several years, particularly concerned about the effects of increasing fuel costs on fixed incomes, who may be willing to invest a few thousand additional dollars for protection.

What price will the market pay? This, of course, is going to depend on fuel costs. Solar heating systems probably are not going to be competitive with natural gas prices until those prices increase by a factor of three or four. It is my belief that solar heating systems can be built into mobile homes or modular houses at prices competitive with current costs of electric or propane heating.

Cost competitiveness depends not only on the initial cost but on the financing arrangements that can be made. If the unit is financed as a mobile home with fifteen-year financing at interest rates of 12 percent, the situation is quite different than if it is financed as a modular house over a period of thirty years at interest rates of 9 percent—the monthly payment is 49 percent greater in the first case. If the add-on costs of the solar heating system are financed along with the house, the effective cost of solar heat to the home-owner would also be 49 percent greater, assuming those costs were the same in both cases. It is usually the double-wide homes that can qualify for longer-term, lower-interest financing.

The first effort sponsored by the federal government to investigate the application of solar energy to mobile home heating and cooling was by the General Electric Co. Funding was initiated under the National Science Foundation and has been continued under the Energy Research and Development Administration. The approach used by General Electric was to modify a 12 by 60-foot mobile home built by the Skyline Corp. The solar heating system uses a liquid-heating collector added to the roof, and water in tanks located in the frame support beneath the mobile home for thermal storage. A solar cooling system was also added by incorporating a lithium-bromide absorption chiller into the mobile home and rejecting the heat from that chiller through a cooling tower on the outside. The heat supply for the chiller was solar-heated hot water.

Continued

The way it works: Mode I heats house directly from the collector. Air is drawn from living room into return air plenum (2) and collector through intake grille (1) and, warmed, returns through duct (3), finned tube heating coil (4), blower (6) and furnace (7) to house distribution duct (8) and overhead air supply register (9) all set in wooden truss space (12). The cycle (10) is then repeated. In Mode II, the system charges storage. Air is drawn from thermal storage (11) to collector inlet duct (2), through the collector cycle and the furnace, and is returned to storage where it flows among water-filled jars stacked roughly ⅔-inch apart. Back draft dampers prevent air flow to house. In Mode III, house is heated from storage. Living space air is drawn into plenum (2), through storage, then blower and furnace to house distribution system. Mode IV heats domestic hot water, primarily in the summer. Air follows the Mode I path through the finned-tube heating coil (5), then is shunted out through an exhaust fan (13). Warmed water goes to 52-gallon solar hot water preheating tank. For those who wonder: (14) is insulation.

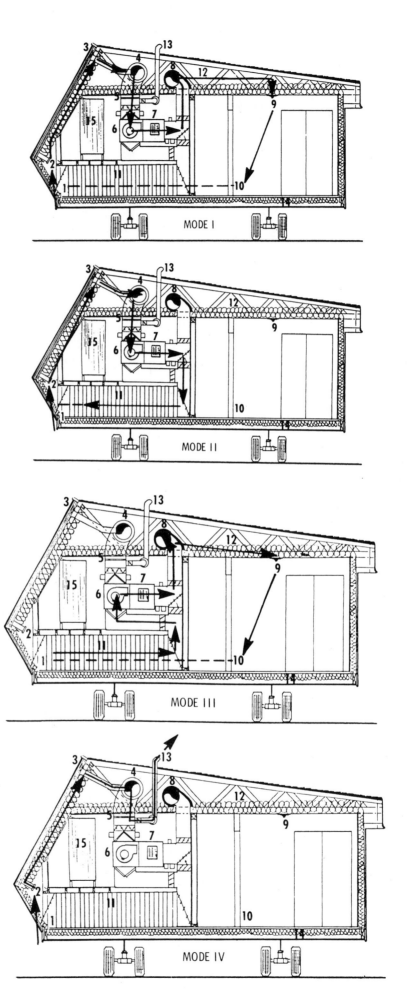

117

MODULAR HOUSES *continued*

The project demonstrated the technical feasibility of the concept. I have serious reservations about the economic feasibility of that particular approach because the total add-on cost of the solar heating and cooling system may well be comparable to the initial cost of the mobile home itself.

The mobile home system has undergone a series of successful tests at several locations. It has also been shown extensively at mobile home fairs, and was recently modified to incorporate a Rankine-cycle type of cooling unit designed by General Electric. This Rankine cycle unit is being tested as a possible alternative to the absorption-type of unit initially installed.

LASL is investigating the application of solar energy to mobile/modular house heating and cooling by testing and evaluating a series of prototype solar heated and cooled homes. Project funding is from the Division of Solar Energy of ERDA. As the country's first solar energy system designed especially for mobile/modular houses the project may have far-reaching effects on the housing industry. The market for this type of housing is expanding rapidly and is projected to account for nearly one-third of the total U.S. single-family dwellings in the near future. The small number of manufacturers and the ease with which a solar heating and/or cooling system can be factory-integrated with these units offer a unique opportunity for immediate and large-scale reduction of domestic fuel consumption at a low cost.

LASL's development plan calls for the design of four prototype units, each unit to emphasize a different technique for a different climate to be designed in conjunction with mobile home manufacturers. Each prototype for testing, evaluating, and demonstration, will accomplish the technical objectives of developing cost-effective solar heating and/or cooling for manufactured houses. The first unit, already built, incorporates an active solar heating system into a double-wide mobile/modular house. The second unit will also be solar-heated but will use a passive system to operate automatically without external energy inputs, using direct energy transfer and natural convection.

Combined solar heating and cooling systems will be built into the last two units. Each will also include the solar heating of domestic hot water.

The project's first unit is an attractive building located at a Los Alamos technical area site, a small three-bedroom house with one and a half baths, a living room, kitchen, furnace room, utility room, and several closets. The dimensions are 24 by 44 feet, for a total of 1,056 square feet. Designed to be relocatable, the unit can be called either a mobile or a modular house depending on its foundation. As a modular house, it will conform to the Uniform Building Code and qualify for normal home financing.

The basic structure is conventional wood frame construction. A special truss structure was designed to provide for the 60° sloping collector array on the south side of the building. To minimize heat loss and air leakage, the house has 4 inches of fiberglass insulation in the walls, about 10 inches in the ceiling, and 8 inches in the floor. Solid-core exterior doors, double-pane windows, and double-pane sliding doors also reduce energy consumption. Five narrow windows are interspersed between collector panels on the south walls to provide additional daylight to the living areas. The building's physical layout reflects the special requirements of highway transportability and of solar energy use.

The solar heating system uses a 340-square-foot seventeen-panel array of air-heating collectors, each 2 by 10 feet, and a bin of glass jars filled with water for thermal energy storage. This should provide approximately 80 percent of the space and domestic hot water heating for the house in the 6,500-Degree-Day Los Alamos climate.

Factory-assembled in accordance with LASL's specifications, the solar panels are fastened directly to the sloping south wall of the house and attached together with a steel cap strip to form a weather-tight roof/wall structure. They are insulated from the house by a 2-inch fiberglass mat, the plywood deck fastened to the trusses, and a 6-inch layer of fiberglass within the truss space. The collectors are made of galvanized steel and have a double-strength

single-pane glass cover sealed within a silicon rubber gasket. For sunlight absorption, the exposed metal surface is painted flat black. The 60° slope of the array is a reasonable choice for any location within the United States.

A unique aspect of the design is the thermal storage system, constructed specifically to fit in a minimum of space in the 4 by 10-foot furnace room. The thermally-insulated storage bin occupies the bottom section of this furnace room to a height of 30 inches. It is filled with 1,536 pint glass jars containing ordinary tap water, sealed, and stacked 2/3-inch apart in four rows. Heat is transferred from the solar-heated air, as it flows around the jars, to the water. The complete package fits into a space that a conventional freezer might occupy, and will store sufficient energy to provide six to ten hours of useful heat on a cold night. An electric furnace provides the backup system for periods when there is a series of cloudy days.

Domestic hot water is integrated, with a hot water finned-copper coil in the collector hot air return. Water preheated in the coil is stored in a 52-gallon solar preheater tank. From there it goes to a 30-gallon tank in which it can be raised to higher temperatures by the electric furnace if necessary. On sunny days, the water in the solar preheater tank ranges from 120° to 150°F in temperature. With a design capacity of 22 gallons per person per day at 125°F, this system should produce 85 percent of the hot water requirement (based on a national average family size of 3.7 persons).

Unit One was designed as part of the on-going solar research directed by the Solar Energy Group of Q-Division at LASL. Industrial and Systems Engineering, Inc. (ISE), of Albuquerque was the prime contractor for the project and designed and integrated the system under the direction of the Solar Energy Group. Designs for the unit were executed by Burns and Peters, an Albuquerque architectural firm, and Aztech International Ltd., Albuquerque, constructed the air collectors as designed by LASL. The unit was manufactured by Albuquerque Western Industries, Inc. (AWI), under a subcon-

The General Electric demonstration mobile home has been shown in locations from Kentucky to Illinois, to California

tract to ISE. AWI manufactures mobile/modular houses for the New Mexico market.

The thermal performance of any solar system depends on its location and the building's thermal load. Within the constraints imposed by a relocatable structure, the thermal load of the first mobile/modular unit was minimized by adequate insulation, double-glazed windows, control of infiltration, and some passive solar gains. At present, this is located at a latitude of 35.8° north and and an altitude of 7,000 feet, where the winter climate is cold but with appreciable snowfall. The measured heat load for the unit is 6.3 Btus per Degree Day, per square foot. This means that for an outside temperature of 0°F and an inside temperature of 70°F the building heat load will be 6.3 x 1,054 x 70 = 466,000 Btus per day, or 19,400 Btus per hour.

A comprehensive instrumentation and data monitoring system was installed in the home to measure performance. Energy transported by air is calculated from air flow and temperature data and the auxiliary furnace and fan energy requirements are measured. During the six-month period from October, 1976 through March, 1977 the heating system delivered 26.7 million Btus to the house. Sixty-six percent of this was from the solar collectors. 21 percent from the auxiliary furnace and 13 percent was electrical energy required by the fan. Over this period 6,562 heating degree-days were measured. House electricity, passive solar gains, and heat losses from the water tanks also added significantly to heating the building.

Based on a detailed analysis of these results, some small modifications in the design would be recommended if future units were to be built. Thermal storage should be made somewhat larger, which could be done by increasing the height of the storage compartment without using additional floor area. Also some modifications in the fan and air ducting should be made to reduce air and heat leakage and fan power.

Estimated total cost of the prototype solar-heated unit is approximately $39,000. Based on a production rate of one unit per week, it is believed that the unit could be produced for around $23,000. Without solar heating, a mass-produced unit would cost around $19,000 with the same insulation as the prototype. Solar add-on costs are therefore approximately $4,000. The collector panels, at about $6.50 per square foot, account for about $2,000 of that.

A second solar heated mobile/modular home has been designed and is currently under construction. It is a totally passive solar heating concept based on solar gains through windows. There are twenty-four windows measuring 28 inches high by 76 inches long that are located in four rows on the south-facing slope of a sawtooth roof. Sunlight passing through the windows is absorbed in water bags that cover nearly all of the flat metal ceiling deck. Small, hinged insulation panels located inside the four roof peaks can be used to close off the ceiling above the water bags either to reduce unwanted solar gain or reduce night time thermal loss through the windows. Aluminum reflectors on both the top and bottom surfaces of the north facing, insulated roof slopes are used to increase the solar gain into the thermal storage bags. The building is heated primarily by radiation from the ceiling.

Performance predictions indicate that this second unit should have a somewhat higher solar heating fraction than the first unit. A conventional forced air heating system is installed as the auxiliary backup. ☼

Money and solar hot water: THE ECONOMICS OF CHOICE

By Richard Livingstone

How do you decide whether or not to buy a solar water heater?

You can make a decision on economics—how much the heat will cost you and how much money it may save you. Or you can make your decision entirely or partly on non-economic grounds.

On economic grounds, the cost of getting hot water from solar energy must compete with the cost of getting it from electricity, natural gas, and heating oil. Today, the cost of a solar water heater ranges from around $1,000 installed up to around $1,400, depending on how much water you may use or want and also the section of the country you live in. These systems typically provide from 50 to 65 percent of a family's hot water needs; since 1 square foot of collector area is usually required for each gallon of water used per day, and the average family uses from 15 to 20 gallons (with dishwasher and washing machine) per person per day, then a 50-square foot collector system would provide about 65 percent of a family's hot water needs.

Each square foot of collector will deliver heat equal to 180,000 Btus (British Thermal Units) per square foot per year. This is on the high side, at least in New England, where a more realistic delivery is 150,000 Btus per square foot per year and probably no more than 170,000 Btus.

With these figures we can begin to examine the economics of the solar water heater. One of the most widely advertised heaters sells at $1,055 if you install it yourself and the system includes two 25-square-foot collectors, or a total of 50 square feet. Now we know that each kilowatt hour output of electricity is equal to 3,413 Btus; when we divide 3,413 into 180,000, or the number of Btus each square foot of collector will deliver per year, we come up with a comparable energy savings of electric heat per square foot of collector area per year of 52 kilowatt hours. If we pay 5¢ per kilowatt hour, we save $2.60 per square foot per year, or a total of $130 per year with a 50 square foot collector. If our system costs $1,055 we can figure a payback, without considering anything else, of about eight years; if we assume an annual fuel price increase of 4 percent per year, averaged savings rise to $195 a year and the payback shortens to a little over seven years. With an 8 percent per year fuel increase, payback drops to a little over six and one-half years.

But how do solar water heater costs stack up against oil costs? A gallon of fuel oil used in a 65 percent efficient furnace (you can of course increase the furnace's efficiency) delivers some 90,000 Btus. So the total heat from 2 gallons of oil, or 180,000 Btus, is equal to the heat output from each square foot of collector per year. If the fuel oil sells at 50¢ a gallon (in the Boston area it is currently around 46¢ a gallon), then you are saving exactly $1 per square foot of collector per year, or $25 for a 25-square-foot collector and $50 a year for two 25-square-foot collectors.

Mr. Livingstone is advertising director for Solar Age *magazine, and author of a solar stock newsletter.*

These savings vary slightly depending on the efficiency of a manufacturer's collectors. One leading manufacturer, for example, claims a delivery for its two 20-square-foot collector system which costs $1,245 installed, or 12 million Btus of heated water per year. This works out to a total system savings of around $56 a year, based on 90,000 Btus from a gallon of fuel oil and current prices of heating oil.

On our original $1,055 system cost, and oil savings that amount to $1,000 in twenty years, the solar system pays for itself in a little over twenty-one years. But we should also assume maintenance costs (there will be minimal operating costs and insurance costs as well) which some figure at 5¢ per square foot and others higher (for solar space heating systems, which are more complex, the government assumes a maintenance cost of 2 percent per year of the initial system cost). At the first figure, maintenance costs come to $2.50 per year, or $60 in twenty years, and for the second, $20 per year or $400 over twenty years.

But whatever figure you use—and without operating and insurance costs, the payback period now moves to twenty-two or twenty-six years and may easily be beyond the lifetime of the unit.

We have not considered, as we did with electric heat, the year-to-year increase in conventional fuel costs. If we assume, as some do, a 4 percent annual fuel increase averaged over twenty years, then the savings amount not to $1 per square foot per year but to around $1.50, or $75 per collector per year. Now the payback is reduced to around twelve years and at the end we would have the equivalent of eight years of free fuel, assuming the system has a twenty-year lifetime, for a savings during those years of around $600.

But if we chose instead to deposit the original capital cost of the solar water heater, or $1,055, in the bank, we would earn at 5 percent interest over twenty years a total of $1,835. So if saving money is your only aim you would do better to put your money in the bank.

But let's assume an 8 percent annual fuel increase. You would then be saving $3 per square foot per year, or $150 per year averaged over twenty years. The payback becomes about seven years and you have the equivalent of thirteen years of free fuel, for a total savings of $2,100. So, if you compare this with putting your money in the bank, you make a profit with the solar water heater, after twenty years, of $265.

However, we have forgotten one thing: the cost of borrowing the money to buy the solar water heater. Borrowing $1,000 at 8½ percent interest for thirty-six months comes to $136.15. Our profit, in this case, comes to $128.85.

There are at least three reasons why a solar water heater is probably a much better buy, even at current prices, than these figures seem to indicate. One is that, if you put your money in the bank you do pay taxes on profits, like the interest from your bank account. You don't, on fuel savings. Second, what will the interest on your savings really be worth twenty years from now, or even ten or five years from now? Fuel savings do not depreciate in value as money does.

But the third and most important reason is that annual increase in the costs of electricity, gas, and oil could be much higher than 8 percent a year. Fred Dubin, president of Dubin-Bloome Associates, Inc., has predicted that energy costs will continue to increase about 10 percent annually. Over the next few years, the increase could be sharper. The annual average could rise to higher than 10 percent; a study by the government's Department of Housing and Urban Development projected a 55 percent increase in fuel oil costs over the next three years, which works out to about an 18 percent a year increase.

Both government incentives and lower costs for solar water heaters will of course improve the economics. And for heavy users of hot water, such as car washes, bottling plants, motels, hotels, universities and other institutions, the economics are not only much better than for private homes but will continue to improve.

Some will decide to buy solar water heaters on non-economic grounds. Conserving energy for the national good is a motive. So also is the desire to be independent from rising fuel costs, however much a system costs originally.

Solar Hot Water
LOOKING BEFORE YOU LEAP

The "best" system is a matter of climate, efficiency/economy, the shape of the building, and the inhabitants' needs.

By Michael Silverstein

Solar water heaters have been commercially manufactured and sold in this country since the 1920s. They achieved wide-spread distribution in Florida and the Southwest for several decades, but lost favor in the 1940s when cheap oil and natural gas became readily available in most parts of the country. With the era of cheap fossil fuels over, however, the demand for this product is again proliferating rapidly.

Literally scores of solar water heating systems are now being sold. How does one decide which to buy? The answer to that question is surprisingly complex. Design and quality considerations aside, the right system depends very much on the climate where it is used; on the particular efficiency versus cost trade-offs appropriate to that climate; on the amount of hot water needed by a homeowner or business; and on the type of structure that will use the system. It is perfectly logical, for example, for a family with a year-round house and a summer house to find good reasons to install a different solar water heater in each.

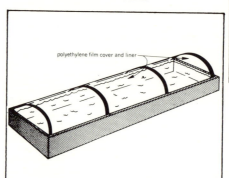

polyethylene film cover and liner

A simple trough of water is the most basic of heaters— a wooden box lined with plastic will do. A transparent cover improves its efficiency, but it won't work on a sloped roof.

There is no "best" solar water heating equipment. The best product is the one that most closely matches a buyer's needs at a price he is prepared to pay.

Materials and efficiency

As in any solar heating system, there are three basic elements in a typical water heating system: a flat plate collector, usually fix-mounted and facing south, that can absorb solar radiation all year long; a circulation pump that sends a heating medium to a storage tank; and the tank itself, which may already be in place for an oil, natural gas or electric system, and can be slightly modified when solar is added. In thermal siphoning solar systems, no

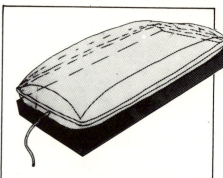

In Japan, water heaters are essentially plastic bags set on level platforms. Some have reflectors below them to reflect additional solar energy up to the bottom sides.

circulation pump is necessary, but these systems are presently not very common in most parts of the country.

In general, the heat collecting medium used in climates where freezing doesn't occur is water. This goes directly from the collector into the storage tank. Where freezing temperatures are common, the preferred medium is a non-toxic antifreeze solution. This must be pumped to a heat exchanger where its heat is transferred to the building's storage tank. Heat sensors and safety valves, that prevent excess pressures and temperatures from building up, are also features of most units.

Michael Silverstein is president of Energy Marketing Associates, in Boston.

Except for its collectors and control devices, a solar water heating system is mostly standard plumbing parts. For this reason, manufacturers concentrate their selling efforts on their collectors.

Collectors

There are some general criteria one can use in evaluating a collector. Some expensive grades of glass used in collectors allow more sunlight in and keep

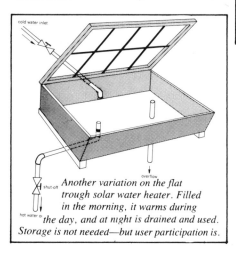

Another variation on the flat trough solar water heater. Filled in the morning, it warms during the day, and at night is drained and used. Storage is not needed—but user participation is.

more gathered heat from escaping. Two glass layers retain more heat than one, but also have a tendency to reject (reradiate) more sunlight. The translucent materials, various forms of plastics, used in some less expensive collectors often have satisfactory solar admittance and retention qualities, but some can't stand up to very high temperatures and may be subject to thermal expansion problems. Aluminum absorbers can corrode if the antifreeze medium they

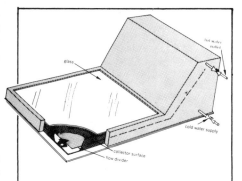

Storage integrated into a solar water heater might look like this unit from the West Indies. The collector unit is flat metal divided by a baffle to encourage thermal siphoning. Storage is at the upper end.

heat isn't blended very carefully, and for that reason aluminum is now generally out of favor in absorbers. A poor insulating layer under the absorber panel can lead to heat loss. All these considerations effect the efficiency of a collector.

Efficiency, however, is an engineering goal that has mixed appeal to a consumer. A typical buyer wants to pay only for the amount of efficiency that saves him money. The most Btus per buck, up to the total amount of Btus required and not much beyond. If your collector gathers more heat than you need to heat your water, it may end up releasing it back into the atmosphere.

Usually, high efficiency collectors are more desirable in very cold climates. Here they repay their higher initial costs in the extra usable heat they gather and retain.

Far more important than the efficiency of a collector, from the consumer's point of view, is its durability. Durability is of paramount importance in any solar system. They are expensive

front-end investments that justify the cost over a long period by saving on fossil fuel or electric bills. If they don't function properly for many years, they are rarely good investments.

The plumbing that goes with a collector in a solar water heating system should be as well insulated as possible to lessen heat loss. This applies to valves, heat exchangers, and especially to storage tanks. In fact, it pays for the owner of any kind of water heating system to make sure his storage tank is well insulated. There is almost always a savings to be realized here.

Installation

Making sure your system is correctly installed may be almost as important today as choosing the right system. In many places, traditional heating system contractors have had little experience with solar units. Manufacturers are only now beginning to establish distribution networks, and your chance of being a contractor's first or second installation is still high. It may be wise to ask your dealer/contractor how many installations he has made.

A solar unit that is shaded by neighboring buildings during part of each day or in the winter season is obviously sited incorrectly. But the situation with trees may be different. In a colder climate, trees on the south side of a house may be bare in winter and not obstruct much solar radiation. In the summer, when the sun is more directly overhead, the trees may not block the collector's access to sunlight. Common sense is the rule.

In fixed mount installations (where the tilt of the collector is constant

Continued

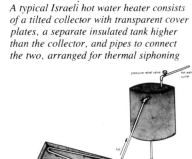

A typical Israeli hot water heater consists of a tilted collector with transparent cover plates, a separate insulated tank higher than the collector, and pipes to connect the two, arranged for thermal siphoning

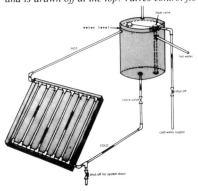

Water-heating collectors have two basic patterns of circulation, either sinusoidal or parallel. Here, parallel pipes are arranged vertically. Cold water enters the bottom, and is drawn off at the top. Valves control flow.

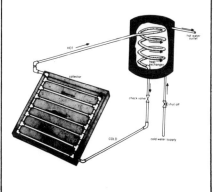

A water heater can also use freeze-proof fluids for transporting solar heat, by keeping them inside a closed loop. Heat is transferred to domestic water through coils inside the water storage tank.

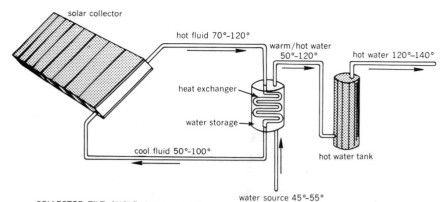

solar collector

hot fluid 70°–120°

warm/hot water 50°–120°

hot water 120°–140°

heat exchanger

water storage

cool fluid 50°–100°

hot water tank

water source 45°–55°

COLLECTOR TILT ANGLE: Latitude ±15°

COLLECTOR FLUID: Non-freezing solution, such as anti-freeze or glycol, circulates when storage temperature is more than 5° less than collector temperature. at a rate of about 6 lbs/ft /hr.

WATER STORAGE: Warm/hot fluid from the collector warms cold water supply. Warmed water circulates to conventional hot water heater and is heated additionally, if necessary.

LOOKING

continued

throughout the year) the actual angle with respect to the sun is less important than the snow shedding ability of that angle in climates that get snow. Where you don't get snow, the tilt may differ.

The parts of most solar systems are guaranteed by manufacturers for a year or more. But check to see how easily a repair can be accomplished if one is needed in that period. You don't want to be unnecessarily inconvenienced—even at no charge. Does your installer offer a service contract similar to those available for gas and oil burners? It's a good investment if he does. External surfaces sometimes require paint; controls need occasional adjustments; and the glass in collectors can develop cracks or break.

Proper sizing of the unit to be purchased is a basic element of any installation. A two-collector system may satisfy all the hot water needs of a small family in Arizona—and have some heat left over for the family livingroom in winter. The same unit might supply only 60 percent of the hot water requirements for a Massachusetts household.

Additional Considerations

Consider the following points before making a purchase:

1. Is insurance available for your solar system? The glass in collectors is susceptible to natural damage and vandalism. Solar use is so new in some states that insurance companies may not even have actuarial charts to base rates on. Some insurance companies, on the other hand, may include a solar installation as part of your present coverage at no additional charge.

2. What sort of bank loans are available to finance your solar purchase? Spreading payments as part of a second mortgage, for example, may make sense for many people. See what arrangements your local supplier has with banks and check your own.

3. Before buying, make sure there are no problems with local building codes for your solar installation.

4. Before even considering a solar unit, carefully examine the southern exposure of your home. Any major and some seemingly minor obstructions above the level of the installed unit, could rule it out.

5. Since a fair percentage of your solar water heater's cost is installation, check to see that the links with your present plumbing are accessible. If they aren't, it could cost a bit of extra money.

6. The same collectors that gather heat for wash water can also provide heat for interior spaces. Even if you only want to go solar for domestic hot water at present, discuss with your supplier the possibilities of stepping up to space heating at a later date.

7. Figuring the savings to be realized from solar water heating is a complicated equation. Solar's costs may be mitigated to some extent by state legislation, and will certainly be cut in the near future by federal tax write-offs. Currently, a consumer might think of the economic value of solar as being related to the cost of the energy it replaces. If you have a large water heating bill and a good southern exposure, it is worth considering. If your water bill is low, you might want to wait.

The sun will be around for a long time. ☀

Drawings from *Solar Energy: Fundamentals in Building Design*, by Bruce Anderson (McGraw-Hill. 1977; $17.50)

How to do it

The mathematics for
SIZING A SOLAR HEATING SYSTEM

By Dan S. Ward

The sizing of a residential solar space and domestic hot water (DHW) heating system includes both technical and economic aspects. The technical sizing is accomplished by providing a relationship (in either graphical or algebraic form) between the solar collector area and the fraction of the annual heating load to be carried by solar energy. Other components and subsystems of the solar heating system are then sized *on the basis of the solar collector area*. The economic sizing determines the optimum solar collector area yielding the minimum cost of the integrated solar and conventional heating system over the expected life of the system. This "life-cycle" cost calculation incorporates amortization of the capital cost of installing a solar heating system, annual operating and maintenance costs of the system, the costs of the required conventional heating fuels, and the predicted effects of future inflation on recurring expenses.

The procedures outlined below are intended to provide a standardized method of sizing residential solar heating installations. The method is based on the work of Klein et al. [1,2], who established the basis for the technical sizing procedure and justified its use for virtually all points in the continental United States. Ward [3] extended Klein's work to allow for a straightforward computation of minimum-cost sizing.

Fraction of annual load

Klein et al. [1,2] have presented the results of numerous computer simulations of a solar heating system, and have estimated the fraction of the annual heating load carried by solar (denoted by f) as a function of two variables that incorporate the collector characteristics and area, the building's heating load, and the local climate.

Dr. Ward is associate director of the Solar Energy Applications Laboratory at Colorado State University.

Table I

Solar collector characteristics				Parameters	
Type	F_R	$(\tau\alpha)$	U_L (Btu/(hr)(ft^2)(°F)	a	b
Liquid, flat-plate, copper-tube type (double glazed)	0.93	0.76	0.80	0.819	0.281
Liquid, evacuated tube type	0.75	0.72	0.20	0.871	0.346
Air, flat-plate, black surface, with duct under plate (double glazed)	0.67	0.73	0.77	0.918	0.432

Equation (1) and Table I are based on three assumptions: (1) there is no heat exchanger between the collector and thermal storage; (2) the tilt of the collector is latitude plus 15 degrees; (3) the orientation of the collector is due south. For installations which do not meet these criteria, corrections to the collector area obtained using Equation (1) must be obtained. This is discussed in the section on "Sizing calculation corrections".)

These two variables are: (1) the ratio of a reference solar collector absorber-plate's energy-loss to the total heating load during a specified time period; and (2) the ratio of total absorbed energy to the total heating load during the same time period.

Using these variables and weather data from nine cities, Ward [3] obtained a relationship between the fraction of annual load carried by solar (f), and (AS_1/L_1), where A is the total solar collector area, S_1 is the total incident solar radiation on a horizontal surface during the month of January, and L_1 is the total January heating load of the building. This is given in Equation (1). (**Note:** While the *January* values of S_1 ad L_1 are used, the fraction of heating load carried by solar is the fraction of *annual* load.)

$$f = a + b \ln(AS_1/L_1) \qquad (1)$$

a and b are dimensionless constants for a given solar collector. The values of a and b depend on the performance of the solar collector, as represented by the "collector heat removal factor", F_R, its transmissivity-absorptivity product, $(\tau\alpha)$, and its overall-heat-loss coefficient, U_L. Table I gives typical examples of the values of a and b for three main types of collectors.

Minimum-cost sizing

The size of the solar collector array that will yield the minimum cost sizing over the life of the solar heating system is given by Equation (2):

$$A = \frac{bH(X - gC_EL)}{(C_C + C_T)} \qquad (2)$$

where b is the solar collector parameter from Equation (1), or Table I, and the other

Continued

SIZING A SYSTEM *continued*

variables are given by Equations (3) through (6).

H (energy inflation factor)=

$$\frac{1}{R_e}[\exp(R_e t_c)-1] \qquad (3)$$

where R_e=Expected energy inflation rate for the lifetime of the system, %/year

t_c=Expected lifetime of the system, years

X (conventional-fuel cost factor)=

$$L_A E_A C_A + L_D E_D C_D \qquad (4)$$

where L_A=Annual space heating load, Btu/year

L_D=Annual DHW heating load, Btu/year

E_A=Efficiency of conventional space heating unit, %

E_D=Efficiency of conventional DHW heating unit, %

C_A=First-year unit cost of conventional energy used for space heating (delivered to furnace), $/(Btu)(year)

C_D=First-year unit cost of conventional energy used for DHW heating (delivered to DHW heater, $/(Btu)(year).

and where [referring again to Equation (2)]

g=Percentage of solar-supplied space heating and DHW loads required in the form of electrical power to operate the solar heating system, %/year

C_E=First-year unit cost of electricity used to operate the solar heating system, $/(Btu)(year)

L=Total annual heating load, Btu

C_C=Capital cost of solar heating system per unit area of collector (includes principal, interest, and down payment), $/ft²

C_T=Insurance and property taxes on installed solar heating system (both of which are subject to general inflation), less savings due to interest and property tax deductions on federal and state income tax returns (per unit area of collector), $/ft²

$$C_C=[DC_s A+(1-D)t_L IC_s A]\frac{1}{A} \qquad (5)$$

where D=Percentage of total cost of installed solar system required as down payment on loan

C_s=Unit cost of installed solar system, $/ft² of collector

t_L=Lifetime of loan, years

I=Amortization rate of loan, %
A=Solar collector area, ft²

$$C_T=[(I_s+t_p-Ft_p)G-F(1-D)(It_L-1)]C_s \qquad (9)$$

where I_s=First-year annual insurance rate percentile, %

t_p=First-year annual property tax percentile, %

G= General inflation factor, i.e.,

$$G=\frac{1}{R_g}[\exp(R_g t_c)-1]$$

where R_g=the expected general inflation rate for the lifetime of the system, %/year

and $F=F_1+F_2-2F_1F_2$

where F_1=Average federal income tax percentile of owner during lifetime of system, %

F_2=Average state income tax percentile of owner during lifetime of system, %

Sizing calculation corrections

The collector area given by Equation (2) must be corrected for variations in tilt and/or orientation, and for the use of a heat exchanger between the collector and storage, if one is used. To correct for the insertion of a heat exchanger between collector and storage, the collector area must be increased by one percent (0.01) for every °F temperature drop across the exchanger. Figures 1 and 2 show what happens when the tilt and/or orientation of the collectors is changed.

Note that each of the corrections is cumulative. For example, a system with a heat exchanger having a $\Delta T=12°F$, at a tilt of 25 degress with respect to the horizontal (at a latitude of 40°N), and facing southeast would require a revised collector area of:

$$A'=A(1.12)(1.11)(1.11)$$

in order to obtain the same fraction of the load.

An additional correction to the collector area must now be obtained by approximating the corrected collector area (A') to the available dimensions of a variety of collector modules. This is accomplished by dividing the corrected collector area (A') by the area of a single collector module, rounding off the quotient to a whole number, and multiplying by the collector module area. Note that some collectors may require installation in pairs and/or groups of three. In this case the number of modules obtained by dividing A' by the collector module area should be rounded off so that it is divisible by two or three, dependent upon the collector module grouping requirements.

Note that the available dimensions of a solar collector array (due to the size of the collector modules) limits the accuracy of obtaining the minimum-cost size of the solar system. In addition, the inaccuracies of knowing the January solar radiation at the particular building site, and of calculating a precise January heating load, also limit the precision of predicting a particular f for a given minimum-cost size of the collector.

Component and subsystem sizing

The other components and subsystems of the solar heating system are sized on the basis of the solar collector area, as corrected for tilt, orientation, use of a heat exchanger, and the collector module dimensions and ganging configurations. Table II gives some typical values for the principal components of a solar heating

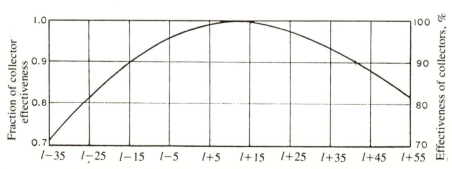

Figure 1. Effect of solar collector tilt on annual heating performance [*l*=local latitude (°N)]

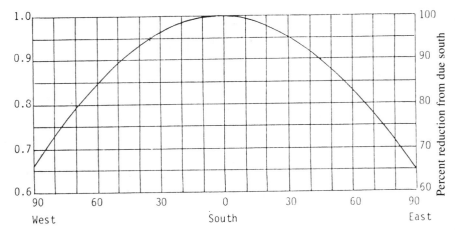

Figure 2. Effect of solar collector orientation on annual heating performance

References

1. Klein, S.A., Beckman, W.A., and Duffie, J.A., "A Design Procedure for Solar Heating Systems," *Solar Energy*, Vol. 18, No. 2, pp. 113-127 (1976).

2. Klein, S.A., Beckman, W.A., and Duffie, J.A., "A Design Procedure for Solar Air Heating Systems," Proceedings of the International Solar Energy Society Conference (U.S. Section), Winnipeg, Manitoba, Canada, Vol. 4, pp. 271-279 (1976).

3. Ward, J.C., "Minimum Cost Sizing of Solar Heating Systems," Proceedings of the International Solar Energy Society Conference (U.S. Section), Winnipeg, Manitoba, Canada, August 1976.

system. In addition, manufacturers' recommendations should also be sought, particularly in sizing such components as heat exchangers, pumps, and blowers.

Life-cycle costs

The total cost savings of using a solar system with a particular collector area is just the cost of fuel for a conventional system less the life-cycle costs of the installed solar system. Equation (7) gives the total cost of *adding* a solar space and DHW heating system, over the lifetime of the system:

$$C = A(C_C + C_T) + gC_ELH(f) + M_SG + (1-f)XH \quad (7)$$

where M_S = first-year maintenance cost of solar system, \$/year; all other variables have been previously defined.

Thus for a particular collector area, A, f may be obtained from Equation (1), and the total cost of *adding* the solar heating system can be obtained from Equation (2). The savings in using a solar system with a collector area A and fraction load f is thus C' [where C' is given by Equation (7) for the case of A, f, and M_S being zero], less C, as given by Equation (7). Thus:

$$\text{Savings} = XH - C \quad (8)$$

Summary

Based on the above discussion, a solar space and DHW heating system may be sized in the following manner.

1. From Table I, find the parameters *a* and *b* which apply to your type of collector. Although *a* and *b* will vary somewhat among different examples of a particular type of collector, the variation is usually quite slight, and the values given in Table I will not significantly affect the outcome of your calculations—especially since factors such as the projected inflation rate, etc., may vary widely from current estimates. If you wish to perform exact calculations for *a* and *b*,

consult references 1, 2, and 3. Hopefully these calculations will be standardized, so that manufacturers can provide values of *a* and *b* for their particular collector. For now, it is convenient to use the approximations given in Table I.

2. Determine and/or estimate all other variables—i.e., those listed following Equations (2) through (7).

3. Calculate the minimum-cost size of the collector area, using Equation (2). (A negative value of A implies that the solar installation cannot be considered cost-effective with the assumed values of inflation rates, etc.)

4. Correct the value derived in step 4 for tilt, orientation, and the use of a heat exchanger between collector and storage. Also divide the corrected collector area by the area of a single module, round off the answer to the nearest whole number (or the whole number divisible by 2 or 3, depending upon the collector module grouping requirements), and multiply by the collector module area.

5. Size the other components of the solar system on the basis of the final collector area obtained in step 5, by utilizing Table II and/or the manufacturer's recommendations.

6. (Optional) Calculate f in Equation (1).

7. (Optional) Calculate the cost savings using Equations (7) and (8). ☼

Table II
Rules of thumb for sizing

Solar air heating systems

Collector slope	Latitude + 15°
Collector air-flow rate	1.5 to 2 cfm/ft² of collector
Pebble-bed storage size	½ to 1 ft³ of rock/ft² of collector
Rock bed depth	4 to 8 feet in air-flow direction
Pebble size	¾" to 1" concrete aggregate
Duct insulation	1" fiberglass, minimum
Pressure drops:	
Pebble-bed	0.1 to 0.3 in. w.g. (inches water guage)
Collector (12-14 ft lengths)	0.2 to 0.3 in. w.g.
Collector (18-20 ft. lengths)	0.3 to 0.5 in. w.g.
Ductwork	approx. 0.08 in w.g. per 100 ft. of duct length

Solar hydronic heating/cooling systems

Collector slope	Latitude + 15°
Collector flow rate	approx. 0.02 gpm/ft² of collector
Water storage size	1.5 to 2.5 gallons/ft² of collector
Pressure drop across collector	0.5 to 10 psi per collector module

Solar domestic-hot-water heating systems

Preheat tank size	1.5 to 2.0 times DHW auxiliary tank size

SOLAR POOL HEATING:
some answers to some practical questions

By Freeman A. Ford

Why is solar pool heating a relatively easy solar application?

It is a basically low-temperature application requiring remarkably simple equipment that can, if properly designed, operate with exceptionally high efficiency. Because solar equipment for pool heating operates at or near ambient air temperature, glass covers, selective surfaces, back insulation, and the like are not required and, in fact, in general are both a needless expense and actually detract from efficiency. Losses from the collector are directly related to its temperature. Thus, keeping the collector at (and frequently below) ambient air temperature allows for convective gains rather than losses, thus obviating the need for a glass cover. The need for heat generally occurs when the solar energy is available and storage is provided by the pool itself. Because all swimming pools have circulation pumps, the purchase of additional pumping equipment is unnecessary.

What can be expected from a well-designed solar pool heating system?

A well-designed, properly installed solar pool heating system can be expected to extend the unheated pool season by approximately 25 percent in the spring and 15 percent in the fall—almost regardless of location. A pool of average size and proper design can be kept at a temperature approximately 10°F above the average unheated temperature. A solar heater cannot be expected to give 80°F temperatures all through the winter unaided in temperature climates like northern California's. While 80°F has been achieved by some pool owners, this was done with 4 to 6 inches of Styrofoam insulation—not considered practical for the average pool owner.

Mr. Ford is president of FAFCO, manufacturer of solar heat exchangers.

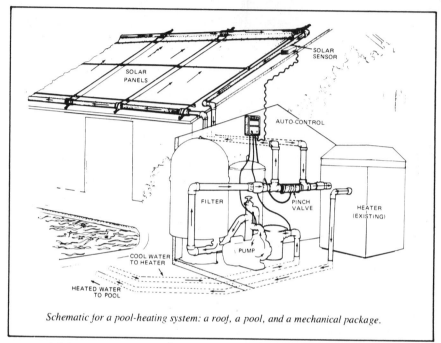

Schematic for a pool-heating system: a roof, a pool, and a mechanical package.

In addition, the solar pool heater will not provide much, if any, heat during cloudy weather. It cannot be expected to provide the 20° and 30°F temperature differentials that can be achieved with the use of conventional oil or gas heaters, unless exceedingly large collector areas are employed.

What constitutes the optimum site?

The critical factor is to have sufficient collector area to collect enough of the weak solar energy to compensate for the large losses inherent in the heating of any pool. A rule of thumb is to have a *minimum* of one-half the swimming pool area.

You should also choose a south-facing exposure and avoid northerly or easterly exposures. However, this is not as critical as is often thought. In general, adding additional panels is preferable to building rack structures to achieve a better angle, because the angle that the panels make, both in azimuth and elevation, is not especially critical. An ideal site would also include a pool house and plumbing within 50 to 100 feet of the panel installa-

tion. Lastly, it is helpful to have a well-designed pool hydraulic system which includes 2-inch suction lines and a 1½-inch return for a high rate of flow with little heat loss.

What comprises an optimum solar pool heating system?

The solar pool heater should be a complete system. While the panel is the critical ingredient, there must be appropriate hardware and automatic controls. A closed or pressurized system is desirable to insure high flow rates with minimum back pressure on the pump. The cooler that one can keep the solar panel, the higher its efficiency. To achieve a high flow rate, some pressure is needed. In addition, a pressurized system eliminates the need for a separate return line to the pool. It reduces the burden on the pump by creating a vacuum so that the static pressure of pumping up to a one or two-story collector bank is compensated in part by the siphoning action of the water falling down on the return side. It is also important to have all panels receive virtually equal flow with a minimum of

A south-sloping roof orientation is best, but west-facing or flat roofs will work too.

Christmas tree-like plumbing—both cumbersome and expensive to install.

The panels should be modular for easy installation and repair, and light in weight when fully loaded with water to avoid crushing the roof.

Is metal better than plastic as a panel material?

Although a well-designed plastic panel is generally more efficient than its metal counterpart, each has its own advantages. Aluminum should be avoided, but copper can be used, though its price tends to be substantially higher than that for a plastic panel.

What is meant by efficiency of the collectors?

Basically, the high efficiency of a properly-designed solar swimming pool heating panel is achieved because of its low temperature. Since the swimming pool heater is designed to operate within 10 to 15°F of ambient temperature, it is reasonable to expect well-designed pool collectors to show exceptionally high efficiency. Maximum efficiency depends on wetting the entire surface of the panel to minimize losses. Efficiency is relatively independent of conductivity, assuming that water flows under each square inch of exposed panel. In addition, a high rate of flow keeps the surface as cool as possible with a resultant very low delta T (difference between input and output temperature).

Many consumers are confused by the difference between heat energy and temperature. They feel that a trickle of hot water is more likely to heat their pool than 50 gallons per minute of water that is only 4 to 5°F warmer than the average pool temperature. This is not the case. If the temperature of the panels goes above the ambient air temperature by more than 10 to 15°F, then the panels are less efficient, and conduction and even radiation losses occur.

Can I use my solar pool heater for hot water house heating?

Generally speaking, the two applications are not compatible. The reason is the pool only needs to be about 75 to 80°F. Most other applications are much higher in temperature, requiring covered insulated panels whose expense would be prohibitive for pool heating.

If I wait, will the price go down?

The likelihood of decreasing price for pool heaters is remote because the cost of plastic and metal will go up with time. History bears this out: the price of solar pool heaters has gone up over the past few years, not down. It is, however, likely that the cost of more complex high temperature panels will come down somewhat, though not as dramatically as some people expect.

What questions should I ask the salesman?

When considering the purchase of a solar heater, the most important question concerns the nature of the company with whom you are dealing. How long have they been in business? How many units have they installed in your area? Can you talk with other owners of the comparable equipment?

One should be especially sensitive to claims that exceed actual performance. The guarantee also deserves caution because it is only a piece of paper. The company standing behind the guarantee is critical. Should the company go out of business, the customer has no hope of satisfaction in case of failure. Therefore, it is very important to establish the competency and reliability of the company from whom you wish to purchase your solar heater. ☼

Solar panels don't have to be on roofs, though roofs are often most convenient. Here, bathhouse and arcade are part of the system.

Complete systems: heat

SOL-PAC III: RESIDENTIAL SPACE HEATING & HOT WATER SYSTEM/ SP-4-0011 THROUGH SP-4-0014 0014

manufactured by:
Solargenics, Inc.
9713 Lurline Ave.
Chatsworth, Cal. 91311
Rowen Collins; phone: (213) 998-0806

Sol-Pac III is a total space heating system for use in new construction and retrofit projects. The system is completely factory assembled and delivered to the job site ready to run. The system is complete and contains all pumps, tanks, controls, collectors, back-up heater, and fan coil units. Installation takes approximately 1 day. **Features and options:** the Sol-Pac III can be fitted with a heat pump to provide air conditioning (up to 5 tons) as well as back up heating. The heat pump replaces the back up boiler and fan coil units. **Installation requirements/considerations:** one day installation is possible with 4 man crew. No excavation is required; may be installed in a garage or outside on concrete pad. **Guarantee/warranty:** 5 year limited warranty with extension on a year to year basis available with service contract. **Maintenance requirements:** minimum service throughout 20 year service life. **Manufacturer's technical services:** complete installation and service with factory trained, licensed contractors. **Regional applicability:** world wide. **Availability:** distributors: **Price:** contact factory.

TITAN 460/RESIDENTIAL SPACE AND COMMERCIAL HOT WATER HEATING

manufactured by:
Energy Converters, Inc.
2501 N. Orchard Knob Ave.
Chattanooga, Tenn. 37406
David Burrows; phone: (615) 624-2608
This solar collection system consists of 8 to 10 copper collectors (models 201GC), a heavy galvanized steel storage tank of 460 gallon capacity, pump, electronics (E10 differential thermostat), and assorted valves. **Features and options:** available

with drain-back operation or can be used with non-freezing liquids. Optional backup electric heating element installed in tank that can eliminate need for other backup heating equipment. **Installation requirements/considerations:** system consists basically of solar loop for collection and storage. Optional equipment (heating coils for ducting, blowers, etc.) is available for distributing the solar energy into the building. **Guarantee/warranty:** 5 years. **Manufacturer's technical services:** project engineers and technicians available upon request for installations. **Regional applicability:** particularly designed for freezing climates. **Availability:** through dealers and direct from factory. **Price:** $4,300-$5,200.

CHAMPION SOLAR FURNACE

manufactured by:
Champion Home Builders Co.
5573 E. North St.
Dryden, Mich. 48428
Henry Leck; phone: (313) 796-2211
A self-contained solar furnace for space heating, capable of retrofit to existing site-built homes, mobile homes, and commercial buildings. Interfaces with conventional forced air heating systems. Unit is placed on ground outside structure to be heated. **Features and options:** complete retrofit capability; residential, as well as small commercial application. **Guarantee/warranty:** five year limited warranty on structure; one year on motor, blowers, and controllers. **Maintenance requirements:** minimal (motors, blowers, controllers). **Manufacturer's technical services:** computer modeling capability, as well as technical staff. **Regional applicability:** national. **Availability:** distributors. **Price:** consult factory.

COMPLETE HEATING SYSTEMS

manufactured by:
Solaron Corporation
300 Galleria Tower
720 S. Colorado Blvd.
Denver, Colo. 80222
Ray T. Williamson
Solar air heating system comprised of the following components: Solaron series 2000 air heating collectors, Solaron air handling unit, necessary controls and sensors, storage design, installation hardware. **Features and options:** can be designed for space heating with domestic hot water preheat. **Guarantee/warranty:** a limited 10 year collector performance warranty and/or 1 year limited warranty on equipment supplied by Solaron. **Maintenance requirements:** adjust fan belts and lubricate motor as recommemded by the manufacturer. **Regional applicability:** all climates and geographic regions. **Price:** consult manufacturer.

SUNPOWER SOLAR HEATING SYSTEMS

manufactured by:
Sunpower Industries, Inc.
10837 S.E. 200th
Kent, Wash. 98031
Al Abramson; (206) 854-0670
Complete air space heating system for commercial and residential application. 115 VAC blower delivers 275 cfm of air to dry storage. Energy sensor control system delivers desired voltage to blower (rock bin and ducting not included by manufacturer). **Installation requirements/considerations:** a minimum of 320 square feet of southern exposure on the average home. **Guarantee/warranty:** any Sunpower component installed in accordance with drawings and instructions, will, if it fails for any reason within one year of installation, be replaced at no charge when such component is returned to an authorized Sunpower dealer. **Maintenance requirements:** several drops of SAE20 oil on the bushings of the 115VAC blower at six month intervals. **Availability:** contact home office for list of manufacturers representatives and dealers and distributors in your area. **Price:** $1,880 for Model 20 kit including: twelve Selector panels, blower and controls, drawings and installation instructions.

WARM™(WALL AND ROOF MOUNT SOLAR HEATING SYSTEM)

manufactured by:
Solar Energy Research Corp.
701B South Main Street
Longmont, Colo. 80501
James B. Wiegand, president; (303) 772-8406
Multiple panels 2 feet wide by any length incorporate the Thermo-Spray™technique. Pure water used for heat transport and storage. Spherical tank of Thermo-Mate™ controls with adjustable differential and digital data temperature read out. Compatible with heat pump equipment and most heating systems. **Features and options:** NLA (no lost area) clear mullion strip to reduce size of collector required. Thermo-Spray™ high efficiency. Removable spray tubes. Selective lacquer. Silicone sealants. Lexan cover for damage resistance. **Installation requirements/considerations:** collector mounting hardware adapts to roof or wall quickly and easily. Gravity return on flow to heat storage tank. **Guarantee/warranty:** one year. **Maintenance requirements:** durable copper used throughout collector, heat exchangers, and manifolding. No valves. All solid state controls with digital temperature readouts and manual overrides. Life-time tank and insulation. Damage-proof collector. **Manufacturer's technical services:** complete system sizing, payback analysis, design and specification, collector and controller repair. **Regional applicability:** no limitations. **Price:** Approximately $22 per square foot.

SOLAR PAK SYSTEM

manufactured by:
Columbia Chase Solar Energy Div.
55 High St.
Holbrook, Mass. 02343
Walter H. Barrett; phone: (617) 767-0513
This system is designed to provide approximately 50 percent of the space heat for a 1,200 square foot house. It operates on a closed loop heat exchanger principle. Electrical backup units are built into the tank; when needed, they operate only during off-peak hours. The tank also has extra hot and cold water inlets for wood burning stoves, fireplaces, etc. **Features and options:** collectors can be double glazed, selectively coated, and various colors are available. **Installation requirements/considerations:** collectors should be mounted toward true south at latitude plus 10 degrees. They can be placed on a roof or on the ground. **Guarantee/warranty:** one year. **Maintenance requirements:** minimal, periodic inspection, oil circulator once a year. **Manufacturer's services:** installation instructions, schematic designs furnished. **Regional applicability:** shipping world wide. **Availability:** see distributors section. **Price:** $4,056.

SOLAR HEAT SYSTEMS

manufactured by:
Solar Comfort Systems Mfg.
4853 Cordell Ave., Suite 606
Bethesda, Md. 20014
Dick Heller; phone: (301) 652-8941

"The Collector"

Complete solar heat and air conditioning systems that can be added to new or existing structures without requiring structural changes. **Installation requirements/considerations:** installed by factory authorized installation distributors, or by homeowner's own HVACP contractor after training. **Guarantee/warranty:** all components in system are warranted for a full five years for material and workmanship. **Maintenance reqirements:** circulation pump and fan motor bearings serviced once every five years. **Manufacturer's technical services:** are available from factory, factory representatives or installation distributors throughout U.S. **Regional Applicability:** unlimited. **Availability:** within two weeks—delivery from factory. Installation within 10 days after delivery. **Price:** $3.50 per square foot of floor area, installed.

INTEGRATED WARM AIR SOLAR HEATING SYSTEM

manufactured by:
Contemporary Systems, Inc.
68 Charlonne St.
Jaffrey, N.H. 03452
John Christopher; phone: (603) 532-7972

An integral roof or wall mounting, air circulating, solar space heating system; all components are compatible and designed to function together, including CSI supplied

DETACHED THERMO-SPRAY™ SOLAR HEATING SYSTEM, PLANS

maufactured by:
Solar Energy Research Corp.
701B South Main Street
Longmont, Colo. 80501
James B. Wiegand, president; phone: (303) 772-8406

Plans for construction on site of a free-standing solar collector with an A-frame structure and south-facing solar wall built over an insulated heat storage tank. Water is sprayed on inside of the absorber. Submersible pump, optional high-performance heat exchanger. Needs no antifreeze; self-draining. **Features and options:** may be used to collect heat, re-radiate heat or cool evaporatively. Use with reversible heat pump for highest efficiency and flexibility. **Installation requirements/considerations:** level site is best. Can be integrated with concrete block or stud wall. Absorber angled 45 to 90° with horizontal. No return piping from collector. No special skills to build. **Guarantee/warranty:** design prepared by registered engineers and architects who certify that design is suitable to intended use. **Maintenance requirements:** periodic replacement of plastic absorber and/or transparent cover plastic when low-cost materials are used. Maintain tank water level. **Regional applicability:** no limitations. **Price:** Plans are $750 per set plus $1 per square foot of collector royalty on patent. $14 per square foot complete unit with material, labor, plans and royalty.

collectors, USU air handling unit, LCU system logic control unit, and architectural drawings for the rock storage sub-system. **Installation requirements/considerations:** collectors are structural units mounted between roof or wall framing members (or external for retrofit); air handling and control units are site applied and compatible with standard trade practices; customer must supply ducting, wiring, storage materials, and installation labor. **Guarantee/warranty:** limited one year warranty. **Maintenance requirements:** collecters require biannual visual inspection and external cleaning if necessary; mechanical components require normal annual inspection and/or lubrication. **Manufacturer's technical services:** computerized project analysis, system integration, site supervision, follow up analysis. **Regional applicability:** designed specifically for temperate Northeast latitudes but suitable for use in all temperate regions. **Availability:** currently available. **Price:** see individual component listings.

SUNPUMP/SOLAR THERMAL ENERGY SYSTEM

manufactured by:
Entropy Limited
5737 Arapahoe Ave.
Boulder, Colo. 80303
Henry L. Valentine; phone: (303) 443-5103
A system for solar thermal energy collection that uses the latent heat of vaporization of water as the energy transport medium. Fixed focusing collectors contain heat pipes to vaporize water, that recondenses on heat exchanger coils transferring its latent heat to the storage medium. **Features and options:** Can be used to complement heating and/or cooling systems in commercial, residential, and agricultural applications. Also, can operate with no auxiliary electrical devices. **Installation requirements/considerations:** collectors can be mounted on horizontal, sloping, or vertical surfaces—either on or remote from the building. **Guarantee/warranty:** collectors and storage tank/heat exchangers are warranted for one year to be free of defects in material and workmanship. System has a minimum design life of twenty years with periodic maintenance. **Maintenance requirements:** periodic inspection and cleaning of outer glazing. **Manufacturer's technical services:** will provide technical training to its distributors/dealers. System description, installation, maintenance and operation literature available upon request. **Regional applicability:** anywhere. **Price:** consult factory.

SKYTHERM SYSTEMS

manufactured by:
Skytherm Processes and Engineering
2424 Wilshire Blvd.
Los Angeles, Cal. 90057
Harold Hay; phone: (213) 389-2300

A system of roof ponds and movable insulation. Exposed ponds are heated by the sun; at night they are covered by insulation to keep heat in. During cooling seasons the process is reversed: ponds are covered during the day, rejecting heat, but are uncovered at night to release heat to the night sky. **Features and options:** works with climate, not against. **Availability:** consult Harold Hay. **Price:** consult, but generally on a par with conventional heat.

"SOLARIS" SOLAR HEATING AND AIR CONDITIONING

manufactured by:
Thomason Solar Homes, Inc.
609 Cedar Ave., Oxon Hill
Fort Washington, Maryland
Washington, D.C. 20022
Harry E. Thomason; phone: (301) 292-5122

Solar heat collectors, storage systems, domestic water heaters, kits to convert to solar heat, air conditioning systems—sized for homes, apartment buildings, factories, office buildings, churches, etc. **Guarantee/warranty:** most licensees guarantee for five years. **Maintenance requirements:** minimum. **Manufacturer's technical services:** inspections and maintenance, if needed, by most licensees. **Regional applicability:** all regions. **Availability:** coast to coast, border to border. **Price:** $2,500 to $6,000 for heating and air conditioning for a house.

HEATING AND COOLING SYSTEMS/ AQUARIUS, CORONA AND ORION

manufactured by:
Independent Living, Inc.
Suite 2200, Two Exchange Place
2300 Peachford Rd.
Atlanta, Ga. 30341
Phone: (404) 455-0927

Heating, heating and cooling, and solar-assisted heat-pump systems for different applications. Consult manufacturer for details. **Prices:** consult manufacturer.

SUNWAVE 420 SYSTEM

manufactured by:
Acorn Structures, Inc.
P.O. Box 250
Concord, Mass. 01742
Stephen V. Santoro; phone: (617) 369-4111

Complete solar system compatible with retrofit or new home construction; includes all required components for solar collection, storage, and delivery for a forced warm air heating system and domestic hot water. Includes six 70 square foot collectors (Sunwave) for a total of 420 square feet, a 2,000 gallon storage tank, 30 gallon stainless steel tank, controls, pumps, and hardware. **Features and options:** self-draining, no antifreeze required. **Installation requirements/considerations:** storage tank may be installed in basement or crawl space. **Guarantee/warranty:** two years. **Maintenance requirements:**(every six months) check water level in storage tank and check for growth of algae. Wash face of collectors as required by local conditions. **Regional Applicability:** anywhere the sun shines.
Price: $6,500 FOB Concord, Mass.

SOLAR HOUSES/PRE FABRICATED

manufactured by:
Acorn Structures, Inc.
P.O. Box 250
Concord, Mass. 01742
Stephen V. Santoro; phone (617) 369-4111

Two models of prefabricated Cape Cod style house including complete solar system and energy-conserving design are available. Model 1950 is 1,964 square feet and model 2300 is 2,295 square feet; both feature Acorn Sunwave solar system. Double insulating glass is used throughout; windows and doors are tightly weather-stripped; minimum north facing windows, maximum south facing windows with protective overhangs for summer sun.
Price: model 1950—app. $70,000; model 2300—app. $90,000.

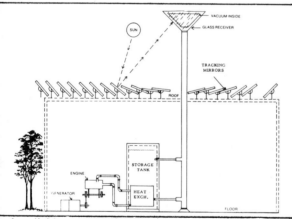
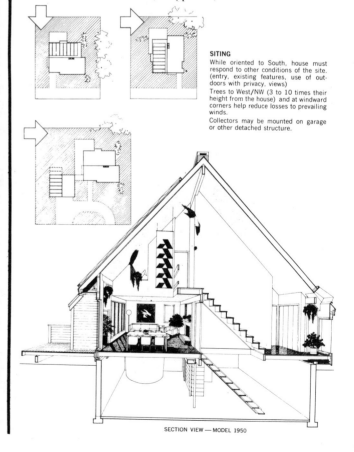

SITING
While oriented to South, house must respond to other conditions of the site. (entry, existing features, use of outdoors with privacy, views)
Trees to West/NW (3 to 10 times their height from the house) and at windward corners help reduce losses to prevailing winds.
Collectors may be mounted on garage or other detached structure.

SECTION VIEW — MODEL 1950

SOL-PAC I: INDUSTRIAL SPACE HEATING & HOT WATER SYSTEM/ DP-1-3400 THROUGH SP-1-6500

manufactured by:
Solargenics, Inc.
9713 Lurline Ave.
Chatsworth, Cal. 91311
Rowen Collins; phone: (213) 998-0806

The Sol-Pac I Industrial Space Heating System is factory prepackaged and includes: pumps, tanks, controls, heat exchangers with back-up boiler on a steel frame, and also includes fan coil units and high performance structural collectors. Energy/storage ratings range from 1,500,000 to 5,000,000 Btu/day. The system has modular expansion capability. **Features and options:** The system can be fitted to drive absorbtive chillers for cooling applications. **Installation requirements/considerations:** the system requires no special preparation for installation. **Guarantee/warranty:** 5 year limited warranty: **Maintenance requirements:** easily maintained; self test capability; servicing instrumentation connections. **Manufacturer's technical services:** complete engineering design, start up and field service. **Regional applicability:** world wide. **Availability:** distributors. **Price:** contact factory.

ELECTRIC HEATING/COOLING UNIT/FE SERIES

manufactured by:
Northrup, Inc.
302 Nichols Dr.
Hutchins, Tex. 75141
E.C. Ricker; phone: (214) 225-4291

Electric heating and air conditioning unit with hot water solar coil for primary or supplementary use. **Features and options:** low RPM fan. Three to 34 kw heat, 18,000 to 60,000 Btus per hour cooling. Hot water capacity to 97,000 Btus per hour. **Installation requirements/considerations:** none unusual. **Guarantee/warranty:** Eighteen months from shipment or twelve months from installation, whichever occurs first. Limited warranty. **Maintenance requirements:** negligible. **Manufacturer's technical services:** inquire regarding specific jobs. **Regional applicability:** worldwide. **Availability:** contact factory. **Price:** contact manufacturer.

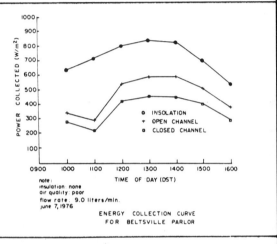

SUN*TRAC/96,128,160

manufactured by:
Future Systems, Inc.
12500 West Cedar Dr.
Lakewood, Colo. 80228
William V. Thompson; phone: (303) 989-0431

Space heating solar furnace consists of three to five collector panels mounted in an A-frame structure together with pebble-bed heat storage and air collection, distribution, and conrol provisions. Aluminum reflector panels supplement direct collection. **Features and options:** absorber vanes on collector absorb sun's rays by multiple reflection. **Installation requirements/considerations:** place in the back or side yard of a residence, connect by air ducts to the cold air return of the forced air furnace. **Guarantee/warranty:** structure, assembly and non-moving parts have a five year warranty; one year on electrical equipment and moving parts; glass excluded. **Maintenance requirements:** periodic belt inspection and annual motor lubrication. **Manufacturer's technical services:** services to homeowners are provided by a network of some 100 dealers and distributors backed up by material and engineering services. **Regional applicability:** U.S. and Canada. **Availability:** forward all inquiries to above. **Price:** manufacturer's suggested retail prices: #96 — $2,695
#128 — $3,195
#160 — $3,595

SOLAR EXPANDER

manufactured by:
Dutcher Industries, Inc.
Solar Energy Div.
7617 Convoy Court
San Diego, Cal. 92111
E.S. Cox; phone: (714) 279-7570

A solar-powered steam engine useable for any shaft power requirements. Develops 5 l.c. at 1,100 rpm using 14 pounds per hour of steam. **Features and options:** is L-head poppet valve engine with push-rod operation from crankshaft driven polydyne cam. Has hydraulic valve lifter, self-relieving valves, and 10:1 expansion ratio. **Installation requirements/considerations:** must have water supply from solar collector at 350°F and up inlet, supply of cooling water, and concrete mounting pad. **Guarantee/warranty:** ninety days on parts and labor. **Maintenance requirements:** crankcase drain and fill every six months. **Manufacturer's technical services:** installation and repair both in field and factory. **Regional applicability:** generally in the southwestern U.S. but useable in other countries of similar insolation rates. **Availability:** factory only. **Price:** consult manufacturer.

SOLECTOR PAK 1000 SERIES

maufactured by:
Sunworks, Division of Enthone, Inc.
P.O. Box 1004
New Haven, Conn. 06508
Floyd C. Perry, Jr.; (203) 934-6301

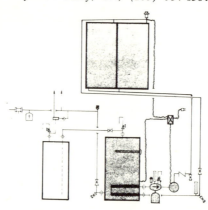

A closed loop forced circulation domestic water-heating system that circulates heat transfer fluid from collector to storage tank. **Features and options:** comes complete with controls, circulator tank, and necessary valves to insure safe operation; available in three sizes: 65 gallon storage tank—Pak 1000; 80 gallon storage—Pak 1001; 120 gallon storage—Pak 1002. **Installation requirements/considerations:** can be installed on any house by any Sunworks dealer, licensed plumber, or by homeowner. **Guarantee/warranty:** five year guarantee on collecors; five years on storage tank; all other items are for one year. **Maintenance requirements:** heat transfer/freeze inhibitor fluid should be replaced every four to five years. **Manufacturer's technical services:** technical staff available for consultation. **Regional applicability:** nationwide. **Availability:** a list of representatives is available on request from Sunworks or by calling Sweets Catalog toll-free Buyline number: (800) 255-6880. **Price:** contact representatives.

SOLECTOR PAK 3000

manufactured by:
Sunworks, Division of Enthone, Inc.
P.O. Box 1004
New Haven, Conn. 06508
Floyd C. Perry, Jr.; phone: (203) 934-6301

An open loop domestic water heating system for use in summer residences and in areas with fewer than thirty freezing days per year. **Features and options:** designed to circulate storage tank water in the collectors to prevent them from freezing, but not recommended for cold areas. **Installation requirements/considera-**tions: can be installed by any Sunworks dealer, any licensed plumber, or any homeowner in the proper geographic areas. **Guarantee/warranty:** five year guarantee; copy available on request. **Maintenance requirements:** In periods of prolonged freezing it is most cost-effective to drain the system at night. **Manufacturer's technical services:** technical staff available for consultation. **Regional applicability:** in areas with thirty freezing days per year or less and in summer only applications in the rest of the country. **Availability:** a list of representatives is available on request from Sunworks or by calling Sweets Catalog toll-free Buyline number: (800) 255-6880. **Price:** contact representatives.

WILCON SOLAR WATER HEATING SYSTEM/120-6

maufactured by:
The Wilcon Corporation
3310 S.W. 7th St.
Ocala, Fla. 32670
Jack W. Rainford; phone: (904) 732-2550

Complete domestic hot water system including solid state control systems using clear vinyl tubing and priority heat transfer fluid. **Installation requirements/considerations:** equipment installed only by licensed dealers with qualified installers. **Guarantee/warranty:** five year warranty on all components manufactured by Wilcon. **Price:** consult factory.

SUNTRAP/SWH401

manufactured by:
Energy Converters, Inc.
2501 N. Orchard Knob Ave.
Chattanooga, Tenn. 37406
David Burrows; phone: (615) 624-2608

A smaller version of the ''Solar Saver'' 80 gallon system except that it uses a 40 gallon tank with no electric element and one 201G collector panel. Components include magnetic drive pump, electronics and drainback tank: they are arranged on 40 gallon tank in compact design. System can supply up to 55 gallons of 150°F water a day. Intended to be added on to an existing hot water heater. **Features and options:** drain back freeze protection; double wall heat exchanger; fill port for adding heat transfer fluid. **Installation requirements/considerations:** install for drain back freeze protection if this mode of operation is desired. **Guarantee/warranty:** 5 years. **Maintenance requirements:** inhibitor in solar loop must be replaced every two years. **Manufacturer's technical services:** available upon request. **Regional applicability:** designed for freezing climates. **Availability:** through dealers and direct from factory. **Price:** $765.

THE NORTHRUP THERMOSIPHON/NSC-TSWH

manufactured by:
Northrup, Inc.
302 Nichols Dr.
Hutchins, Tex. 75141
E.C. Ricker; phone: (214) 225-4291

A self-contained solar hot water system with no moving parts. Storage and collector combined in one unit. **Features and options:** optional frame and mounting apparatus. **Installation requirements/considerations:** installs easily on roof or frame; simple piping. **Guarantee/warranty:** Eighteen months from shipment or twelve months from installation, whichever occurs first. Limited warranty. **Maintenance requirements:** negligible. **Manufacturer's technical services:** inquire regarding specific jobs. **Regional applicability:** suggested for low and non-freeze areas. **Price:** consult manufacturer.

SUN-TRACKING RESIDENTIAL SOLAR HOT WATER SYSTEM/1000

manufactured by:
Energy Applications, Inc.
830 Margie Drive
Titusville, Fla. 32780
Napoleon P. Salvail; phone: (305) 269-4893

A residential, north-south oriented, daily sun-tracking solar collector system that focuses the sun with rotating parabolic reflectors on Pyrex-jacketed stationary absorbers. The 36 square foot unit comes in kit form for homeowner installation. **Features and options:** fully automatic operation; lightweight—2.5 to 3 pounds per square foot; structure of aluminum, plastics, and Pyrex to reduce corrosion and deterioration in humid environments. Add-on tracking modules (12 square feet each) are optional as is support structure for ground and flat roof installations. **Installation requirements/considerations:** collector faces south with a fixed inclination equal to the latitude of the location. **Guarantee/warranty:** one year on parts and labor. **Maintenance requirements:** none. **Manufacturer's technical services:** product technical consultant. **Regional applicability:** not advisable for areas with frequent ice storms at present time. **Availability:** from manufacturer. **Price:** $950 (optional add-on tracking modules of 12 square feet: $150 each).

SUNSPOT SOLAR WATER HEATING SYSTEM

manufactured by:
El Camino Solar Systems
5330 Debbie Lane
Santa Barbara, Cal. 93111
Allen K. Cooper; phone: (805) 964-8676

Sunspot is an energy management system using Cascade autocontrol. Sunspot is compatible with existing hot water systems and ca be installed using the existing tank only or as a dual tank system. **Features and options:** the Cascade autocontrol is a three channel autocontrol system that actuates the circulation pump, the Cascade valve, and an electric or gas override. **Installation requirements/considerations:** no special handling or tools required. The Sunspot system is designed to be installed by a two person crew. Primary skill required is that of an apprentice plumber. **Guarantee/warranty:** five year warranty. **Regional applicability:** Sunspot system incorporates a frost cycle freeze protection system For areas where freezing occurs often a failsafe draindown freeze protection system is available. **Price:** $1,225.

SOLAR TUBE SYSTEMS WATER HEATER KITS

manufactured by:
Heilmann Electric
127 Mountainview Rd.
Warren, N.J. 07060
phone: (201) 757-4507

Kits provide the major components necessary for the installation of a solar water heater in a new or exossting building. System will heat a tank supply by 60°F on a sunny day. Each kit includes Solar Tubes (⅝ inch copper tubing binded to aluminum in a 6 foot by 4 inch Lexan tube with fiberglass insulation), differential thermostat, two sensors, storage tank with heat exchanger, circulating pump, expansion tank, check valve, air release valve, filler valve, and two galvanized straps for each solar tube. **Features and options:** breakage resistant; lightweight—easy to handle and transport. Directional flexibility: tubes may be rotated during and after installation to maximize solar collection. **Installation requirements/considerations:** one man can install tubes for an entire system using ordinary hand tools. **Guarantee/warranty:** tibes guaranteed against defects in material and workmanship for five years.

Price:

storage capacity	number of tubes	
42 gallons:	10	—$497
52 gallons:	13	—$578
66 gallons:	16	—$682
82 gallons:	20	—$784
120 gallons:	30	—$1,094

SUNEARTH DOMESTIC HOT WATER SYSTEMS

manufactured by:
Sunearth Solar Products Corp.
R.D. 1, Box 337
Green Lane, Pa. 18054
H. Katz; phone: (215) 699-7892

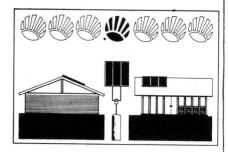

Sunearth domestic hot water systems are available in two types of completely automatic packages. Both systems come complete with collectors, tank, controls, fittings and instructions. The systems are designed to use either motorized drain valves or antifreeze solution to provide freeze protection. **Features and options:** depending on field conditions. **Installation requirements/considerations:** Sunearth's solar domestic hot water manual, available from a dealer or the factory for $3 covers all aspects of sizing, installation and service for collectors and hot water systems. **Guarantee/warranty:** limited five year warranty on defects in material and workmanship from date of installation. **Maintenance requirements:** in case of accidental damage or vandalism glazing and collectors can be easily replaced or repaired with Sunearth Factory Parts and Service. **Manufacturer's technical services:** complete sizing of system to region and domestic needs. **Regional applicability:** freeze protected. **Price:** according to system size.

RAYPAK/DHWS-2-T82

manufactured by:
Raypak, Inc.
31111 Agoura Rd.
Westlake Village, Cal. 91361
H. Byers or A. Boniface; (213) 889-1500
Domestic hot water package consists of two Raypak model SG-18-P Solar-Pak collectors, differential thermostatic controls; high

limit control, freeze protection control, pump, 82 gallon storage tank. **Features:** fully automatic system, simple to install, long life all metal collectors, meets potable water codes. **Installation requirements/considerations:** installation and operating manuals available. **Guarantee/warranty:** one year on materials and workmanship. **Maintenance requirements:** pump motor must be oiled once or twice each year. **Regional applicability:** representation in all states and international markets. **Availability:** from inventory. **Price:** $1,460.

COLUMBIA DOMESTIC HOT WATER SYSTEM

manufactured by:
Columbia Chase Solar Energy Div.
55 High St.
Holbrook, Mass. 02343
Walter H. Barrett; phone: (617) 767-0513
This system operates on the closed loop heat exchanger principle. A non-toxic fluid is circulated through the collectors and a heat exchanger located inside the hot water tank. The fluid is controlled automatically. The system is designed to work only during the efficient collector hours. **Features and options:** hot water tanks of different sizes—65, 120 gallons. **Installation requirements/considerations:** collectors should face true south and be mounted on a 45 degree angle on a roof or on the ground. **Guarantee/warranty:** one year. **Maintenance requirements:** minimal, periodic inspection, oil circulator once a year. **Manufacturer's technical services:** installation instructions, schematic diagrams furnished. **Regional applicability:** shipped worldwide. **Availability:** see distributors section. **Price:** $1,086.

COLUMBIA DIRECT EXCHANGE DOMESTIC HOT WATER KIT

manufactured by:
Columbia Chase Solar Energy Div.
55 High St.
Holbrook, Mass. 02343
Walter H. Barrett; phone: (617) 767-0513
This kit is designed to attach to the existing hot water heater with a minimum amount of modification and labor. The system may be equipped with automatic drains for use in cold climate. **Features and options:** additional collectors may be purchased for added efficiency. **Installation requirements/considerations:** collectors should be mounted facing true south at a 45 degree angle for the best efficiency. They can be placed on the roof or on the ground. **Guarantee/warranty:** one year. **Maintenance requirements:** minimal, periodic inspection, oil circulator once a year. **Manufacturer's technical services:** installation instructions, plumbing and wiring diagrams included. **Regional applicability:** shipping worldwide. **Availability:** see distributors section. **Price:** $478.88.

SERC DOMESTIC HOT WATER SOLAR PRE-HEATER

manufactured by:
Solar Energy Corp.
701B South Main St.
Longmont, Colo. 80501
James B. Wiegand, president; (303) 772-8406
Preheat tank with integral pump, controls, double-wall heat exchanger. Available as a kit with pre-assembled collectors. Sized 2 x 8, x 10, x 12, or x 16 feet. Tanks 40, 80, 120, 200 gallons. **Features and options:** integral electrical heating coil for stand by hot water in cloudy weather. **Installation requirements/considerations:** An eighty page manual explains installation, operation and maintenance; all mounting hardware provided; all CPVC tubing and flexible foam insulation to complete. **Guarantee/warranty:** one year on collector, tank and controls; manufacturer's warranty on stainless steel Grundfos pump. **Maintenance requirements:** maintain water level in closed loop heat exchange. **Manufacturer's technical services:** sizing the system. **Regional applicability:** unlimited. **Price:** $799 and up.

HOT WATER SYSTEMS

manufactured by:
Solaron Corporation
300 Galleria Tower
720 S. Colorado Blvd.
Denver, Colo. 80222
A hot water heating system that uses flat plate air heating collectors with heat exchange through an air to water heat exchanger. **Features and options:** optional collector package (2-4 collectors). **Installation requirements/considerations:** installation hardware is available from Solaron. **Guarantee/warranty:** a limited 10 year collector performance warranty and a 1 year limited warranty on equipment supplied by Solaron. **Maintenance requirements:** adjust fan belts and lubricate motor as recommended by the manufacturer. **Regional applicability:** all climates and geographic locations. **Price:** see distributor/dealer.

SOLAR DOMESTIC WATER HEATING SYSTEM

manufactured by:
Revere Copper and Brass Inc.
Solar Energy Dept.
P.O. Box 151
Rome, N.Y. 13440
William J. Heidrich; phone: (315) 338-2401
Solar domestic hot water system includes virtually all the necessary components except the copper tubing. Offered as a single package in sizes from 65 to 120 gallons. **Features and options:** three types of systems available (internal heat exchanger, external heat exchanger, no heat exchanger) in four different sizes (65, 80, 100, 120 gallons). Solar collectors use

Revere's Tube-In-Strip absorber plate. **Installation requirements/considerations:** system should be properly piped, with collectors correctly oriented, securely fastened and protected against freeze-up in a manner suitable for the area. **Guarantee/ warranty:** five year warranty on major components (except glass on collectors). One year warranty on minor components. **Maintenance requirements:** periodic inspection to see that glass has not been obstructed. **Manufacturer's technical services:** trained staff to provide technical advice related to installation and application of the system. **Regional applicability:** nationwide. **Availability:** nationwide through authorized distributors and dealers. **Price:** available on request.

SOLARCRAFT/I & II
manufactured by:
State Industries, Inc.
Ashland City, Tenn. 37015
or
Henderson, Nev. 89015
A pre-packaged system containing all components for a complete solar installation. Aluminum and Fiberglas collector panels; aluminum heat exchanger. **Features and options:** manual override switch; electric back-up booster element. Closed loop heat transfer. Systems include two collectors and an 82 or 120 gallon storage tank and pump. **Price:** consult factory.

SUNSTREAM®DOMESTIC HOT WATER SYSTEM/50
manufactured by:
Sunstream, Div. of Grumman Houston Corp.
P.O. Box 365
Bethpage, N.Y. 11714
Two collectors in kit form, ready to assemble; storage tank with heat exchanger; circulating pump; controls; expansion tank. **Features:** check valve, Sunstream anti-freeze. **Price:** consult factory.

DOMESTIC HOT WATER SYSTEM/ 12A or 28D
manufactured by:
Cole Solar Systems
440A East St. Elmo Rd.
Austin, Texas 78745
Warren Cole; phone: 512/444-2565
Model 12A comes with one 40 square foot Cole 410AT collector, 82 gallon tank with auxiliary heating coil, pumps, valves, tamperature controller, and pump turn-on freeze protection for southern climates. Model 28D comes with two 32 square foot collectors, automatic draindown freeze protection for northern climates. **Features and options:** 52 gallon tanks for retrofitting with existing system. **Guarantee/warranty:** five years on parts and labor if returned to factory. **Availability:** 3 weeks. **Price:** consult manufacturer.

WATER HEATING SYSTEM
manufactured by:
W.L. Jackson Mfg. Co., Inc.
1200-26 E. 40th St.
P.O. Box 11168
Chattanooga, Tenn. 37401
Ralph L. Braly; phone: (615) 867-4700
Domestic hot water heating system with all aluminum collector panels; closed loop design. Includes 1/10 HP magnetically driven pump to circulate deionized water. **Features and options:** can supply up to 95 gallons of 150°F water per day. Auxiliary heating element. **Guarantee/warranty:** five year warranty, in effect only if distilled water with inhibitor is changed every two years. **Price:** consult factory.

SOLAR HOT-WATER KIT
manufactured by:
SolarKit of Florida, Inc.
1102 139th Ave.
Tampa, Fla. 33612
Wm. Denver Jones; phone: (813) 971-3934
A kit consists of the number of collectors required for the size of job, a circulating pump, a solid state controller with both freeze and upper temperature limit control, all necessary valves, simple installation instructions, and tips for efficient use. **Features and options:** collectors have redwood frames, are lightweight (50 lbs.). **Installation requirements/considerations:** collectors are made with pre-drilled 2-inch-long aluminum leg at each corner and can be bolted directly to roof or any frame. Plumbing is straightforward. Electrical system requires only low-voltage dc connections when a 110 volt outlet is already available. **Guarantee/warranty:** collectors are warranted against defects in workmanship and materials under normal use for three years from date of purchase. Circulation pump and solid state conroller are warranted separately for one year by

their manufacturers. **Maintenance requirements:** minimal; pump motor should be oiled annually; occasional painting or staining of collectors is recommended (redwood is treated with a sealer at the factory). **Manufacturer's technical services:** can custom build. **Regional applicability:** suitable primarily for the southern half of the U.S. **Availability:** direct from factory, or Bob Vijil, Sun-Gard Systems, 528-B E. Brandon Blvd., Brandon, Fla. 33511. **Price:** $375 and up.

SOLAR SAVER/SWH801
manufactured by:
Energy Coverters, Inc.
2501 N. Orchard Knob Ave.
Chattanooga, Tenn. 37406
David Burrows; phone: (615) 624-2608
The Solar Saver potable hot water system is a matched system of two solar panels, storage tank (80 gallon), pump and electronics. These components are arranged into a compact, self-contained system for installation in new or existing homes; can supply up to 110 gallons of 150°F water per day. Twenty-five gallons of hot water are always maintained in the system by the auxiliary heating element. **Features and options:** drain-back freeze protection; double wall heat exchanger; fill port for adding heat transfer fluid; available with one collector panel. **Installation requirements/considerations:** install for drain-back operation. **Guarantee/warranty:** 5 years. **Maintenance requirements:** inhibitor in solar loop must be added every two years. **Manufacturer's technical services:** available upon request. **Regional applicability:** designed for freezing climates. **Availability:** through distributors or direct from factory. **Price:** $1,098. $835 with only one collector.

SOLAR WATER HEATER/SD-5
manufactured by:
Solar Development, Inc.
4180 Westroads Drive
West Palm Beach, Fla. 33407
Don Kazimir; phone: (305) 842-8935
System includes 4 foot x 10 foot or 2 foot x 10 foot collector, roof mounting hardware, del Sol control, March 809 pump, check valve and relief valve. Collectors are series or parallel and either single or double glazed. Drain down freeze protection system available. **Features and options:** variety of

collector sizes and roof mounting hardware available. Series or parallel absorber. **Installation requirements/considerations:** fastened to roof rafters with lag screws, angle brackets and pitch pans; can be mounted horizontally or vertically. **Guarantee/warranty:** 5 year limited warranty on collector; 1 year on pump and controller. **Maintenance requirements:** oil pump once a year. **Manufacturer's technical services:** engineering design services available. **Regional applicability:** unlimited. **Price:** $495 for 4 x 10 foot single glazed system.

SOL-PAC II: MULTI FAMILY-INSTITUTIONAL HOT WATER SYSTEM/SP 36-0011 THROUGH SP36-0018

manufactured by:
Solargenics, Inc.
9713 Lurline Ave.
Chatsworth, Cal. 91311
Rowen P. Collins; phone: (213) 998-0806
The Sol-Pak II Multi—Family/Institutional Hot Water system is factory prepackaged and includes: pumps, tanks, controls, heat exchangers and back-up boiler in a decorative housing on a steel frame. High performance structural collectors are included. The system is suitable for condominiums, apartments and rest homes.

Features and options: system contains special low stack loss boiler and energy economizing valving for use in circulating hot water systems. **Installation requirements/considerations:** does not require burying of tanks and can be installed in 1 day or less with a crew of 2. Can be installed indoors or outdoors (with factory built housing). **Guarantee/warranty:** 5 year limited warranty with extensions on a year by year basis available with service contract. **Maintenance requirements:** minimum service throughout 20 year service life. **Manufacturer's technical services:** complete installation and service with factory trained licensed contractors. **Regional applicability:** worldwide. **Availability:** distributors. **Price:** contact factory.

SOL-PAC II: INDUSTRIAL PROCESS WATER HEATING/SP 96-3400 THROUGH SP 96-6500

manufactured by:
Solargenics, Inc.
9713 Lurline Ave.
Chatsworth, Cal. 91311
Rowen Collins; phone: (213) 998-0806
The Sol-Pac II Industrial Process Water System is factory prepackaged and includes: pumps, tanks, controls, heat exchangers with back-up boilers on a steel frame, and includes high performance structural collectors. Energy storage ratings range from 1½ to 5 million Btus per day. **Features and options:** complete mounting system optional. **Installation requirements/considerations:** the system requires no special preparation for installation. **Guarantee/warranty:** five year limited warranty. **Maintenance requirements:** easily maintained; built-in self test capability; servicing instrumentation connections. **Manufacturer's technical services:** complete engineering design, start-up, and field service. **Regional applicability:** world wide. **Availability:** distributors. **Price:** consult factory.

SOL-PAK IV: RESIDENTIAL HOT WATER SYSTEM/SP 66 THROUGH SP 120

manufactured by:
Solargenics, Inc.
9713 Lurline Ave.
Chatsworth, Cal. 91311
Rowen Collins; phone: (213) 998-0806
The Sol-Pac IV residential hot water system is factory prepackaged and includes: pumps, controls, tanks and high performance collectors. The system is designed as a pre-heater for a standard residential hot water system. **Features and options:** health code/IAPMO approved double wall heat exchanger available in hard freeze areas; system has building code approval; system meets HUD financing requirements; thermosiphon configuration available. **Installation requirements/considerations:** system typically installs in about 4 hours with 2 man crew. No special piping required; installs like conventional hot water heater. **Guarantee/warranty:** 5 year limited warranty with extension on a year by year basis available with service contract. **Maintenance requirements:** minimum service throughout 20 year service life. **Manufacturer's technical services:** complete installation and service with factory trained licensed contractors. **Regional applicability:** world wide. **Availability:** distributors. **Price:** contact factory.

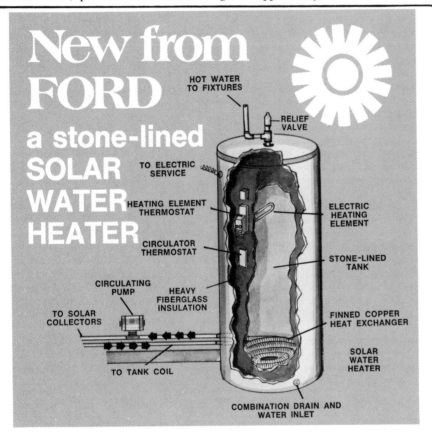

Swimming pool heat

ALCOA SOLAR HEATING SYSTEMS FOR SWIMMING POOLS

manufactured by:
H.C. Products Co.
(subsidiary, Aluminum Company of America)
P.O. Box 68
Princeville, Ill. 61559
W.M. Foster; phone: (412) 553-3185

A pool heating system designed to be used either as a sole source or supplement to a conventional heater. Modular design suitable for most pool sizes, incorporates weather-resistant ABS plastic panel approximately 4 feet square in size. **Features and options:** automatic venting and drain-down for freeze protection plus pressure relief valve as standard. Automatic temperature solid state control optional. High wind load capability. **Installation requirements/considerations:** may be either roof or ground-mounted. Typically installed in one or two days; 15° minimum slope for drainage. **Guarantee/warranty:** one year on materials and workmanship. **Maintenance requirements:** occasional washing with garden hose. Field repair kit for minor leaks. **Manufacturer's technical services:** contact factory representative—Robert Hartnett at address above or phone: (309) 385-4323. **Regional applicability:** no restrictions. **Available from:** Bicknell, Inc., Continental Mfg., Kubiak Builders. **Price:** $1,800-$3,600.

SWIMMING POOL HEATING SYSTEM

manufactured by.
Fafco Inc.
138 Jefferson Drive
Menlo Park, Cal. 94025
Larry Hix; phone: (415) 321-6311

Solar equipment designed and engineered as a complete system, specifically for heating swimming pools. Solar panels and auto control, interconnecting and hold-down hardware insure safe operation and quick installation. **Features and options:** a closed system: no separate pipes needed to return the water to the pool; the panels can withstand all the pressure a pool pump can exert. **Installation requirements/considerations:** factory-trained crew can usually install the entire system in less than a day. **Guarantee/warrantee:** five years on materials and labor. **Maintenance requirements:** panels should be drained in the winter in areas where the temperature drops below 25°F. **Maufacturer's technical services:** factory trained dealers-installers. **Price:** consult manufacturer.

SDI SWIMMING POOL HEATER

manufactured by:
Solar Development, Inc.
4180 Westroads Drive
West Palm Beach, Fla. 33407
Don Kazimir; phone: (305) 842-8935

Two systems now available. One system depends on high flow (all pool water) through 4 foot x 10 foot parallel copper panels—no box, no glazing. System is operated by a differential controller or hand bypass. Other system consists of standard 4 foot x 10 foot SD-5 collector with either bypass or full flow depending on number of panels. **Installation requirements/considerations:** bare plates mounted on roof using brackets and pitch pans; 1½ inch pvc supply and return manifolds. For standard SD-5 system, insulated copper pipes required. **Guarantee/warranty:** one year limited warranty. **Maintenance requirements:** none. **Manufacturer's technical services:** engineering services available for design schematics and sizing. **Regional applicability:** unlimited. **Price:** standard SD-5 panel—$347 plus $28 for roof mounting hardware. Absorber plates—$120—4 foot x 10 foot parallel flow with ¾ inch headers.

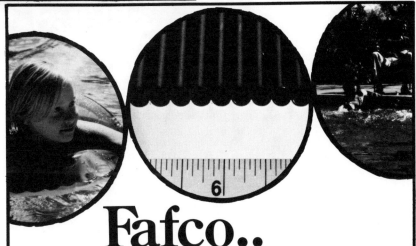

SOLAR POOL HEATER/SPH-0001

manufactured by:
Solar Home Systems, Inc.
12931 West Geauga Trail
Chesterland, Ohio 44026
Joseph Barbish; phone: (216) 729-9350
50 square foot low cost, low temperature

plastic solar collector with integral manifold for heating swimming water. **Features and options:** Tedlar-coated vinyl for strength and long life is standard, Differential thermostat and control valve is optional. **Guarantee/warranty:** limited warranty to original purchaser that the collector shall be free from defects in material and workmanship for two years. **Manufacturer's technical services:** design and engineering. **Regional applicability:** national. **Price:** $2.25 per square foot.

BURKE SOLAR HEATER

manufactured by:
Burke Industries, Inc.
2250 S. Tenth St.
San Jose, Cal. 95112
Larry R. Schader, Mgr.
Phone: (408) 297-3500
Flexible flat-plate collector using water as transfer medium, for house or commercial swimming pool heating, either alone or with conventional heat sources, or for similar large-volume medium-flow-rate systems operating to temperatures of 140°F. **Features and options:** collection surface of Dupont Hypalon [r], gravity flow, U.S. Patent #3,991,742. Sizes: 8x8 feet, 8x12 feet, 8x16 feet, 8x20 feet, 8x24 feet. **Installation requirements/considerations:** uniform surface for mounting panel, minimum flow of 2-20 gallons per minute

per panel, proper orientation of system to sun for maximum effectiveness. **Guarantee/warranty:** ten year limited. **Maintenance requirements:** none. **Manufacturer's technical services:** project planning, installation training. **Regional applicability:** no limitations. **Price:** $2.50-3 per square foot retail.

SOLAR-PAK/SK800 AND SK1000

manufactured by:
Raypak, Inc.
31111 Agoura Road
Westlake Village, Cal. 91361
R. Dominguez or A. Boniface; phone: (213) 889-1500
Liquid collector designed for swimming pool heating consisting of all copper waterway with aluminum heat-absorbing surface. Designed to withstand stagnation temperatures of 250°F. and working pressures up to 125 psi. Weighs approximately 1 pound per square foot. **Features and options:** long life non-combustible all-metal construction, lightweight, simple to install. **Installation requirements/considerations:** installation and operating manuals available. **Guarantee/warranty:** one year on materials and workmanship. **Regional applicability:** representation in all states and also marketed internationally. **Availability:** from inventory. **Price:** SK-800: $169. SK1000: $178.

DEARING SOLARCAP™ ENERGY SAVING MODULAR BLANKET SYSTEM FOR LARGE POOLS

manufactured by:
L.M. Dearing Associates, Inc.
12324 Ventura Blvd.
Studio City, Cal. 91604
LeRoy M. Dearing; phone: (213) 769-2521

The sun's heat is transmitted through transparent polyethylene and absorbed by the pool water. Thousands of air-filled bubbles insulate pool while eliminating evaporative losses of heat, chemicals, and water. Pool temperatures are raised 10 to 15° F. **Options:** reel systems and storage benches. **Installation requirements/considerations:** no special requirements — system comes on reel core with winding wheels, detachable leaders, positioning devices to hold module blanket sections in

position at the edge of the pool and adjacent to each other. **Guarantee/warranty:** one year pro-rated warranty. **Price:** 84¢ per square foot for the system.

SOLAR POOL HEATER/SPH-0001

manufactured by:
Solar Home Systems, Inc.
12931 West Geauga Trail
Chesterland, Ohio 44026
Joseph Barbish; phone: (216)729-9350
50 square foot low cost, low temperature plastic solar collector with integral manifold for heating swimming water. **Features and options:** Tedlar coated vinyl for strength and long life is standard. Differential thermostat and control valve is optional. **Guarantee/warranty:** limited warranty to original purchaser that the collector shall be free from defects in material and workmanship for two years. **Manufacturer's technical services:** design and engineering. **Regional applicability:** national. **Price:** $2.25 per square foot.

SWIMMING POOL COLLECTOR

manufactured by:
Cole Solar Systems
440A East St. Elmo Rd.
Austin, Texas 78745
Warren Cole; phone: (512)444-2565
Solar collector for swimming pool heating, 47×122 inches, with copper tubes and spot welded aluminum fins. Flat black surface. **Guarantee/warranty:** five years on parts and labor. **Maintenance requirement:** repaint every five to ten years. **Availability:** 3 weeks delivery. **Price:** consult manufacturer.

AQUASOLAR

manufactured by:
Aquasolar Inc.
1232 Zacchini Ave.
Sarasota, Fla. 33577
G.J. Zella or John Pickett, phone: (813) 336-7080 or 958-5660

Heater manifold is made from 1½ inch internally finned black Cycolac pipe. One hundred feet of pipe per 1,000 gallons of pool water is used to complete a unit. A temperature differential controller is used to make the unit fully automatic. **Features and options:** the temperature differential controller is adaptable to separate pump or solenoid valve. Cycolac manifold mounting brackets for pitched roof installations. **Installation requirements/considerations:** no special installation requirements other than those outlined in the installation manual. **Guarantee/warranty:** ten years on Aquasolar pipe and fittings. One year on components such as temperature controller and valves. **Maintenance requirements:** no maintenace requirements except the normal freeze protection procedure. **Manufacturer's technical services:** installation manuals and an owner's manual is provided. **Regional applicability:** unit has a freeze protection feature to make it adaptable to any climate. **Price:** $1250 retail.

Solenoid valves

ELECTRIC SOLENOID VALVE 1½ INCH/R219N/O AND R221N/C

manufactured by:
Richdel, Inc.
1851 Oregon St.
Carson City, Nev. 89701
P.O. Drawer A
Dale Soukup; phone: (702) 882-6786
The model 219 normally open and 221 normally closed are 2 way, diaphragm, pilot operated, solenoid valves with 1½ inch NPT female inlet and outlet connection. The valve body is molded from high strength plastic materials resistant to corrosion. **Features and options:** designed to operate at minimum pressure loss through the valve in systems where low pressures and GPM are encountered, such as swimming pool solar heat systems. **Installation requirements/considerations:** valve is not position sensitive and can be mounted in any position. Adaptable for either straight or angle installation. **Guarantee/warranty:** one year limited warranty on defective workmanship and materials. **Maintenance requirements:** none—high strength corrosion resistant PVC. **Manufacturer's technical services:** all valving and control engineering analysis for any solar application. Richdel has the capability to design and manufacture all types of valves for specific or OEM utilization. **Regional applicability:** worldwide. **Availability:** factory direct. **Price:** R219 N/O 1½ inch: $56 (list). R221 N/C 1½ inch: $48 (list). Oem pricing on request.

ELECTRIC SOLENOID VALVE 1½ INCH/R216N/C AND R216 N/O

manufactured by:
Richdel, Inc.
1851 Oregon St.
Carson City, Nev. 89701
P.O. Drawer A
Dale Soukup; phone: (702)882-6786
The model R216N/C normally closed and the R216N/O normally open are 2 way, diaphragm, pilot operated, solenoid valves with 1½ inch NPT female inlet and outlet connection. The valve body is molded from high strength plastic materials resistant to corrosion. **Features and options:** designed to operate at minimum pressure loss through the valve systems where low pressures are encountered, such as swimming pool solar heat systems **Installation requirements/considerations:** the valves are not position sensitive and can be mounted in any position. Adaptable to either a straight or angle installation. **Guaranty/warranty:** 1 year limited warranty on defective workmanship and materials. **Maintenance requirements:** none. High strength corrosion resistant PVC. **Manufacturer's technical services:** all valving and control engineering analysis for any solar system application. Richdel has the capability to design and manufacture all types of valves for specific or OEM utilization. **Regional applicability:** worldwide. **Availability:** factory direct. **Price:** $216/C $45.50 (list), R216N/O: $53 (list). OEM pricing on request.

ELECTRIC SOLENOID VALVE 2 INCH/R220N/O AND R222N/C

manufactured by:
Richdel, Inc.
1851 Oregon St.
Carson City, Nev. 89701
P.O. Drawer A
Tom Erwin; phone: (702) 882-6786
The model 220 normally open and 222 normally closed are 2 way, diaphragm, pilot operated, solenoid valves with 2 inch NPT female inlet and outlet connection. The valve body is molded from high strength plastic materials resistant to corrosion. **(See listing above for further information). Price:** R220 N/O $67.50 (list). R222 N/C $60 (list).

"SOLAIRE 36"
manufactured by:
Arkla Industries Inc.
P.O. Box 534
Evansville, Ind. 47704
Thomas M. Helms, phone: (812)424-3331
"Solaire 36" is Arkla's second generation 3-ton absorption unit designed specifically for solar cooling applications. Production began in mid-December, 1976 and the unit is now available for immediate shipment. Absorption chillers are the only commercially produced equipment for solar cooling applications. **Features:** full cooling capacity is produced from 195° F. solar heated water. The unit can operate on temperatures as low as 170° F. **Installation requirements/considerations:** the unit is for indoor installation, is water cooled, and requires a maximum draw of 250 watts of 115V, 1 phase, 60 hz electrical current. **Guarantee/warranty:** five year limited warranty. **Maintenance requirements:** Springtime check-out. **Manufacturer's technical services:** application manual, installation and start-up manual, service manual and a factory service school for distributors, dealers and service com-

panies. **Regional applicability:** wherever air conditioning is desired.
Price: from distributors.

"SOLAIRE 300"
manufactured by:
Arkla Industries Inc.
P.O. Box 534
Evansville, Ind. 47704
Thomas M. Helms, phone:(812) 424-3331
"Solaire 300" is Arkla's second generation 25-ton absorption unit designed specifically for solar cooling applications. Production began in December 1976 and the unit is now available for immediate shipment. Absorption chillers are the only commercially produced equipment for solar cooling applications. **Features:** full cooling capacity is produced from 195° F. solar heated water. The unit can operate on temperatures as low as 160° F. **Installation requirements/considerations:** The unit is for indoor installation, is water cooled, and requires a maximum draw of 150 watts of 115v, 1 phase, 60 hz electrical current. **Guarantee/warranty:** one year limited warranty is standard. An optional limited

warranty is offered for an additional four years on the sealed refrigeration cycle. **Maintenance requirements:** springtime check-out. **Manufacturer's technical services:** application manual, installation and start-up manual, service manual, and a factory service school for distributors, dealers and service companies. **Regional applicability:** wherever air conditioning is desired.
Price: from distributors

ABSORPTION LIQUID CHILLERS/ES
manufactured by:
York, Div. of Borg-Warner
P.O. Box 1592
York, Penn. 17405
Automatic heat powered cooling systems that may be used with liquid-cooled solar panels. Applicable to installations requiring 40 tons or more of full load cooling. **Features and options:** available in 21 sizes; hot water temperatures as low as 150°F can effectively supply energy required for cooling. **Availability:** local or nearest York office. **Price:** consult York.

Photovoltaic converters

SOLAR CELL MODULE/1002
manufactured by:
Solar Power Corp.
5 Executive Park Drive
North Billerica, Mass. 01862
Arthur Rudin; phone: (617) 667-8376
A 2 volt, .6 amp sealed, durable, and easily replaceable photovoltaic building block. It offers protection against both human and natural environments by using an optically bonded ultraviolet stable polycarbonate case. **Features:** impact resistant; only four screws need be removed for installation or replacement. **Installation requirements/considerations:** arrays must face south and be adjusted to recommended tilt angle. **Guarantee/warranty:** one year warranty against defects in material and workmanship. **Maintenance requirements:** general annual inspection; cleaning of module surface. **Manufacturer's technical services:** design complete systems for specific power loads and provide complete engineering and service capabilities. **Regional applicability:** world wide. **Price:** consult factory.

SOLAR CELL MODULES/SERIES E
manufactured by:
Solar Power Corp.
5 Executive Park Dr.
North Billerica, Mass. 01862
Arthur Rudin; phone: (617) 667-8376
The series-E solar module is offered in 4, 6, and 12 volt building blocks. Protection is provided to the solar cells by incorporating a dual layer of silicone rubber encapsulant. **Features:** dual layer of silicone rubber gives superior resistance to impact while providing a slick surface to prevent buildup of dirt, dust, snow, etc. **Installation requirements/considerations:** solar arrays must face south and be adjusted to recommended tilt angle. **Guarantee/warranty:** one year warranty against defects in material or workmanship. **Maintenance requirements:** annual general inspection and possible wiping of module surface. **Manufacturer's technical services:** design of complete systems for specific power loads, complete engineering and service capabilities. **Regional applicability:** world wide. **Price:** consult factory.

PHOTOVOLTAIC CONVERTERS
manufactured by:
Sensor Technology, Inc.
21012 Lassen St.
Chatsworth, Cal. 91311
Kees van der Pool; phone: (213) 882-4100
Complete solar electric power in arrays to supply from a few Watts to tens of kilowatts of continuous power output. Also indivi-

dual solar modules to charge batteries, etc. **Features and options:** modules are easily interconnected in arrays. Integral Lexan or integral glass covers, built-in Schottky diode, grounding terminal are all optional **Installation requirements/considerations:** dependent on local circumstances. **Guarantee/warranty:** five years. **Maintenance requirements:** dependent on local circumstances. **Manufacturer's technical services:** installation and checkout, periodic maintenance. **Availability:** dependent on size of system. **Price:** consult factory.

PHOTOVOLTAIC DEVICES
manufactured by:
International Rectifier
Semiconductor Div.
233 Kansas St.
El Segundo, Cal. 90245
Phone: (213) 322-3331
Silicon photovoltaic converters, silicon power modules, silicon card/tape sensors, selenium light sensors, and other devices. **Price:** consult manufacturer.

SUNCHARGER™
manufactured by:
Solar Energy Co.
Drawer 649
Gloucester Point, V. 23062
G. H. Hamilton
A battery charger that converts light to electricity by using silicon solar cells. Available in several sizes and current outputs. **Price:** write manufacturer.

Data and Measurement Collection

CHANNELYSER RECORDER MULTIPLEXER/CHL-1, CHL-1R

manufactured by:
Bailey Instruments
515 Victor St.
Saddle Brook, N.J. 07662
Inez C. Sims; phone: (201) 845-7252

An electronic commutating device that accepts signals from up to four transducers, samples them consecutively, and feeds them to the recorder, where a dot appears on the chart each time a transducer is sampled. **Features:** solid state; available battery operated or with floating batteries and a built-in charger for AC and battery use. **Availability:** four weeks. **Price:** $475.

SUNPUMP/SOLAR INSOLATION METER/SIM-100 AND SIM-200

manufactured by:
Entropy Limited
5735 Arapahoe Ave.
Boulder, Colo. 80303
Henry L. Valentine; phone: (303) 443-5103

Meters measure and display real time solar radiation in Langleys per minute as well as the integrated insolation in Langleys. Model SIM-100 is designed for permanent installation; SIM-200 is a portable (suitcase) model. Both units are solar powered. A battery provides operational power for night time integrated read-outs. **Features and options:** converts incident solar radiation to provide an automatic digital display in Langleys per minute. Also provides a digital display of integrated solar insolation up to 99,999 Langleys before resetting to zero. **Installation requirements/considerations:** normal operating temperature range - 20° to + 45° C. **Guarantee/warranty:** freedom from defects in material and workmanship for ninety days. **Maintenance requirements:** none. **Regional application:** anywhere. **Price:** consult factory.

PYRANOMETER/RS 1008

manufactured by:
Rho Sigma, Inc.
15150 Raymer St.
Van Nuys, Cal. 91405
Robert Schlesinger; phone: (213) 342-4376

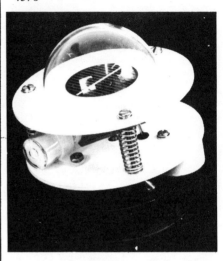

Photovoltaic pyranometer manufactured under license from John I. Yellott. The RS 1008 can drive a strip chart recorder directly to record continuous values of insolation. **Features and options:** Rho Sigma can provide strip chart recorders calibrated to pyranometer. Full scale on recorder will be 400 Btus per square foot per hour or 1,000 Watts per square meter. **Installation requirements/considerations:** easy mounting of RS 1008 on camera tripod allows flexible and rapid changes in orientation. **Guarantee/warranty:** one year warranty against defects in parts and materials. **Maintenance requirements:** recharge dessicant as needed. **Manufacturer's technical services:** documentation that accompanies the RS 1008 details the response pyranometer as a function of solar input. **Regional applicability:** national.
Price: $129.50 in quantities of one.

MICROPROCESSOR FOR DATA ACQUISITION

manufactured by:
Rho Sigma, Inc.
15150 Raymer St.
Van Nuys, Cal. 91405
Robert Schlesinger; phone: (213) 342-4376

The RS 5080 is an advanced microprocessor designed for energy-related engineering studies, to monitor energy-related parameters and make cumulative energy, average, and instantaneous efficiency calculations. **Features and options:** programmed to customer specifications to answer questions of interest to engineers. **Installation**

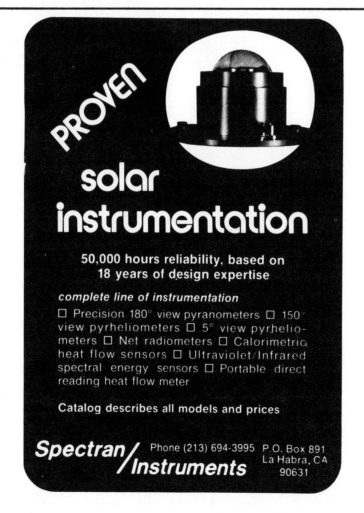

requirements/considerations: Rho Sigma field engineers help select compatible sensors and install the complete system to assure turn-key operation. **Guarantee/warranty:** one year warranty against defects in parts and materials. **Maintenance requirements:** replace tape or paper as needed. **Manufacturer's technical services:** experienced instrumentation and field engineers specializing in solar energy systems available to obtain the data and answers of interest in engineering studies. **Regional applicability:** national.
Price: varies with requirements of the task.

THERMATRACE
manufactured by:
Barnes Engineering Co.
30 Commerce Rd.
Stamford, Conn. 06904
Jayme Prins; phone: (203) 348-5381

A thermal scanning instrument that depicts temperature distribution by superimposing a thermal A-trace over a visual image of a remote object or area, and measures temperature along a single scanning line with a display amplitude that varies with the temperature along that line. Can be used for non-destructive testing and thermal inspection in industry, in the laboratory, and in air conditioning/heating, and energy-related applications. **Features and options:** the instrument operates on batteries rechargeable for four hours of continuous operation, and is hand-held. Scans a field 25° wide, with the visual observed through the viewfinder. Turning it 90° produces a vertically scanned line. An optional photographic

recording system using the Polaroid SX-70 is available. Has five selectable temperature ranges: 10°, 30°, 100°, 300°, and 1,000°F. **Guarantee/warranty:** workmanship and material guaranteed for one year. **Price:** $6,950; add $765 for photographic recording system.

INSTATHERM
manufactured by:
Barnes Engineering Co.
30 Commerce Rd.
Stamford, Conn. 06904
Jayme Prins; phone: (203) 348-5381
A compact temperature measurement instrument that indicates the temperature of any object or area found in its viewfinder. Is used by sighting through the viewfinder at the object or area whose temperature you want to measure; the temperature will register on a built-in meter. A pushbutton on the top of the instrument will hold the reading. Small temperature differences above and below a present value are measured by switching the meter from absolute to differential. **Features and options:** An optional audible seeker that indicates increasing temperature by increasing frequency of tone. A calibrator is built in, and standard replaceable batteries (life: 200 hours) power the unit. Available in three temperature ranges: $-10°$ to $60°$ C; $0°$ to $200°$ C; and $0°$ to $600°$ C. **Price:** $695.

LI-100 LEVEL INDICATOR
manufactured by:
Solar Innovations
412 Longfellow Blvd.
Lakeland, Fla. 33801
Phone: (813) 688-8373
Instrument for measuring the tilt angle of a collector to within ½° accuracy. Eliminates requirement of measuring roof pitch and figuring mathematical calculation. **Price:** $9.95.

NORMAL INCIDENCE PYRHELIO-METER/NIP
manufactured by:
The Eppley Laboratory, Inc.
12 Sheffield Ave.
Newport, R.I. 02804
Goerge L. Kirk; phone: (401) 847-1020
For either total or spectral measurements of the direct solar intensity. Instrument was designed for measurements of solar radiation at normal incidence; it may be considered a thermoelectric version or variation of the Smithsonian Silver Disk pyrheliometer, as it incorporates in its design some of the basic features of that instrument. **Features and options:** the sensitive element of the instrument is an E 6 type wire-wound thermopile with a termistor temperature-compensating circuit if required, embedded in heat sink of thermopile. The receiver is coated with 3M Velvet Black. **Installation**

requirements/considerations: the resistance matches all standard recording potentiometers very well but is not suitable for use with microammeter recorders. For periodic readings, the instrument should be attached to a mount with provision for varying the elevation and the azimuth settings. If a continuous record is required, the instrument must be mounted on a power-driven equatorial mount such as the Epply Model EQM. **Price:** $890.

BLACK AND WHITE PYRANOMET—ER/8-48
manufactured by:
The Eppley Laboratory, Inc.
12 Sheffield Ave.
Newport, R.I. 02840
George L. Kirk; phone (401) 847-1020
For the measurement of global-total sun and sky-radiation, a development of the Eppley 10-and 50-junction 180° pyrheliometer originally introduced by Kamball and Hobbs in 1923. The detector is a differential thermopile with hot-junction receivers blackened and the cold-junction receivers whitened. **Features and options:** the innovations include replacement of a two concentric-ring detector or gold-palladium—platinum-rhodium alloys—by one of radial wirebound-plated construction, and a precision-ground optical glass envelope instead of the former blown-glass bulb. **Installation rquirements/considerations:** the cast aluminum case carries a circular spirit level and adjustable levelling screws. Also supplied is a desiccator that can be inspected readily. **Maintenance requirements:** the hemispherical envelope has a weatherproof seal but is readily removable for instrument repair.
Price: $590 for model 8-48; add $20 for model 8-48A (lower sensitivity).

EQUATORIAL MOUNT/EQM
manufactured by:
The Eppley Laboratory, Inc.
12 Sheffield Ave.
Newport, R.I. 02840
George L. Kirk; phone: (401) 847-1020
Designed to accomodate one to four normal-incidence pryheliometers. electrically driven and geared to solar time, it prmits continuous measurements. The unit can also accomodate other instruments used to measure direct solar radiation. **Features and options:** latitude setting (any value from $(0°$ to $90°$); declination setting $(+23.5°$ to $-23.5°$); weatherproof housng. **Availablity:** when ordering, the number of pyrheliometers or other measuring devices should be specified. Unless otherwise noted the unit wll be supplied to accomodate three Eppley model NIP Pyrheliometers. **Price:** $1,275, for 220V 50 Hz add $60. Noted the unit will be supplied to accomodate three Eppley model NIP pyrheliometers.

ANGSTROM PYRHELIOMETER ANG

manufactured by:
The Eppley Laboratory, Inc.
12 Sheffield Ave.
Newport, R.I. 02804
George L. Kirk; phone: (401) 847-1020

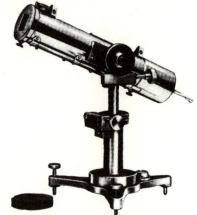

Electrical compensation pyrheliometer introduced by K. Angstrom, well-known and reliable instrument for measuring the intensity of the sun's radiation, is used as a primary working standard for the calibration of secondary pyrheliometers and pyranometers. The Eppley model is a development introduced in collaboration with Dr. Anders Angstrom. **Features and options:** principal improvements over earlier European models are: element is of permanent soldered construction and the leads to readout instrumentation are continuous; conical shield behind the manganin strip receivers minimize reflection and emission effects; strips and shield are coated with Parsons' optical black lacquer. **Price:** $1,425. Control unit with digital meter (model 402) $1,525.

SPEEDOMAX H AND W RECORDERS

manufactured by:
The Eppley Laboratory, Inc.
12 Sheffield Ave.
Newport, R.I. 02840
George L. Kirk; phone: (401) 847-1020
The L&N compact Speedomax H recorder or wide-chart W both complement all types of Eppley pyranometers and other radiometric instruments for such applications as agricultural research where local weather conditions directly affect plant growth, in manufacturing where weather exposure tests confirm the quality of products such as paint and textiles, and in meteorology where the amount of solar radiation is an important factor of the total environment. **Features and options:** the recorder consists of a sensitive electronic servo-mechanism with solid-state amplifier, operating with an integral null-type d-c potentiometer measuring circuit. **Installation requirements/considerations:** for operation on 120 volts, 60 hertz. For 50 hertz operation, specify "−5" instead of "−6" in list number. (Speedomax H: 300-113-000-1822-6-001-601; W: 500-113-000-1822-6-001-601.) **Price:** consult manufacturer.

SOLAR TRACKER/ST1

manufactured by:
The Eppley Laboratory, Inc.
12 Sheffield Ave.
Newport, R.I. 02840
George L. Kirk; phone: (401) 847-1020

Designed to accomodate one normal-incidence pyrheliometer. Electrically driven and geared to solar time, it permits continuous measurements. The unit can also accomodate other instruments used to measure direct solar radiation. **Features and options:** pointing accuracy of plus or minus .25° daily; latitude (any value from 0° to 90°); declination setting (plus 23.5 to −23.5°); leveling screws; waterproof housing. **Price:** $740. For 220 V 50 Hz add $30.

ULTRAVIOLET RADIOMETER PHOTOMETER/TUVR

manufactured by:
The Eppley Laboratory, Inc.
12 Sheffield Ave.
Newport, R.I. 02840
George L. Kirk; phone: (401) 847-1020

For the measurement of sun and sky ultraviolet radiation. This instrument is a rugged, relatively simple detector for the measurement, on a continuous basis, of the solar UV radiation. Ease of operation combines with performance accuracy comparable to meteorological pyranometers intended for the recording of total short-wave radiation received from the sun and sky on a horizontal surface (i.e. global radiation). **Features and options:** hermetically sealed selenium barrier-layer photoelectric cell protected by a quartz window. **Installation requirements/considerations:** the unit is of brass construction, assembled to be completely weatherproof; adjustable leveling screws and a circular spirit level are provided.
Price: $1150.

ABSOLUTE RADIOMETER/KAR

manufactured by:
The Eppley Laboratory, Inc.
12 Sheffield Ave.
Newport, R.I. 02840
George L. Kirk; phone: (401) 847-1020
For the precise measurement of thermal radiation in the spectral range 0.2-50 cm at intensities of 10 to 200 mw cm^{-2}. The Eppley model of this Kendall (JPL) radiometer is a modification (approved by J.M. Kendall) of those described in the Kendall-Berdahl article *(Applied Optics 1970, 9, 1802).* Construction is somewhat different but the basic physical principles are identical. Self-checking instrument es-

sentially consists of a cavity-type radiation receptor in a massive copper body inserted n a gold-plated case. **Features and options:** Eppley also supplies the control unit with a built-in power supply and readout including a 4½ digit voltmeter and null detector. **Price:** consult manufacturer.

PRECISION INFRARED RADIOMETER/PIR
manufactured by:
The Eppley Laboratory, Inc.
12 Sheffield Ave.
Newport, R.I. 02840
George L. Kirk; phone (401) 847-1020

A development of the Eppley Precision Pyranometer, this radiometer is intended for unidirectional operation in the measurement, spearately, of incoming or outgoing terrestrial radiation as distinct from net long-wave flux. Comprises the same type of wire-wound plated, nonwavelength-selective thermopile detector, and chromed brass desiccated case, as in the pyranometer model. **Features and options:** temperature compensation of detector response is incorporated. Instrument calibration is fundamentally referred to a precision low-temperature blackbody, supplemented by a group of working standard pyrgeometers. **Price:** $1,190.

ELECTRONIC INTEGRATOR/410, 411,411-6140
manufactured by:
The Eppley Laboratory, Inc.
12 Sheffield Ave.
Newport, R.I. 02840
George L. Kirk; phone: (401) 847-1020
This electronic integrator accepts an analog DC input signal from a typical Eppley Radiometer and converts it to an output pulse rate proportional to the input signal. Output pulse train is then counted and totaled to produce the integral of the input signal. The total count is displayed on a digital counter. **Features and options:** the basic integrator employs a mechanical counter that can be manually reset, as well as an analog output proportional to the count rate to drive a standard millivolt strip chart recorder if a trace is desired. Also offered is a complete system with a digital printer. Here the mechanical counter is replaced by an electronic unit with a BCD output. **Price:** model 410 with mechanical counter, $550. Model 411 with electronic counter and BCD, $950. Model 411-6140 complete with printer, $1,800.

PRECISION SPECTRAL PYRANO—METER/PSP
manufactured by:
The Eppley laboratory, Inc.
12 Sheffield Ave.
Newport, R.I. 02840
George L. Kirk: phone (401) 847-1020
For the measurement of sun and sky radiation totally or in defined wavelength bands, an improved smaller model of an earlier instrument introduces in 1957. It comprises a circular multi-junction Eppley thermopile of the wire-wound type. **Features and options:** the thermopile can withstand severe mechanical vibration and shock. Its receiver is coated with Parsons' black lacquer (nonwave length; selective absorption). **installation requirements/considerations:** spirit level is provided; instrument has a cast bronze body with white enamel guard disc and comes with a transit or storage case. **Maintenance requirements:** instrument is supplied with a pair of removable precision ground and polished hemispheres of Schott optical glass; desiccator is supplied and is readily inspectable.
Price: $990.

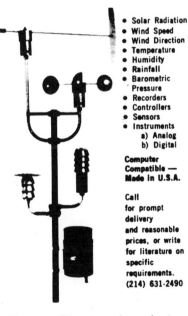

COMPARATIVE DATA

Glazing

Product identification manufacturer and/or trade name	Type	Light transmittance (after aging 10 years or longest test available)	Tensile strength	Impact strength	Coefficient of thermal conductivity	Coefficient of thermal expansion
DUPONT CO. PLASTIC AND RESINS DEPT. TEFLON	FEP Flourocarbon film	96% (1 mil film)	3000lbs./in^2	n.a.	1.7	6×10^{-5}/°F
KALWALL CORP. SUN-LITE PREMIUM	Fiberglas reinforced polyester	80% after 20 years	11,000 psi	10ft.-lb./in. (Izod Impact; D-638, D-256)	.87 Btu-in./hr/ ft.^{2}/°F (C-177)	1.4 in./in./ °F = 10^{-5} (D-696)
KALWALL SUNWALL^r	Two layers .040 in. thick, separated by ½ or 1½ in. aluminum beams	77%	11,000 psi (cover sheet)	5ft. lbs. (SPI Method B)	.46 to .55 Btu-hr./ft.^{2}/°F	.40 Btu-in./hr. ft.^{2}/°F
KALWALL SUN-LITE GLAZING PANEL	Two layers .040 in. thick, separated by 1 or 1½ in. aluminum beams	77%	11,000 psi (cover sheet)	5 ft. lbs. (SPI Method B)	.46 to .55 Btu-hr./ft.^{2}/°F	1.24×10^{-5}in./ in./°F
3M COMPANY FLEXIGARD	Transparent composite film	Not specified	14,000 psi-60% elongation (across web); 11,000 psi-65% elongation (down web)	Not specified	.014 cal./cm^2/ sec./0°C	5.0 to 9.0× 10^{-5}/in./°C
MARTIN PROCESSING, INC. UX-V	Polyester film. Contact manufacturer for further specifications					
ROHM AND HAAS TUFFAK	Plastic polycarbonate sheet and film	82 to 89%, depending on thickness; long-term data not yet available	9,500 psi	Izod notched. ⅛ in.; 16 ft. lb./in.	1.35 Btu/hr./ ft.^{2}/°F/in.	3.8 in/in./ °F=10^{-5}
ROHM AND HAAS CO. PLEXIGLAS G.	Plastic acrylic sheet	91% after 10 years	10,500 psi	8.0 ft. lb.(charpy unnotched-½ × 1 in. section)	1.3 Btu/hr./ ft^2/°F/in.	.000041 in./ in./°F

TABLES

Flammability	Chemical resistance	Surface weathering	Maximum operating temperature	Heat shrink capability and temperature	Maximum span width for each thickness	Longevity	Transparent or translucent
Non-flammable	Highly resistant	No measurable change after 20 years	400°F	Yes, 300°F	4 ft. (1 mil film)	Data not yet available	Transparent
Less than 1½ in. per minute (ASTM: D-635)	Resistant to most acids and alkalides	Color stability—exterior face will not change more than 4 units	200°F	Material will not shrink	.040 in.: 36 in.: .025 in.: 30 in.	Estimated 20 years	Translucent
Less than 1½ in./min. (ASTM D-635)	Resistant to most airborn chemicals	20 year life expectancy; prolong by refinishing every five years with Kalwall Weatherable Surface	200°F	n.a.	½ in.: up to 24 in.; 1½ in.; up to 48 in.	20 years with minimal maintenance	Translucent
Less than 1½ in./ min. (ASTM D-635)	Resistant to most airborn chemicals	20 year life expectancy: prolong by refinishing every five years with Kalwall Weatherable Surface	200°F	n.a.	½ in.: up to 24 in.: 1½ in. up to 48 in.	20 years with minimal maintenance	Translucent
Yes, no specification	Available upon request	Highly resistant	250°F (framed): 180°F (unsupported)	30% unsupported: 0% framed	48 in. wide	7 to 10 years	Transparent
Combustible, self-extinguishing	Resists water, weak acids, aromatic solvents. Attacked by alkali, strong acids, and chlorinated solvents	Resistant	250° to 270°F	.005 to .007% at 500 to 600°F	Not specified	5 to 7 years	Transparent: tinted grades of lower transmission available
Combustible	Resists, water, alkali, dilute acids, gasoline, oils. Dissolved by ketones, esters, aromatic and chlorinated solvents	Little or none	180° to 200°F	.002 to .008% at 350 to 500°F	Not specified	10 to 20 years	Transparent; translucent also available

Absorber Plates

Product Identification	Dimensions	Applications	Plate and tubing materials	Plate design	Method of bonding tube to plate	Dimensions and wall thickness of tube
BARKER BROTHERS ABSORBER PLATE	46×76×.0135 in.; 48×120×.0135 in.; special sizes available upon request	Dom. water heating, space heating, industrial process water heating, pool heating	Copper	Not specified	50/50 solder (for medium temp. models)	.5 in. O.D.; wall thickness .025 in; length variable
BERRY SOLAR PRODUCTS ABSORBER PANELS	24×96 in.; or custom	Dom. water heating; space heating, etc.	Copper and stainless steel	Tube-in-sheet (copper); and two-panel sandwich (stainless steel)	Solder, clip, heat transfer compounds	⅜ in. O.D.
BURTON INDUSTRIES, INC. SUNSTRIP	2 in. wide, up to 20 ft. long	Dom. hot water, space heating, cooling, pool heating	Aluminum extrusion, copper tube liner	Tongue-and-groove extrusions with imbedded tubes	Impulse formed, heat treated	.653 in. O.D., .513 in. I.D.; liner: .018 in. thick
DELTA T COMPANY	33 5/16 in. × 74½ in.; area: 17.138 ft.^{2}	Dom. hot water space heating, air conditioning, etc.	Galvanized steel, copper, aluminum	Steel plate; copper tubes with bonded aluminum fins	Silver solder	Not specified
ILSE ENGINEERING INC. SANDWICH PANEL™ Information available from manufacturer						
OLIN BRASS ROLL-BOND^R	Aluminum: Maximum size 36×110 in.; Copper: maximum size 34×96 in.	All solar applications	Copper alloy-122; Aluminum alloy-1100 (other alloys available)	Two sheets of metal bonded and expanded to form channels	Silk screen process followed by inflation and trimming	.120 and .140 in. height; ⅜, 7/16, ½ in. thickness
THE SOLARAY CORP. C-18 AND C-6	33×73¾ in. and 17½×47¼ in.	Absorber plates supplied to manufacturers to assemble complete panels for heating water or for pool heating	All copper	Flat plate formed to fit runner pipes	50/50 solder with added copper-graphite compound for heat transfer	Copper 1⅛ in. OD and ⅝ in. OD header pipes, both with ⅜ in runner pipes
SOLAR DEVELOPMENT, INC.	4×10 ft and 2×10 ft. are standard; other lengths available up to 20 ft.	Hot water, and pool heating	.012 in. thick plate, ½ in. ID copper tubing	Series or parallel flow; parallel flow available with ½, ¾, or 1 in. (or larger) headers	Plate is grooved to accept ½ of pipe circumference. Pipe is clamped to plate and then soldered	½ in. ID copper water tube
TRANTER, INC. 3482S	34¼×82¼ in.	Various, water heating for space and surface water heating	Carbon steel or stainless steel	Two sheets of steel seam-welded together and pressure formed fluid circuitry	Not applicable	Not applicable

Flow rate recommended and pressure drop per absorber at recommended flow rate	Maximum operating temperature and pressure	Absorber surface (e/a)	Recommended heat transfer fluid (s)	Materials recommended to interface collectors into fluid supply and return lines	Is an' external manifold necessary to connect individual absorbers to each other as well as to supply and return lines?	Location of the connections to supply and return lines	Limitations
1.8 GPM; .5 psi	250° F; 200 psi	e= .95; a= .95	Water	Copper fittings or dielectric unions for iron pipe	No	All four corners	400°F
2 to 3 gpm	220°F	e= .10; a= .92	Water and anti-freeze	Copper tubing	No	Ends of absorbers	Not specified
.0175 gpm/ft^2; P= .003 in. H$_2$O/ft of legnth with water as coolant	600°F; 150 psig	e= .98; a= .89 (for 3M Nexel)	Water, glycol, etc.	Standard plumbing materials	Yes	Both ends of manifolds; mid-points optional	None
Not specified	400°F; 200psig	Not specified	Not specified	Soft copper tubing	Not specified	Not specified	Refer to max. operating temp. and pressure
Not specified	Temperature not to exceed melting point of base material; stock panels 100 psi (can be designed in excess of 500 psi)	Not specified	Aluminum: non-aqueous solutions recommended	Aluminum: non-dielectric connectors; copper: same as copper plumbing tube	No	To customer specification	None specified
One gallon per hr/ft^2	400°F and 150 psi working (tested pressure 300 psi)	Nickel black chrome selective service (e/a: .1/.93)	Any liquid compatible with copper	Copper tubing (supply lines)	No	Parallel header constuction allows connection from either end of headers. (header ends extend beyond plate sides at either end of plate.)	None
Flow rate depends on application and header size	360°F and 150 psig	Flat black lacquer base auto primer or black chrome on nickel flash	Various	Copper tube	Depends on header size	Connector locations optional	Not specified
App. .6 gpm/plate flow rate and app. 2 psi pressure drop	50 psi for 20 gage and 75 psi for 18 guage	Not normally supplied	Water with glycol or other heat transfer fluids compatible with carbon or stainless steel	Plastic connectors may be used; in many cases the connestions are direct steel to copper	2 or 3 units may occassionally be connected in parallel without a manifold; otherwise manifold is required	In the center of each end	The absorbers should not be used at excessive pressures without inhibitors or without freeze protection. Prefer to sell to collector manufacturers or similar firms rather than to individuals.

Absorber Coatings

Product identifications-manufacturer and/or trade name	Type	Absorptivity	Emmissivity	e/a	Application mode	Temperature limitations
AMERICAN SOLARIZE, INC.	Aluminum honeycomb	.90 to .98	.2 to .4	Not specified	Not specified	600°F
BERRY SOLAR PRODUCTS SOLARFOIL AND SOLARSTRIP	Black chrome	.92	.10	11%[1]	Electroplated; formable for tube-in-sheet type plate; applicable as spiral pipe wrap; or adhesive backed	750°F
DESOTO, INC. ENERSORB	Flat black, non-selective	.97	.92	Not specified	Not specified	Not specified

Flat Plate Collectors: Liquid

Product identification: manufacturer and/or trade name	Collector dimensions and net absorbing area	Applications	Absorber plate material and specifications	Glazing	Insulation	Case material	Sealants and adhesives
ACORN STRUCTURES, INC. SUNWAVE 420	4×20 ft; net area 70 ft^2	Space heating dom. hot water	020 aluminum, painted black, with copper tubes mechanically attached	Filon #548 acrylic-fortified, glass-reinforced polyester. .032-in.-thick, Dupont PVF Tedlar coating, guaranteed for 2 years against thermal shock and degradation	2-in.-thick fiberglass blanket	1½-in. redwood glazing frame, painted black; 1½-in.-thick fir box-frame with ⅜-in. plywood back and #15 felt	GE#1200 construction silicone sealant
ALTEN CORP. 200 G	92¾×46¾×4¾-in.; net area 78 ft^2	150 to 200°F solar heating and cooling requirements	Extruded aluminum finned plate, copper tubes embedded; coating on copper, plus high-temperature-resistant hermetic seal at interfaces	Double 3/16-in. tempered glass	Furnace-grade high-temperature fiberglass, 2½-in. back, 1-in. sides	Extruded aluminum frame	Silicone, neoprene
AMETEK, INC. POWER SYSTEMS GROUP	96×22-in.; net area 14.67 ft^2	Absorptive chiller, dom. hot water, space heating	Copper, Olin Roll Bond	Double ⅛-in. tempered, low-iron glass	4-in. fiberglass and rockwool	Cor-Ten steel, welded at exposed joints	Various sealants compatible with the glazing system
BURKE INDUSTRIES INC.	8-ft width×8, 10,12,16,20 or 24 ft	Home or commercial pool heating; medium-flowrate systems operating to 140°F	30-mil Dupont Hypalon	n.a.	n.a.	n.a.	n.a.

152

Longevity	Durability and outgassing potential	Materials which attack coating	Other limitations		
Not specified	Withstands any atmospheric conditions	None	Not specified		
Not specified	No outgassing properties	Oxygen, moisture, and some corrosive gasses under long-term high temperature exposure	Not specified		
Not specified	Not specified	Not specified	Not specified		

Spacer materials and desiccants	Temperature limitations on entire collector	Method of installation to roof (are components for roof installation supplied or available?)	Are instantaneous efficiency curves (in accordance with NBS guidelines) and day-long efficiency testing available?	Are there any high-temperature limitations on particular components within the collector that require the system to contain a cooling mode or heat-dump capability?	Can the collector stagnate indefinitely without damage?	Limitations	Does the guarantee specifically cover any potential for thermal shock or chemical breakdown of the components due to temperature?
None	250°F shutdown	Directly over roof sheathing; screws supplied with insulation kit	Instantaneous curve	No	Yes, in any climate where glazed collectors are appropriate	See literature	No
None	350°F (continuous)	Various methods available through manufacturer	No	No	Yes	350°F	No
Silica gel desiccant	No stagnation temperature limitations within ASHRAE 93-77	Bolts to support rack	Instantaneous curve	No	Yes	40 psi @ 250°F	No
n.a.	Surface limitations: —40 to 250°F	Installation kits and instructions available	Instantaneous curve not yet available; day-long testing results available	No	Yes, if properly installed	Applications to 140°F	Warranted not to crack or leak due to normal weathering

Continued

Product identification manufacturer and/or trade name	Collector dimensions and net absorbing area	Applications	Absorber plate material and specifications	Glazing	Insulation	Case material	Sealants and adhesives
BURTON INDUSTRIES, INC.	40½×80-in.; net area 18 ft^2; 52½×104-in., net area 32 ft^2	Domestic or industrial hot water, space heating and cooling, pool heating	Finned-tube aluminum extrusions with internal thin-walled copper tube. 2-in. net width per strip. Assembled laterally. See "Sunstrip" on Absorber Plates comparative data chart	2 sheets of Kalwall Sunlite Premium (top: .040-in.; bottom: 0.25-in.) Can be replaced in place	Back: 3-in. of fiberglass batt; edges: rigid fiberglass, organic bonded	Aluminum extrusion, 1/16-in.-thick anodized black	Adhesive: foam tape; sealant: silicone
CALMAC MFG. CORP. SUNMAT	4-ft width, any length up to 125 ft	Space, hot water and pool heating	Plastic tubing bonded with adhesive cement	.025-in. fiberglass reinforced plastic, resurface every 5 to 7 years, easily replaced. No guarantee	1-in. or 2-in. 3 lb/ft^3-density foil-faced Owens-Corning Fiberglas #703	1-in. 3lb/ft^3-density foil-faced fiberglass, sealed with black mastic cement	Cal-Zorb adhesive cement and 3M 1300 contact cement
CHAMBERLAIN MFG. CORP.	82.24×34.3 in. (19.6 ft^2) or 94.24×34.3 in. (22.4 ft^2)	Space heating or cooling, domestic hot water (with intermediate heat exchanger)	Cold rolled steel (mfg. by Tranter, Inc.; see Absorber Plates data chart)	⅛-in. tempered low-iron glass, field replaceable, single or double	5-in. blanket of low-binder-content fiberglass, compressed to 3 in. Rated to 1,000°F face temperature	18-guage galvanized steel	EPDM glazing channel and EPDM foam tape (sealants)
COLUMBIA CHASE SOLAR ENERGY DIV REDI-MOUNT 77-3376 AND 76-3496	40×104 in.; net area 22.6 ft^2	Dom. hot water, space heating, pool heating	Copper plate, Kennecott Terra-Light Absorber	Kalwall Sunlite Premium, Filon, Tuffak-Twinwall or tempered glass	2½-in. fiberglass	⅛-in.-thick, one-piece fiberglass frame-and-back ass'y various colors	Silicone
DAY STAR CORP. '20'	72¾×44½× 5½ in.; net area 21 ft^2	Dom. hot water, space heat, absorption cycle air conditioning, pool heating	Solid copper sheet with ½ in. type L copper serpentine soldered to back of absorber	3/16 in. tempered low iron glass in proprietary heat trap (pleated polycarbonate 7½ mil.)	1 in. sides, 2 in back Isocyanurate (foamed in place); tan	.032 aluminum, baked enamel finish, white	GE 1200 sealant
EL CAMINO SOLAR SYSTEMS SUNSPOT	48¾×96¾ in.; net area 30.3 ft^2	Hot water up to 212°F	Copper tubes with extruded aluminum fins, baked siliconized polyester coating on fins	.040 fiber acrylic with .001 Tedlar UV filter	½-in. urethane	.060 extruded aluminum side and cove sections, bronze or red, sheet aluminum back	RTV neoprene. closed-cell vinyl
ENERGY CONVERTERS, INC. SERIES 200	34×96 in.; net area 22.5 ft^2	Open or closed loop system, max. 125-psi	Copper Roll-Bond	5/32-in., fully tempered water white crystal #76 glass	4-in. high-temp., low-binder fiberglass, compressed to 3.5 in.	26-guage galvanized steel, grey primer coating	Butyl and silicon rubber, etc.
FAFCO INC.	4×10 or 4×8ft; net areas 40 and 32 ft^2, respectively	Swimming pools, other low-temperature applications	Polymer (plastic)	None	n.a.	n.a.	n.a.

Spacer materials and desiccants	Temperature limitations on entire collector	Method of installation to roof (are components for roof installation supplied or available?	Are instantaneous efficiency curves (in accordance with NBS guidelines) and day-long efficiency testing available?	Are there any high-temperature limitations on particular components within the collector that require the system to contain a cooling mode or heat-dump capability?	Can the collector stagnate indefinitely without damage?	Limitations	Does the guarantee specifically cover any potential for thermal shock or chemical breakdown of the components due to temperature?
Spacers: rigid fiberglass, organic bonded; desiccant: silica gel	350°F	Roof-surface or free-standing mounting. Loads taken by header tubes. Installation components not supplied.	NBS instantaneous curve	No. Optional venting available if alternate components are to be used	Yes, but lower cover could degrade optically during extended stagnation periods	No significant limitations	No, but can be included with optional over-temperature venting provision
Silica gel	400°F	Built on and cemented directly to roof or other mounting surface	ASHRAE 93-P testing done. Day efficiency data available	No	Yes	None	Yes, covers breakdown of tubing for any reason when installed according to directions
Silica gel	None	Hold-down points provided on corners	NBS instantaneous curve	No	Yes	None	Yes
Fiberglass	400°F	Internal full-perimeter mounting flange	NBS instantaneous curve	None. Collector will stagnate at 190°F	Yes	Yes. Unless tempered glass is used, double-glazed collector or selective coating should have dump system bypass	Yes
Stainless steel trap support		Unistrut channel (optional) or direct	NBS efficiency curve	Heat dump panels recommended and supplied where necessary	Yes	see max. temp.	Yes
None	450°F on absorber, 250°F on remaining components	Lagged through side frames	NBS instantaneous curve; day-long test data available	Yes	No	See temp. limitations	Yes
Wood spacers, with or without desiccant	Stagnates at 310°F	Installation components available	No	No	Yes	None	No
n.a.	210°F at 5 psi	n.a	No	n.a.	Yes	20 psi up to 100°F, 5 psi at 200°F. Burst pressure: 60 psi at 80°F	No guarantee implied

Continued

155

Product identification: manufacturer and/or trade name	Collector dimensions and net absorbing area	Applications	Absorber plate material and specifications	Glazing	Insulation	Case material	Sealants and adhesives
GULF THERMAL CORP. CUS30 AND CUP30	98.5×48.5×2.57 in.; net area 30 ft^2	Dom. hot water, space heating, pools	½-in. copper tubes within extruded aluminum fins. Tubes are brazed to ¾-in. headers. Assembled plate is chemically treated and coated	Triple ⅛-in. or single 3/16-in. tempered or annealed low-iron glass. Designed to withstand 150-mph winds	1¼-in. polyisocyanurate foam	Anodized aluminum extrusion. .032 mill-finish aluminum backplate	Silicone
HALSTEAD & MITCHELL SUNCEIVER 35775	35⅜×77⅜×4 in.; net area 17.2 ft^2	Dom. hot water, space heating, pools, tropical fish breeders	Copper tube with aluminum fins	2⅛-in. tempered water white glass	1.5-in. Foam Glas—UL File #R2844	.051 aluminum	JS-780 butyl, RTV-108 silicone
LIBBEY-OWENS-FORD GLASS CO. LOF SUNPANEL	Net area 19.47 ft^2	Dom. hot water, space heating, air cond., process heat	Copper, with brazed copper manifold	⅛-in. tempered glass; can be replaced	3 in. high-temp. fiberglass, prebaked	Extruded aluminum frame with two handles on each side	Silicone
MARK M MANUFACTURING MODELS# 16, 24 and 32	Not specified	Dom. hot water, space heating, pools	Aluminum or copper plates with flat black or selective coatings, copper manifolds. Polycarbonate available on request (for pool heating)	Kalwall Premium Lexan polycarbonate, glass, or Rohm and Haas Twinwall polycarbonate, in all available thicknesses	2 in. special water-blown sealed urethane foam protected by aluminum foil on both sides	Aluminum, stainless steel, and galvanized steel	Not specified
NORTHRUP, INC. NSC-FPIT	Net area 22 ft.	Dom. hot water and space heating	Extruded aluminum mechanically connected copper circulating tubes; coating: 3M Nextel[R]	.004 in. DuPont Tedlar[R]	3½ in. bottom and 1 in. sides Fiberglass	20 guage galvanized steel; baked enamel finish	Not specified
O.E.M. PRODUCTS, INC., SOLARMATIC DIV.,	Contact manufacturer for details						
PAYNE, INC.	84×24×2½ in.; net area approx. 14 ft^2	Single glazed: pool heating; double glazed: dom. hot water and space heating	Single-piece, vacuum-cast lightweight concrete, steel reinforced, PVC ducts and headers. Aluminum insert for glazing. Trickle-type collector	Tedlar or Teflon film, single or double glazing	n.a.	n.a.	n.a.
PAC, DIV. OF PEOPLE/SPACE CO. PAC OPEN FLOW COLLECTOR	Net area 27 ft^2	Dom. hot water, space heating, preheating water	Conductive-plastic (graphite) and polyester resin) integral absorber plate	Kalwall Sunlite	2 in. urethane foam	Polyester resin, thickness averages ⅛ in., colors available	Silicone
RAYPAK, INC. SOLAR-PAK SG-18-P AND DG-18-P	37.5×79.5 in.; net area 17.3 ft^2	Dom. hot water, pools, central heating in moderate climates	Copper tubes with aluminum surface plate	⅛-in. low-iron glass	3-in. fiberglass	20-gauge galvanized metal with baked enamel brown finish	Silicon and butyl rubber seals

Spacer materials and desiccants	Temperature limitations on entire collector	Method of installation to roof (are components for roof installation supplied or available?)	Are instantaneous efficiency curves (in accordance with NBS guidelines) and day-long efficiency testing available?	Are there any high-temperature limitations on particular components within the collector that require the system to contain a cooling mode or heat pump capability?	Can the collector stagnate indefinitely without damage?	Limitations	Does the guarantee specifically cover any potential for thermal shock or chemical breakdown of the components due to temperature?
None specified	300°F on plate surface, 400°F for all other components	Components for flat or pitched-roof mounts, fixed angle or adjustable, available from manufacturer	NBS instantaneous curve	No	Not known	See temp. limitations	Yes
½-in. galvanized spacer. Desiccant: 10/20 mol-siv beads	400+°F	Attaches to frame with sheet-metal screws (not available from manufacturer)	Instantaneous curve	No	Yes	None specified	Yes
Teflon standoffs, aluminum spacers, etc.	400°F	Bracket fixtures available	NBS instantaneous curve; day-long data to be available	Excursions to 400°F	Yes	Corrosive or reactive collector fluids should not be used	Yes
Not specified	Not specified	Standard procedures	No	None implied	Not specified	Not implied	Yes, on plates using copper manifolds
	Not specified	Framing and mounting apparatus available as an option; copper tie-in tubing not included	Not specified	Not specified	Not specified	Not specified	Not specified
n.a.	None. Do not, however, exceed PVC plumbing pressure limits	Not specified. External manifold required	No	No	Yes	None	Yes
None	180°F for normal use, excursions to 300°F tolerated	Bolt holes on collector body are fastened to support structures of various types	Instantaneous curve; in-house testing	180°F for manifold pipe, can endure brief excursions to 200°F	No	See temp. limitations	No. Water must circulate in collector, or unit must be covered
Kiln-dried wood spacers and silica gel desiccant	400°F	Operating and installation manuals available	NBS instantaneous curve	Will withstand 400°F	Yes	None implied	Yes

Continued

	Collector dimensions and net absorbing area	Applications	Absorber plate material and specifications	Glazing	Insulation	Case material	Sealants and adhesives
REVERE COPPER AND BRASS INC., SOLAR ENERGY DEPT. SUNAID	35×77 in.; net area 17.2 ft^2	Dom. water, space heating and cooling, pool heating	Copper-Tube-In-Strip^R, grid flow pattern with ½-in. (nom.) header and six ¼ in. (nom.) flow tubes	One or two layers tempered low iron or water white glass	2½ in. fiberglass, approximate R value of 10	6063-T5 aluminum extrusion, .086 in.	Silicone channel used at glass, silicone sealant at corners
REVERE COPPER AND BRASS, INC., SOLAR ENERGY DEPT. SUN ROOF	14.66 ft^2 net area	Dom. water heating, space heating and cooling, pool heating	.010 in. copper laminated to a substrate of ⅜ or ¾ in A-C, Group Exterior Grade Fire Retardant plywood; .010 in. aluminum backing	⅛ in. tempered glass	6 in. fiberglass (R-19 suggested)	mill finish copper	Butyl rubber sealing tapes; copper filled epoxy conductive sealant
SOLARAY CORP. AP-18 AND AP-6	34¼×76¼ in., net area 17 ft^2, 18×48 in., net area 5¾ ft^2	Water heating	Copper plate and tubes, nickel-black chrome selective surface. Silver-solder brazed pipe connections	Single glass or plastic-laminate (Filon) glazing. Attached with silicone to resist thermal shock	2-in. glass wool for large collector, 1 in. for small collector, reflective aluminum foil between plate and wool	Anodized aluminum.	Silicone
SOLAR CORP. OF AMERICA MARK III, MARK V	4×8 ft×5 in. (Mark III); 3×6.5 ft×5 in. (Mark V)	Dom. hot water, space heating, etc.	Copper tubes, aluminum fins. Manifolds attached	Low-iron tempered glass	Insulated to an R-15 factor	Galvanized steel	High-temp rubber
SOLAR ENERGY PRODUCTS, INC. CU 30	45×91.5 in.; net area 29.65 ft^2	Water, pool, space, or process heat; water preheating; cooling; cooking	½-in. copper tubes, extruded aluminum fins, ½-in. copper headers	⅛-in. standard lime glass or 3/16-in. water white glass, tempered	1¼-in. isocyanurate foam board	3/16-in. extruded aluminum alloy, clear anodized finish. Colors optional	Silcone (UV-stable)
SOLAR ENERGY RESEARCH CORP.	2×8,10,12, or 16 ft; net area 30 ft^2 for 2×16 ft panel	Medium temp. hot water for space heating, pools, car wash, heating greenhouses	.010 half-hard copper	.040 or .060 Lexan or Kalwall. 15-year life. Easily replaced. Guaranteed 2 years against degradation or thermal shock	6-in. fiberglass	Redwood, primed and painted, or galvanized steel, primed and painted	Butyl and silicone
SOLARGENICS, INC. SERIES 76-77	3×10,15, or 20 ft; net areas 27.2, 41, and 55 ft^2, respectively	Dom. hot water and space heating, absorptive cooling, boiler preheating	Copper tubes, bonded to aluminum that has a black chrome selective plating	⅛-in. tempered low-iron glass. Optional water white and other glazings	Foam and fiberglass. R-12 rating or better	Structural aluminum channel sides and ends, heavy aluminum back. Sealed	Exterior-grade silicone

Spacer materials and desiccants	Temperature limitations on entire collector	Method of installation to roof (are components for roof installation supplied or available?)	Are instantaneous efficiency curves (in accordance with NBS guidelines) and day-long efficiency testing available?	Are there any high temperature limitations on particular components within the collector that require the system to contain a cooling mode or heat-dump capability?	Can the collector stagnate indefinitely without damage?	Limitations	Does the guarantee specifically cover any potential for thermal shock or chemical breakdown of the components due to temperature?
Silicone spacers	450°F	Aluminum bracket systems available from manufacturer	NBS efficiency curve	450°F	Yes	None	Yes
Not specified	300°F stagnation temperature	Install on roof trusses, or on previously placed roof deck	NBS efficiency curve	325°F	Undetermined at this time	Same as above	Yes
None	350°F (subject to glazing used)	Straps or brackets for mounting to custom installation frame	Not specified	No	Yes, subject to type of glazing used	None	No
n.a.	Not specified	Wood or steel supports. Collectors should be installed with a 1-in. drop per 20 ft of horizontal run, to take advantage of gravity drain	Instantaneous curve	None specified	Not specified	None specified	Yes
Neoprene spacers	300°F	Mounting package includes hinges, stand-offs, brackets, hardware. Available from manufacturer	NBS instantaneous curve	No	Yes	None	Yes
Butyl silica gel	300°F	Installation components available for wall or roof. Manual available	NBS instantaneous curve	No	Yes	Gravity-return to tank	Yes
Replaceable desiccant bags	500+°F	May be installed as part of roof or as carport/patio cover. Mounting hardware available from manufacturer	NBS instantaneous curve; day-long data available	No, if proper heat-transfer fluids are used	Yes	None	Yes, if installation is done by authorized contractors

Continued

Product identification: manufacturer and/or trade name	Collector dimensions and net absorbing area	Applications	Absorber plate material and specifications	Glazing	Insulation	Case material	Sealants and adhesives
SOLARGIZER CORP.	48×96 in.	Dom. hot water, space heating, pools	Copper panels	Not specified	Insulated	Aluminum frame, fiberglass case and cover	Not specified
SOLAR HOMES SYSTEMS INC. SHS-00L1	94×46 in.; net area 30 ft²	Space and dom. hot water heating	½-in. copper tube in extruded aluminum fin, mechanically bonded; tubes spaced 6 in. on center	Single 3/16-in. tempered glass	5/8-in. Celotex TF 400, foil on both sides	Aluminum extrusion	GE silicone
SOLAR INNOVATIONS SC-200	3¼×22¾× 96¼ in.	Dom. and process hot water, space heating and cooling	Copper Roll-Bond, 3M Black Velvet coating	Single Kalwall Sunlite Premium	1 in. polyurethane	Aluminum	None specified
SOLARKIT OF FLORIDA, INC.	3×5 ft×3½ in.	Pool and dom. hot water heating	Aluminum sheet grooved to accept ½-in. copper tubing spaced 6-in. on center, coated with flat black enamel	⅛-in. tempered glass	1-in. closed-cell Technifoam isocyanurate with reflective aluminum	1×4-in. redwood backed with .015-in. aluminum sheet	None specified
SOUTHERN LIGHTING MFG. UNIVERSAL 100 THERMOTUBE	4×10 ft, net area 38.8 ft²; or 4×5 ft, net area 19.2 ft²	Space heating, dom. and commercial hot water	.008 copper sheet painted with Farboil Solar Black	Outer: 3/16-in. tempered glass; inner: .040 Kalwall Sunlite Premium	1 in. of Celotex isocyanurate and ½ in. of homasote	.090 aluminum, painted black	Sodium silicate and contact cement
SUNWORKS, DIV. OF ENTHONE SOLECTOR	3×7 ft, net area 18.68 ft²; or 3×5 ft 4 in.; net area 14.09 ft²	Space heating and cooling, dom. hot water, drying, process heat	Copper sheet and tubes with Enthone selective black surface	Single: 3/16-in. tempered no-iron-glass, swiped edges; double: ⅛-in. tempered no-iron glass, swiped edges	1 in. compressed glass fiber over 1 in. foil-faced isocyanurate (R-10)	Extruded aluminum sides, sheet-aluminum back, standard mill finish; anodized clear or black enamel available	Neoprene "U" gasket for glazing, closed-cell elastomer, compressible high-temp. silicone seal for absorber sheets
SOLAR DEVELOPMENT, INC. STANDARD SD-5	4×10 ft; net area 34.84 ft²; or 2×10 ft, net area 17.42 ft²	Space heating and hot water	.012-in. grooved copper with ½-in. copper tubing soldered into groove and spaced 4.6 in. on center; series or parallel flow	Kalwall Sunlite Premium (.025 in. thick, or .040 in.)	Celotex Technifoam TF 400, 1 in. thick, 1 or 2 layers	1/16-in. reinforced 6063T5 aluminum for 30 lb/ft² wind load, or 6061T5 aluminum for 48 lb/ft² load	Aluminum tape sealant for glazing
SUNEARTH SOLAR PRODUCTS CORP. 3296	Net area 19.26 ft²	Space heating and dom. hot water	Copper waterways, aluminum fins; selective paint, integral headers	Replaceable .080-in. acrylic, 20-year life, 5-year guarantee	1-in. rigid fiberglass	Mill-finish aluminum	None
SUNEARTH SOLAR PRODUCTS CORP. 3597A	Net area 19.78 ft²	Space heating and dom. hot water	Copper waterways, aluminum fins; selective paint, integral headers	Replaceable .080-in. acrylic, 20-year life, 5-year guarantee	1½-in. thermal wool	Mill-finish aluminum	Not specified

Spacer materials and desiccants	Temperature limitations on entire collector	Method of installation to roof (are components for roof installation supplied or available?)	Are instantaneous efficiency curves (in accordance with NBS guidelines) and day-long efficiency testing available?	Are there any high-temperature limitations on particular components within the collector that require the system to contain a cooling mode or heat-dump capability?	Can the collector stagnate indefinitely without damage?	Limitations	Does the guarantee specifically cover any potential for thermal shock or chemical breakdown of the components due to temperature?
Not specified	Not specified	Not specified	Not specified	Not specified	Not specified	None specified	Yes
Silica gel desiccant	400°F	Bolt, screw, or nail to roof	NBS Instantaneous curve	None	Yes	None specified	No
None specified	Not specified	Universal mounting bracket	Not specified	Not specified	Not specified	Not specified	Not specified
None specified	None specified	Not specified	Not specified	Not specified	Not specified	None specified	Not specified
Neoprene gasket around glass	350°F	Pitch pans, roof flashings, mounting hardware available; bolts purchased locally	NBS instantaneous curve	Not in normal use	Yes	n.a.	
Condensation trough directs moisture to weep-holes. Double glass collector has desiccant between layers	400°F	Various methods	NBS instantaneous curve; all-day data available	None required	Can undergo long periods	pH must be maintained between 6.5 and 8	No
Vented collector	n.a.	Hardware provided	NBS tests in progress	Kalwall is limited to 300°F, but collector is vented	Yes	None	No
Neoprene and EPDM seals	Not specified	2×6-in. curb for skylight mount, counter-flashing integral to unit	NBS instantaneous curve	No	Can stagnate in all U.S. climates	Not specified	Yes, for defective material
Neoprene and EPDM seals; filtered condensate drain	Not specified	Direct roof-mounting system	NBS instantaneous curve	No	Can stagnate in all U.S. climates	Not specified	Yes, for defective material

Continued

Product identification: manufacturer and/or trade name	Collector dimensions and net absorbing area	Applications	Absorber plate material and specifications	Glazing	Insulation	Case material	Sealants and adhesives
UNIT ELECTRIC CONTROL, INC. SOL-RAY DIV. SOL-RAY 1-6-50M	35½×77¼× 2¾ in.; net area 17.2 ft^2	Space heating, domestic water and pool heating	.016-in. copper sheet soldered to ½-in. i.d. copper tubing	3/16-in. replaceable tempered glass	1-in. urethane foam edge insulation; ½-in. air space plus ½-in. urethane foam rear insulation.	.063-in. aluminum, heat and weather resistant flat black finish	Silicone sealant for glass and top frame

Flat plate collectors: air

Product identification: manufacturer and/or trade name	Collector dimensions and net absorbing area	Applications	Absorber plate material and specifications	Glazing	Insulation	Case material	Sealants and adhesives
AMERICAN SOLARIZE, INC	3×7 ft×7in.; net area 20.17 ft^2 or 4×8 ft×7 in.; net area 31 ft^2.	Up to 300°F	.004 in. aluminum with aluminum honeycomb over it; black crystaline surface.	Top layer: glass or fluorinated ethylene propylene; lower layer; double skinned polycarbonate. Retains 95% of total transmission after 10 years.	3½ in. of fiberglass or 2 in. foamglass.	Extruded aluminum or reinforced fiberglass	Polyurethane
CONTEMPORARY SYSTEMS, INC. SERIES IV	2 ft × any length: net area .90 of gross area	Space heating	7 mil optically blackened metal	Outer: Kalwall Sunlite Premium 40 mil; Inner: Tedlar 4 mil film with integral spacers; one year limited warranty	Site installed to need; normally fiberglass batt; exterior to collector chassis	Extruded aluminum, unpainted, .065 in. thick	GE Clear Silicone #GE2567-012 (or equivalent product).
FUTURE SYSTEMS, INC. SUN*TRAC MODEL 232	4×8 ft: net area 29.33 ft^2 (larger models available)	Residential and commercial heating	Black-painted aluminum veins, horizontal, with ¾-in. spacing	Replaceable 3/16 in. tempered glass, guaranteed against thermal shock	4-in. urethane foam (R-29) with 2 lb/ft^3 density	20-mil aluminum with embossed brown baked-enamel finish, and ⅛-in. fiberglass in several colors	Glass: butyl tape; metal: acrylic and silicone caulking
KALWALL CORP. SOLAR-KAL AIR HEATER	47⅜×95½ in.: net area 30.08 ft^2	Space heating	.032 in. flat aluminum; black coating; corrugated (optional)	.040-in.-thick Sun-Lite Premium II	Apply as necessary	6063-T6 aluminum I beam frame: mill finish aluminum fiberglass back	Kalwall proprietary sealing adhesive compound
NORTHRUP, INC. SUNDUCT™ SDC-8 & SDC-12	96×33×6⅛ in. and 144×33×6⅛in.	Space heating, make-up, drying or process air	Not specified	Double cover of DuPont Tedlar^R	Not specified	Galvanized steel	Not specified

Spacer materials and desiccants	Temperature limitations on entire collector	Method of installation to roof (are components for roof installation supplied or available?)	Are instantaneous efficiency curves (in accordance with NBS guidelines) and day-long efficiency testing available?	Are there any high-temperature limitations on particular components within the collector that require the system to contain a cooling mode or heat-dump capability?	Can the collector stagnate indefinitely without damage?	Limitations	Does the guarantee specifically cover any potential for thermal shock or chemical breakdown of the components due to temperature?
n.a.	350°F	Roof mount, certified to withstand hurricane force wind is available	n.a.	No	Yes	None	Yes, except for glass, paint and freezing

Spacer materials and desiccants	Temperature limitations on entire collector	Method of installation to roof (are components for roof installation supplied or available?)	Are instantaneous efficiency curves (in accordance with NBS guidelines) and day-long efficiency testing available?	Are there any high-temperature limitations on particular components within the collector that require the system to contain a cooling mode or heat-dump capability?	Can the collector stagnate indefinitely without damage?	Limitations	Does the guarantee specifically cover any potential for thermal shock or chemical breakdown of the components due to temperature?
None specified	300°F	Attaches to supports or roof; components available	Instantaneous curve	Stable to 325°F	Yes	Max. air flow 4 CFM/ft^2	Yes
n.a.	180°F max. operating temp	Integrated with roof frame; no special hardware	Tests in progress	Automatic thermal vent system incorporated	Continuous stagnation without use of vent will degrade glazing	Installation to manufacturer's specifications required	Yes, if used as specified
Wood sash spacers; desiccants unnecessary	250°F	Structural supports at 4-ft intervals	NBS Instantaneous curve; day long testing	Yes	No	Aluminum reflectors, when closed, limit temperature	No
6063-T6 aluminum spacers.	200°F	Attaches to south facing walls or roof; hardware available	NBS efficiency curve	200°F	No, unless automatic vent option specified	See max. temp.	No, but thermal shock cannot occur
Not specified	Not specified	Attaches to roof or wall; components available	Not specified	Not specified	n.a.	Not specified	

Continued

Product identification: manufacturer and/or trade name	Collector dimensions and net absorbing area	Applications	Absorber plate material and specifications	Glazing	Insulation	Case material	Sealants and adhesives
SOLARON CORP. SERIES 2000	3×6½ ft; net area 17.46 ft^2	Space and process heat; service and dom. hot water	Porcelain-enameled steel	⅛-in. tempered low-iron glass, easily replaced (standard patio-door size)	Fiberglass batt with 1 in. rigid fiberglass in internal manifold	20-guage steel sheet, painted black	Silicon-rubber gasket between ports. EPDM gasket around glass. silicon silastic between absorber and case.
SOLAR DEVELOPMENT, INC.	4×8 ft×5⅜ in. nominal; net area 29.5 ft^2	Space heating	Vitreous screen (details proprietary)	Double .025-in. Kalwall Sunlite Premium	1-in. Celotex Technifoam TF 400	1/16-in. extruded aluminum 606T5 alloy	Aluminum tape
SOLAR HOMES SYSTEMS, INC. SHS-00A1	94×46 in.; net area 30.02 ft^2	Space heating, greenhouses, grain drying	Corrugated 1100 aluminum	Single 3/16-in. tempered glass	⅝-in. Celotex TF 400, foil on both sides	Aluminum extrusion, natural color	GE silicone
SUNPOWER INDUSTRIES, INC. SELECTOR PANEL	4×8 ft; net area 31 ft^2	Space heating	Black, multi-coated, thermoformed diathermic material	Outer: acrylic modified 4-oz. greenhouse fiberglass; inner: 1-mil. DuPont Tedlar	n.a.	Galvanized steel	Silicone adhesive on inner glazing.
SUNWORKS, DIV. OF ENTHONE SOLECTOR	3×5 ft; net area 18.68 ft^2	Space heating and cooling; drying; process heat	Copper sheet with corrugated aluminum inner channel, selective surface	Single: 3/16-in. tempered no-iron glass, swiped edges; double: ⅛-in. tempered no-iron glass, swiped edges	½-in. fiberglass over ¾-in. foil-faced isocyanurate. (Glass fiber: R-9, 1.2 lb/ft^3 density)	Extruded aluminum sides, sheet-aluminum back, standard mill finish; anodized clear or black enamel available	Neoprene "U" gaskets for glazing: closed-cell elastomer. compressed hig-temp. silicone seal for absorber sheet

Controls

Product identification	Type	Dimensions	Applications	Power input and output	Electronic or Mechanical design	Principle of operation
AERODESIGN CO. AD-101 AND AD-102	Differential controls or temperature limit switches	AD-101: 9½×5×2½ in. (surface mount); AD-102: 7×5½×2½ in. (shelf mount)	Differential thermostat; high-temperature shut-off; freeze protection activator; auxiliary cut-in; thermostat	Input: 120 VAC, 50/60 Hz, 6 watts; output: 10 amps resistive; 1/6 HP at 120 VAC. Switching of higher currents available on order	Electro-mechanical relay contacts enclosed. SPDT standard, DPDT or 3PDT contacts available order	Two sensors change in resistance in response to temperatures being sensed. Resulting voltage levels are amplified, and actuate relay
CONTEMPORARY SYSTEMS, INC. LCU-110	Logic control for air circulating systems	8×10×4 in.	Most warm air systems	Input: 110 VAC, 60 Hz; triac/relay outputs o.k. with standard HVAC practices	Electronic digital logic	The LCU logic circuits compare collector, storage room temperatures with set values, determine the mode of operation, and switch proper damper actuators and blowers

Spacer materials and desiccants	Temperature limitations on entire collector	Method of installation to roof (are components for roof installation supplied or available?)	Are instantaneous efficiency curves (in accordance with NBS guidelines) and day-long efficiency testing available?	Are there any high temperature limitations on particular components within the collector that require the system to contain a cooling mode or heat-dump capability?	Can the collector stagnate indefinitely without damage?	Limitations	Does the guarantee specifically cover any potential for thermal shock or chemical breakdown of the components due to temperature?
Polysulfide between glazings as spacer, silica gel desiccant	350° to 400°F	Hold-down hardware is available	N.B.S. Instantaneous curve	None	Yes	n.a.	Not specified
None specified	None, vented	Aluminum angle clips	n.a.	Is vented	Yes	None	No
Silica gel	400°F	Bolt, screw, or nail to roof	No	No	Yes	None specified	No
n.a.	230°F	Standard construction materials and methods; instructions included	Not specified	Not specified	Yes	Blower is necessary	Yes
Condensation trough in extrusion directs moisture to weep holes	400°F	Integral mounting leg, prefabricated "U" clips supplied; through-bolted anywhere	No	Stable to 400°F	Can stagnate for long periods	See Sunworks guarantee	No

U.L approval	Differentials available if differential thermostat	Sensitivity	Operating range	Degree of adjustment possible	Limitations
On enclosure and switches	Pre-set to turn on at 18°F and turn off at 5°F	1°F	to 110°F	Adjustable within: turn-on: 10 to 25°F; turn-off: 3 to 10°F	Not specified
U.L. testing in progress	Collector and storage differentials incorporated in logic circuits and are user adjustable	±3%	For logic circuits: 0 to 40°C; and for probes: −40 to 125°C	17°C range; hysteresis also setable in 17°C range	Install according to specifications

Product identification	Type	Dimensions	Applications	Power input and output	Electronic or Mechanical design	Principle of operation
DEL SOL CONTROL CORP. 02A	Differential, usually for temperature but may include water level, pressure, etc., depending on type of sensor used	4×2×2 in.	Control of pumps, fans, valves, and transformers	Input: negligible: output; 230 VA max.	Electronic, 100% solid state	Sensitive to the relative resistance of two sensors
ENERGY APPLICA-TIONS, INC. SUN-TRAK 100	Electronic tracking device	1¼×1¼×2 in.	Collectors with north-south axis, motorized shades and shutters, tracking platforms, solar cell panels, etc.	Input: 12 VDC, 20 mA: output: ½ amp, 28 volts, 3 watts (all DC) maximum	Electronic	Two sensors: one detecting day/night conditions and the other the sun's east to west movement
ENERGY CONVER-TERS, INC. E10	Differential thermostat	Not specified	Turn on pumps or other devices	Input: 120 volts	All solid state	Two appropriately placed thermistors
HAWTHORNE IN-DUSTRIES, INC. FIXFLO AND VARIFLO	Differential temperature, on-off and variable flowrate controls	6×3×3 in.	All systems	Input: 117 volts, 4 watts; output: 117 volts, up to 6 amps	Solid state: triac switching circuits	Thermistor sensing at collector and storage activates triac switch
HELIOTROPE GENERAL DELTA-T	Automatic motor control	4 11/16 in. square	Controls pump or blower	Input: 120 VAC, 50/60 Hz: output: ⅓HP at 120 VAC	Solid state electronic and mechanical relay	Sensors indicate when the collector temperature exceeds the storage temperature by preset differentials
RHO SIGMA, INC RS 106	Differential thermostat	6×8×4 in.	Space and hot water heating systems	Input: 120 VAC, 6 watts: one SPDT relay rated at 10 amps (DPDT and 3PDT relays available on request)	Electronic circuits with relay output	Relay energizes pump when the temperature of the collector minus the temperature of the storage equals 20°F (±3°F): relay de-energizes pump when collector minus storage equals 3°F (±1°F)
RHO SIGMA, INC. RS 500S SERIES	Differential thermostat with two independent outputs, high temperature and low temperature override circuits	6×6×3 in.	Domestic water heating systems, including those with drain valves for freeze protection	Input: 115 VAC at 6 amps (240 VAC optional); output: 115 VAC at 6 amps (240 VAC optional)	All solid state	Provides differential control of solar circulating pumps and fans with high temperature turn-off (from storage side) and low temperature turn-on (from collector side). The second optional output can control a second pump or normally open or normally closed valves

U.L. approval	Differentials available if differential thermostat	Sensitivity	Operating range	Degree of adjustment possible	Limitations
Yes	Normally 8 to 10°F, but others may be provided	.05°F	Control box: 0 to 70°C; sensors: −50 to 300°F	None required	Application expertise (contact factory for assistance)
Not applicable to a 12 volt system	Not applicable	Adjustable	−25 to 85°C	Sensitivity to haze conditions and optional tracking accuracy built in	North-south axis of rotation for mounting of tracking device
Not specified	Settings other than standard can be specified	Not specified	Not specified	Optional on-off temperature differential settings available	Not specified
Yes	On-off is factory set at 16°F on, 3°F off. Will set to user's specifications	±.5°F	Control: 32 to 125°F ambient; sensors up to 300°F: control near linear from 30 to 220°F	None required	Virtually unlimited applications
Yes, for most models	Various, many models	Not specified	Various, many models	Varies	Not specified
Yes	To customer's specifications	Stable under line voltage reductions of 15%	Covers the temperatures normally encountered in solar systems	Wide adjustment possible	Output capabilities increased into horsepower range with addition of booster relay into the bottom compartment of the control
Yes	Delta-T on equals 20°F ±3°F: delta-T off equals 3°F±1°F	21°F±3°F	−60°F to stagnation temperatures of collectors	Adjustable over full ranges normally encountered in solar systems	Designed for use for fans and starter configuration. Intended for use with 1/6 horsepower motors or smaller. Controls with higher output drive circuits available for larger permanent capacitor and shaded-pole fans and pumps.

Continued

Product identification	Type	Dimensions	Applications	Power input and output	Electronic or Mechanical design	Principle of operation
RHO SIGMA, INC. RS 500P SERIES	Proportional differential thermostat with two independent outputs. High temperature and low temperature override circuits	6×6×3 in.	Domestic water heating systems, including those with drain valves for freeze protection	Input: 115 VAC 6 amps (240 VAC optional); output: 115 VAC, 6 amps (240 VAC optional)	All solid state	Provides proportional control of solar circulating pump with high temperature turn-off (from storage side) and low-temperature turn-on (from collector side). The second optional output can control a second pump or normally open or normally closed valves
RHO SIGMA, INC. RS 104	Differential thermostat with freeze protection	6×8×4 in.	Space heating, air or liquid systems	Input: 120 VAC, 6 watts; output: two SPDT relays; rated at 10 amps (DPDT and 3PDT relays available on request)	Relay outputs driven by electronic circuits	Relays energize and de-energize according to differences in collector and storage temperatures
RHO SIGMA, INC. RS 240	Differential thermostat	6×8×4 in.	Control of valves and pumps in solar systems used in swimming pool heating	Input: 120 VAC (240 VAC optional) output: capable of driving (a) one 24 VAC. 4watt solenoid. (b) three 24 VAC. 6 watt solenoids or (c), through an auxiliary relay, any load desired. Specify desired output	Relay output with NC and NO contacts for valves. Interior of the box is laid out to accommodate W.W. Grainger's relay 5X846 for direct switching of a horsepower rated auxiliary pump	Circulates water through the solar system when the collectors become 6°F hotter than the pool
ROBERTSHAW CONTROLS CO. COMMANDER SD-10	Differential thermostat	Not specified	Operates circulating pump in a liquid solar heat storage system	Input: 80 to 130 volts, 50 or 60 Hz	Solid state	Provides pump circulation cut-in any time the collector panel temperature is 15°F higher than the storage tank
SIMONS SOLAR ENVIRONMENTAL SYSTEMS	Differential thermostat	8×6×4 in.	Flat plate and low-temperature concentrating systems	Input: 120 V at 60 Hz: output 120 V, 50/60 Hz, 6 or 20 amps	Solid state design	Thermistors are arranged in series and compared to a fixed voltage arranged such that an 18°F difference fires the comparator. On firing, feedback is introduced, lowering the fixed voltage point and thus allowing turnoff at 15°F less than on

U.L. approval	Differentials available if differential thermostat	Sensitivity	Operating range	Degree of adjustment possible	Limitations
Yes	Delta-T (slow) equals 3°F; delta-T (full) equals 12°F; high temperature and low temperature circuits to customer's specs. (140°F/37°F standard)	±2°F	−60°F to stagnation temperatures of collectors	Adjustable over full ranges normally encountered in solar systems	Designed for use with permanent capacitor and shaded pole fans and pumps commonly used in solar systems. Intended for use with 1/6 horsepower or smaller motors. Controls with higher output drive circuits available for larger motors
Yes	Available to customer specs.	Stable to line voltage drop of 15%	Covers temperatures normally encountered in solar systems	Wide adjustment possible	Output capabilities increased to horsepower range with addition of booster relay into the bottom compartment of the control
Yes	T on=6°F, T off=3°F, high temperature limit adjustable from 60 to 115°F	±2°F	Temperatures over the range normally encountered in pool heating applications	60 to 115°F	None specified
Yes	Cut-in: 15°F; ±5°F; cut-out: 5°F±3°F	Not specified	32 to 150°F	Cut-in temperature can be an adjustable option between 8 and 20°F	None specified
No	18°F on, 3°F off, or custom ordered	Not specified	None, factory set	Control: 32 to 120°F; sensors: to 350°F	None specified

Continued

Product identification	Type	Dimensions	Applications	Power input and output	Electronic or Mechanical design	Principle of operation
SOLAR CONTROL CORP.	Differential thermostat	Details available from manufacturer				
SOLAR CONTROL CORP.	Logic and switching control	8×7×4 in.	Home heating and cooling	Varies by model	All solid state	Logic and switching
SOLAR CONTROL CORP.	Solar air mover	Many models; details available from manufacturer				
SOLAR ENERGY RESEARCH CORP. THERMO-MATE™ DELUXE, DC-761	Thermostatic controller	8×5×6 in.	Flexible. Various systems, air or liquid	Input: 12 watts meter off, 20 watts meter on; output: 1 amp, 120 volts	Solid state electronics	Not specified
SOLAR ENERGY RESEARCH CORP. THERMO-MATE™ STANDARD, SC-762	Thermostatic controller, two differential circuits, two sensors per circuit, adjustable	5×4×3 in.	Solar heat collection and distribution	Input: 8 watts; output: 10 amps; optional: 10 amp output relay. 110 volts	Solid state electronic	Not specified
SOLARICS SPC-1000 AND SPC-2000 SERIES	Differential pump or fan temperature controller with hysteresis	6×4×3 in.	Hot water, space or pool heating	Input: 117 VAC, 60 Hz, 3 watts; output: 117 VAC, 60 Hz, 120 watts to 1,755 watts (customer specified)	All solid state	Two sensors: one each in tank and collector; differential with factory-set hysteresis
SOLARON CORP. AU-40-50	Air handling unit	6×6×3 in. and 10×12×4 in. standard NEMA boxes	Water and space heating, space cooling, solar heat pumps	Power in and out: 120 VAC, 60 Hz	Electronic and mechanical	An electronic differential thermostat in the solar system is integrated with a room thermostat

Concentrating Collectors

Company name/product identification	Design	Dimensions	Applications	Materials and absorber surface	Concentrating surface reflectivity and longevity	Day-long efficiency under clear sky conditions	Pressure and temperature limitations
ALPHA SOLARCO SUNTREK-7 SOLAR ENERGY LABORATORY	Parabolic trough, evacuated tube	1 meter^2 concentrator mirror	Research, demonstration, data acquisition	Aluminum mirror; black chrome-finish steel pipe inside Corning glass tube receiver	Reflectance: 85%. Longevity not specified	Not specified	Pressure: not specified. Temperature: to 600°C (1,000°F)
ENERGY APPLICATIONS, INC. MODEL 2000 CONCENTRATING & SUN-TRACKING SOLAR COLLECTOR	Parabolic-trough reflectors rotate around fixed absorbers. North-south axis; tracks sun	10 × 25 ft per array. Nine modules per array, ea. 16 in. ×9.5 ft	Supplies hot water 140° to 220°F for absorption air-conditioning, restaurants (dishwashing), bottling plants (bottle cleaning), laundries, textile mill dyeing operations, plating operations, etc.	Blackened copper absorber pipe with sealed clear Pyrex jacket; aluminized Teflon FEP film on fiberglass-reinforced polyester reflector shells; aluminum supports	Reflectivity: 85%. Longevity: at least 15 years	50 to 60%	Built for pressurized operation. Max. recommended. temp. is 250°F; will withstand 400°F dry condition in event of a power failure

U.L. approval	Differentials available if differential	Sensitivity	Operating range possible	Degree of adjustment	Limitations
None specified	Varies with model	Not specified	Varies with model	Not specified	
Yes	Field adjustable -100 to 130°F	±1°F. Hysteresis selected by buyer, 7°F standard if not specified	0 to 250°F	All functions field-adjustable	
Yes	Adjustable: 3 to 90°F	±1°F, regulated power supply standard. Available with 2 to 30°F hysteresis. 7°F standard	0 to 250°F	Differentials field-adjustable with screwdriver	None
Applied for	Factory set at 7°C on and 1.7°C off. Customer may specify any desired setting	Capable of ±6°C resolution	Controller: 0 to 70°C; sensor: −30 to 150°C	Customer may specify as follows: on and off at 0°C or greater	1 Hp or 15 amps maximum power
Yes	Fixed for air system	±2°F	Appropriate for air heating collector systems	None	Suited for air systems

Pressure drop through absorber	Recommended heat-transfer fluid	Are tracking devices supplied with the collector?	Limitations
Not specified	Accepts a variety of fluids, from water to Therminol-66	Yes	None implied. However this is an eduational or research device, and is not specifically designed for other uses. The manufacturer makes custom units for full scale applications.
Minimal	Water/glycol	Yes	Severe ice storms

Continued

Company name/ product identification	Design	Dimensions	Applications	Materials and absorber surface	Concentrating surface reflectivity and longevity	Day-long efficiency under clear sky conditions	Pressure and temperature limitations
ENTROPY LIMITED SOLAR COLLECTOR MODULE SCM-100	Concentrating collector, non-tracking; partial cylinder with closed ends; heat pipe vaporizer section concentration ratio of 5 to 1	77 in. long, 27 in. wide, 14.5 in.deep; total collection area: 1 meter^2	Heating and/or cooling	Concentrator: fiberglass inner and outer shell, double thickness foam insulation, tempered glass cover. Heat pipe: aluminum with flat black polymide coating	Reflectivity: 83 to 87% (estimated). Longevity: 20 years design life	20% (sunrise to sunset)	Normal operating temp.: 212° F (100°C) at 1 psi. Heat pipe burst pressure: greater than 20 psi
HEXCEL PARABOLIC TROUGH CONCENTRATOR	Parabolic trough	20 ft × up to 9 ft, modules	Heating and/or cooling	Troughs are aluminum honeycomb with aluminum skins; surface is aluminized modified acrylic	Not specified	Not specified	Optional over-temperature and no-flow protection controls available
KTA CORP. SOLAR COLLECTORS	Glass tubes with copper waterways	Width: 24, 36 or 48 in.; length: 52 or 74 in.	Space or hot-water heating	Copper water tubes and semi-cylindrical reflector coating, within glass tubes	Longevity: expected to be at least 15 years	Not specified	Max. operating temp.: 200°F
SOLAR KINETICS, INC. MK-4 AND SKI-MK-4 LINEAR PARABOLIC	Concentrating collector with receiver tube	Collectors measure 3.5 × 10 ft, are assembled in arrays of up to 50 ft long	Heating	Aluminum concentrating collector, glass-encapsulated mild steel receiver tube with black chrome selective surface	Reflectivity 89%. Longevity: unknown	up to 50%	650°F at 200 psi

Complete Systems: Heat

	Type and Dimensions	Collector Trade Name Type Design and Materials	Control Specifications; attachment location sensors	Components and materials	Heat transfer fluid	Storage type and capacity	Heat exchangers
ACORN STRUC-TURES, INC.	Liquid 420 SF with 2,000 gallon tank	Sunwave 420 System with 6 collectors 70 ft^2 each STG tank 30 gallon domestic HW tank, two pumps and controls	Thermal switch to activate collector pump and 2-stage room. State to activate heating pump and interconnective relays and specialties	2,000 gallon STG tank, 30 gallon stainless steel tank, heating coil, collector pump, heating pump, thermalstat 2 STG, 1-sensor SW, and all hardware	Water	2,000 gallon steel reinforced plywood tank with vinyl liner. 3½ in. fiberglass installation, Masinate outer shell	3-row copper tube in plate. Amana

Pressure drop through absorber	Recommended heat-transfer fluid	Are tracking devices supplied with the collector	Limitations
n.a.	Water vapor (steam)	n.a.	Must be used with a heat exchanger (condenser)
	Air, water or oil	Yes	None specified
0.37 psi	Water or water/glycol	n.a.	None specified
3 psi at 10 gpm of water	Therminol-55 or -66	Yes	None specified

Design performance and capacity of system	Based on Btu/hr ft^2 incident on collector and total system efficiency, dollar cost per Btu deliverable at a 100°F delta T	Applications	Options	Limitations
Not specified	Not specified	Space heating with domestic hot water	Not specified	Not specified

Continued

	Type and Dimensions	Collector trade name type design and materials	Control Specifications; attachment location sensors	Components and materials	Heat transfer fluid	Storage type anc capacity	Heat exchangers
CHAMPION HOME BUILDERS CO. SOLAR FURNACE MODELS 96, 128, 160	Vertical vane fixed flat plate 96, 128, or 160 ft^2	Champion solar furnace	100% solid state; switching by triacs; Thermistor temperature sensor. Interfaces Solar Furnace with conventional fueled system. Controlled by house thermostat. Logic determination for on/off switching of collection and distribution motors	Collector and storage are one unit	Forced air	Washed, screened granite and sandstone river gravel	None
COLUMBIA CHASE SOLAR ENERGY DIV. SOLAR PAK SYSTEM	220 ft^2 home heating system	Columbia Redi-Mount Solar Collector	Differential thermostat, sensor on storage tank and on outlet tube of last collector	Circulator pump, heat, exchange fluid, control, 10 collectors	Suntemp	500 gallons hot water	Built into tank
CONTEMPORARY SYSTEMS, INC. INTEGRATED WARM AIR SOLAR HEATING SYSTEM	See component listings	Series IV Solar low-temperature integrated roof wall type	All control functions by LCU 110 logic control unit with 5 sensors, 2 in storage, 1 in collector; 2 sensors in living space	See component listings	Air	Rock storage recommended; sized as required	n.a.
ENERGY CONVERTERS, INC. TITAN 460, MODELS SWH 460 AND CWH 460	Copper collectors (Model 201GC)	Titan 460;.	Not specified	8 to 10 collectors 460 gallon galvanized steel storage tank, pump, electronics (E10 differential thermostat), assorted valves	Water or anti-freeze	Not specified	Not specified
FUTURE SYSTEMS, INC. SUN*TRAC, MODELS 96, 128, 160	Solar air system; pebble bed storage. Model 96: 8×12 ft 9 in.: Model 128; 8×16 ft 9 in.: Model 160: 8×20 ft 9 in. All units 8 ft high	Sun*Trac; tempered 3/16 in., 4×8 ft glass covers, vertical vane absorber plate supported by aluminum structure; 4 in. urethane insulation	Thermistor sensors on absorber plate and in pebble bed; solid state logic control switching of collection, distribution motors	Centrifugal blowers, driven by ½ hp split-capacitor motors; damper assemblies. Aluminum housing with baked-enamel exterior finish; 4-in. urethane foam backing	Air	Pebble bed. Model 96: 9 cu yds; Model 128: 12 cu yds.: Model 160: 15 cu yds	None
SOLAR ENERGY RESEARCH CORP. THERMO-SPRAY™	Liquid collector built on site	4 ft wide, any length	Not specified	A-frame unit built over heat storage tank; submersible pump; stud or concrete block	Water	Water; not specified	None

Design performance and capacity of system	Based on Btu/hr ft² incident on collector and total system efficiency, dollar cost per Btu deliverable at a 100°F delta T	Applications	Options	Limitations
Not specified	Not specified	Residential space heat	n.a.	Not specified
Designed to provide 50% of heating load for 1,200 ft² house	$.13656 per Btu on entire system	House heating; expandable to larger buldings	Double glazing; selective coating on collectors	Tank only useful in unpressurized systems
Standard design sizes correspond to collector areas of 200 ft² to 1,000 ft²; other size requirements by special order	@ 100°F. $T(T_a - T_i)$, $I = 300 Btu/ft^2 hr$ $.043/Therm 47.4:1 energy efficiency ratio	Residential, commercial, industrial space heat	Alpha numeric mode indicator	Installation in accordance with manufacturers specifications
Not specified	Not specified	Residential space heat; commercial water heat. Particularly designed for freezing climates	Optional electric backup heating elements in tank; optional heat distribution equipment	Not specified
Model 96: 120,000 Btu/day with air flow of 500 CFM at maximum temperature rise of 60°F	n.a.	Residential, commercial space heat	Collectors may be ordered individually or as part of integrated furnace	Electric power required for control, blower
Up to 250°F	Not specified	Space heat or evaporative cooling; use with heat pump	Variety of absorber plate and glazing materials	Does not adapt to roof; should not undergo long stagnation

Continued

	Type and Dimensions	Collector Trade Name Type Design and Materials	Control Specifications; attachment location sensors	Components and materials	Heat transfer fluid	Storage type and capacity	Heat exchangers
SOLAR ENERGY RESEARCH CORP. WARM™	Multiple liquid panels; 8×10, 12, or 16 ft	Thermal-Spray™ liquid, copper, plastic, wood	Thermo-Mate™ sensors: 2 in heat storage, 1 in collector, 1 in space to be heated	Collector panels; storage tank; submersible pump; and controls	Water	Spherical plastic tank (buried): 350, 500, 750 1,000, 1250 gallons	Copper tube
SOLARGENICS, INC. SOL-PAK III, MODELS SP-4-0011 THROUGH SP-4-0014	Mechanical package 36 ft × 78 in. × 60 to 150 in.; collectors from 120 to 360 ft², 3×20 ft	Solargenics Series 76 (selective) or Series 77 (non-selective) flat plate, structural, single or double glazed	Solid state differential pump control and failsafe valve controls. Sensors factory-installed at optimum locations	Standard fan coils, air/vent heat pump	Water, or Dow Therm J or equivalent	Glass lined steel with anode, 120 to 480 gallons	Integral jacket type, double metal wall; expansion tank included
SOLARGENICS, INC. SOL-PAK 1: MODELS DP-1-3400 THROUGH SP-1-6500	Industrial space heat, mechanical equipment 9×10×16 to 27 ft; collector array 2,000 to 7,000 ft²	Solargenics Series 76 (selective) or Series 77 (non-selective), flat plate, structural, single or double glazed	Solid state pump controls and failsafe valve controls, sensors factory installed at optimum locations	Supplied complete with coiling, or wall-mount fan coil units; industry standard	Water; Dow Therm J or equivalent	Steel tank, 125 psi ASME rating, glass lined with anodes; 3,400 to 6,500 gallons	Collector loop isolation heat exchanger, tube-in-shell type, external to tank
SOLARON CORP.	Air-heating flat plate	Solaron series 2000: gross area 3×6.5 ft.	2-stage thermostat, first stage solar, second stage auxiliary. Differential collector T=40°F. Storage sensor set point =85°F	Solaron series 2000 collectors, air handling unit, necessary controls and sensors, storage design	Air	Pebble bed, designed size=½ cu ft rock/ft² collector	Water-heating coil required for service water preheat
SUNPOWER INDUSTRY, INC. SUNPOWER SOLAR ENERGY SELECTOR PANEL	Hot air to rock; 320 to 384 ft²	Sunpower Solar Energy Selector Panel; continuous baffle hot air	115 VAC blower delivering 275 CFM of air to dry storage. Energy sensor control system delivers desired voltage to blower	Round galvanized ducting to 5 inch diameter max.	Air	20 to 100 tons washed, round graded gravel	n.a.
THOMASON SOLAR HOMES, INC. SOLARIS	Details available through manufacturer						

Complete Systems: Hot Water

	Type and dimensions	Collector™ type, design, and materials	Control specifications, attachment location for sensors	Pump hp and make	Storage type and capacity	Heat exchangers
COLUMBIA CHASE SOLAR ENERGY DIV DIRECT EXCHANGE DOMESTIC HOT WATER KIT	Hot water kit; 22.6 ft² effective absorber area	Columbia Redi-Mount Solar Collector	Differential thermostat control, sensor collector outlet and storage tank	Teel pump 1/25 hp	Customer's hot water tank	None

Design performance and capacity of system	Based on Btu/hr ft^2 incident on collector and total system efficiency, dollar cost per Btu deliverable at 100°F delta T	Applications	Options	Limitations
Computed for each product.	Computed for each project see performance	Commercial and residential space heat	Reversible heat pump	Not for absorption cooling
Designed to provide up to 80% of heating and hot water, heat storage up to 170°F, hot water to 140°F	Equivalent to $.025/kwh electric, $.57/gal oil, and $3/1,000,000 Btu gas, based on 20 year service life	Residential heat, new or retrofit	Heat pump can be provided for air conditioning and backup heat. Pool heating coupler.	Depends on type of construction
1,500,000 to 5,000,000 Btu/day delivered to fan coil units. Design temperatures up to 180°F; 100% backup	Equivalent to $0.25/kwh electric, $.57/gal. oil, and $3/1 million Btu gas, based on 20 year service life backup	Commercial and industrial space heat	Increased collector array output and dual tankage provide 210°F water for absorption chiller	None within designed range; may be ganged or expanded modularly
At (Tout-Tamb) H=0.30, collector efficiency is approximately 45%	Dollar per Btu, depending on geographic location, = $7 to $15 per million. assuming 30-year mortgage. Installed cost, $22 to $28 per ft^2	Space heat, domestic and service hot water, process heat, drying, etc.	Domestic water preheat, also available separately	n.a.
60% of heat load	$12/ft^2 collector	Commercial and residential space heat	None specified	Minimum of 12 ft^2 southern exposure on average house

Heat transfer fluid	Other components supplied with system	Design performance and capacity of system	Based on 300 Btu/hr/ft^2 incident on collector and total system efficiency, state dollar cost per Btu deliverable at 50°F ΔT	100°F	Applications	Limitations
Water	4 drain solenoids, 1 drain sensor	Approximately 50 gallons 140°F water on sunny day	$.0945 per Btu at 50°F ΔT	$.14189 per Btu at 100°F ΔT	Domestic hot water	Proper sizing required

Continued

	Type and dimensions	Collector ™, type, design, and materials	Control, specifications attachment location for sensors	Pump hp and make	Storage type and capacity	Heat exchangers
COLUMBIA CHASE SOLAR ENERGY DIV DOMESTIC HOT WATER SYSTEM	Domestic Hot Water System: 45 ft² effective absorber area	Columbia Redi-Mount Solar Collector	Differential thermostat control, sensor at bottom of storage tank and on outlet of one collector	Teel pump, 1/25 hp	80-gallon hot water tank	Heat exchange unit built into tank
EL CAMINO SOLAR SYSTEMS SUNSPOT	Direct heating	SUNSPOT Model 3200 flat plate liquid collector, aluminum absorber plate, copper waterways, Tedler-coated acrylic cover.	SUNSPOT CASCADE 3-channel auto-control: 3 sensors. 3 activators	Grundfos 85-watt pump or TACO 102B	30 to 240 gallons	None
ENERGY APPLICATIONS, INC.	Concentrating, sun-tracking collector; three 16 in. × 9.5 ft parabolic reflectors, 7 × 10 ft	Energy Applications' Model 1000 Concentrating & Sun-Tracking Solar Collector; Parabolic trough; reflectors rotate about fixed absorbers oriented in a north-south axis. Blackened copper absorber pipe with sealed clear Pyrex jacket: Fiberglas reinforced polyester reflector shells laminated with reflective aluminized Teflon FEP film (clear Teflon on outside surface): upper and lower bearing and gear assembly fabricated from nylon: upper and lower supports of structural aluminum; drive assembly, stainless steel threaded shaft with engineered plastic rider, aluminum drive link with nylon rack gears: painted aluminum top and bottom end covers	Hawthorne solid state differential temperature control	1/200 hp Teel pump	Existing hot water tank or optional tank	Not required
ENERGY CONVERTERS INC.	Liquid flat plate	Suntrap SWH 401 liquid flat plate	Not specified	Not specified	40 gallon tank with no electric element	Double wall heat exchanger
ENERGY CONVERTERS, INC.	Liquid flat plate	Solar Saver SWH801	Not specified	Not specified	80 gallon tank with auxiliary heating element	Double wall heat exchanger
HEILMANN ELECTRIC, INC.	Flat plate tube collector 10 to 30 tubes by 6 ft length	Solar Tube Systems Water Heater Kits; 5/8-in. copper tubing bonded to aluminum inside insulated 4 in. diameter Lexan tube	Differential thermostat, 2 sensors	Not specified	Water tanks, 42 to 120 gallons	Not specified
RAYPAK, INC. HOT WATER, MODEL DHWS-2-T82	Liquid flat plate collector 37.5 × 79.5 ft × 3.75 in.	2 Model SG-18-P Solar-Pak Collectors	Differential thermostatic control; high-limit control, freeze protection control, sensor locations as per manual	1/20 hp March pump	82-gallon glass-lined storage tank	Not applicable

Heat transfer fluid	Other components supplied with system	Design performance and capacity of system	Based on 300 Btu/hr/ft^2 incident on collector and total system efficiency, state dollar cost per Btu deliverable at 50°F △T	100°F △T	Applications	Limitations
Suntemp	None	Designed to supply approximately 50% of a family of four's requirement	$.0945 per Btu at 50°F △T	$.14189 per Btu at 100°F △T	Residential and commercial hot water	Proper sizing required
Potable water	System includes: 1 CASCADE autocontrol; 1-3 collectors; 1 solar storage tank, 1 85-watt pump, 1 misc. parts kit	See report to NASA, ESC-4, "System Performance Specification"	For model 250C: 50°F △T=$0.1204	100°F △T= $0.1823	Domestic, commercial, industrial hot water	None identified
Water	Hawthorne pump/temperature control; Sun-Track 100 sun tracking device	Maximum continuous operating temperature 250°F; designed to cease tracking when temperature exceeds 160°F	Net efficiency at 50°F △T is 55%	At 100°F △T, 53% efficient	Residential hot water	Severe ice storms
Inhibited water	Tank, 1201GA collector panel, magnetic drive pump, electronics, drainback tank	55 gallons of 150°F water a day; intended to be added to existing hot water heater	Not specified	Not specified	Retrofit hot water	None specified
Not specified	2 solar panels, storage tank, pump, electronics	Up to 110 gallons of 150°F water per day; 25 gallons always maintained by auxiliary heat	Not specified	Not specified	New or retrofit domestic hot water	None specified
Not specified	Tubes, controls, storage tank with heat exchanger, circulating pump, expansion tank, assorted valves, galvanized straps	Storage capacity × 60°F on sunny day	Not specified	Not specified	Water preheater for new or existing building	Not specified
Water	All controls	Not specified	At 50°F △T cost/Btu delivered is $0.09 (collector only) or $0.21 (system)	At 100°F △T Cost/Btu delivered is $0.14 (collector only) or $0.35 (system)	Single family residences	Not specified

Continued

	Type and dimensions	Collector™ type, design, and materials	Control specifications, attachment location for sensors	Pump hp and make	Storage type and capacity	Heat exchangers
SOLAR DEVELOPMENT, INC.	Liquid flat plate collector 4×10 ft or 2×10 ft	Solar Development, Inc. SD-5; copper absorber plate, aluminum box.	del Sol Control Corp.: sensors inside collector and against bottom of tank surface (held by magnet)	March 809 1/220 hp	Electric water heater (solar version), use applicable size	None
SOLAR ENERGY RESEARCH CORP.	Domestic hot water solar preheat tank, 21 ft diameter, 59 in. high; kit, with collectors	WARM™ Thermospray™; copper, redwood, galvanized steel	Adjustable differential, SERC Thermo-Mate™, 1 sensor at tank, 1 sensor at collector	1/12 hp Grundfos	National 40 gallon (80, 120, 200 gallon) tank	Integral with heat transfer fluid (water) reservoir; double wall. Optional automatic make-up water
SOLARGENICS,INC. SOL-PAC IIMODELS SP 36-0011 THROUGH SP 36-0018	Multi-family/institutional hot water system; mechanical package 36 × 78 × 60 to 150 in. L, collectors from 60 ft² to 240 ft², 3×20 ft panels	Solargenics Series 76 (selective) or Series 77 (non-selective) flat plate, structural, single or double glazed	Solid state differential pump control and fail-safe valve controls, sensors factory installed	1/20 hp Grunfos or 1/12 hp TACO stainless or bronze	Steel, glass lined with anode in 120-gal increments to 480 gallons	Integral jacket type, double metal wall separating potable and non-potable water. Expansion tank included
SOLARGENICS, INC. SOL-PAC II, MODEL SP 96-3400 THROUGH SP 96-6500	Industrial process water system, mechanical equipment: 9×10 ft ×12 to 21 ft long; collector array 2,000-5,000 ft²	Solargenics™ series 76 (selective) or series 77 (non-selective) flat plate, structural, single or double glazed	Solid state pump controls and fail-safe valve controls; sensors factory installed	Pump size varies to meet requirements, typically in the ½ to 5 hp range	Steel tank, 125 psi ASME rating, glass-lined with anodes; capacities range from 3,400 to 6,500 gallons	Collector loop isolation heat exchanger. tube-in-shell type, external to tank
SOLARGENICS,INC. SOL-PAC IVMODEL SP 66 THROUGH SP 120	Residential hot water system, factory prepackaged; 66 gallons 28 × 62 in; 82 gallon 32 × 54 in; 120 gallon 34 × 65 in.; collectors 30, 45, 60 ft²; 3 × 10 ft, 3 × 15 ft or 3 × 20 ft.	Solargenics Series 76 (selective) or Series 77 (non-selective) flat plate, structural, single or double glazed	Solid state differential pump control, sensors factory installed	Integral jacket type, double metal wall separating potable and non-potable water, expansion tank included	Steel, glass lined with anode, 66, 82, 120 gallon	1/20 hp Grunfos or 1/12 hp TACO stainless or bronze
SOLARKIT OF FLORIDA, INC.	Liquid flat plate hot water heater	Not specified; redwood frame, treated with sealer	Hawthorne controller	March 809 (Teel 1P760); 30 watt; 110 volt	Not specified	Not specified

Heat transfer fluid	Other components supplied with system	Design performance and capacity of system	Based on 300 Btu/hr/ft^2 incident on collector and total system efficiency, state dollar cost per Btu deliverable at 50°F ET	100 °F ΔT	Applications	Limitations
Water	Check valve, relief valve supplied. Drain-down freeze system available	4×10 ft panel with 82-gal. solar tank, family of 3-5. Three 2×10 ft panels with 120 gal. solar tank, family of 6 or more in Florida of for average family in northern climates	Available on request	Available on request	Domestic, commercial water heating	n.a.
Pure tap water	Collector mounting hardware	Flow per 20 ft^2 panel, 1.25 gal/min. See NBS performance chart of WARM collector	n.a.	n.a.	Residential/ commercial domestic hot water	Gravity return, lower edge of collector must be above top of preheat tank. Provide for sloped return lines to tank
Water in mild freeze areas; Dow Therm J or equivalent in hard freeze areas	High-efficiency/ low stack-loss back-up boiler, 80,000 to 160,000 Btu/hr; all other operating valves and fittings	Up to 80% of user hot water load at 140°F	Equivalent to $.02/kwh electric, $.51/gal oil and $2.80/ 1,000,000 Btu gas based on 20 year service life		Cluster homes, condominiums, apartment buildings, small industrial buildings, nursing homes	None
Water, in mild freeze areas; Dow Therm J or equal in hard freeze areas	Back-up boiler (gas, elect, oil), all valves, strainers, and fittings to monitor performance. Collector mounts optional	1,500,000 to 5,000,000 Btu/day delivered hot water. Design temperatures up to 180°F., 100% back-up	Equivalent to $.02/KWH electric, $.51 Btu/oil, and $2.80/1 million Btu gas based on 20 year service life		Commercial laundries; canning plants; chemical, dairy, textile, paper processing, or other high-level hot water requirement	None within design range; may be ganged or expanded modularly for higher outputs
Water in mild freeze areas; Dow Therm J or equal in hard freeze areas	Collector flow adjustment valves, check valves, and pump isolation fittings. Tempering valve is optional	Up to 80% of a family hot water load at 140°F	Equivalent to $.02/kwh electric, $.51 Btu/oil, and $2.80/ 1,000,000 Btu gas, based on 20 year service life		All types of residential hot water heating	None
Not specified	Circulation pump, controller, assorted valves	Sized to supply hot water needs of one person (2-collector system supplies 2 people, etc.) assuming consumption of 20 gallons per day	Not specified	Not specified	Domestic hot water	None specified

Continued

	Type and dimensions	Collector[TM] type, design, and materials	Control specifications, attachment location for sensors	Pump hp and make	Storage type and capacity	Heat exchangers
SOLARON CORP.	Air heater with air-to-water heat exchanger. Panel dimensions 3 × 6.5 ft gross area	Solaron Series 2000 air heating collector	Differential thermostat between collector outlet and bottom of water preheat tank. Turn on T =40°F; Turn off T=20°F	1/20 hp Grundfos	Stonelined water tank, 60 gal to 120 gal	Water heating coil
STATE INDUSTRIES, INC. SOLARCRAFT I & II	Prepackaged system	Solarcraft: aluminum and Fiberglas collector panels	Not specified	Not specified	82 to 120 gallon storage tank	Aluminum
SUNEARTH SOLAR PRODUCTS, CORP.	Drain-down system, size according to load	Sunearth Model 3296 or 3597A	Frost sensor operates drain valves; plate and tank sensors	1/25 or 1/12 TACO (usually)	66 to 120 gallon glasslined tank	None
SUNEARTH SOLAR PRODUCTS, CORP.	Antifreeze system; sized according to load	Sunearth 3296 or 3597A	Plate and tank sensors	1/25 TACO (usually)	45 to 120 gallon stonelined tank	Coil in tank
SUNWORKS DIVISION OF ENTHONE SOLECTOR PAK 1000 SERIES	Liquid flat plate; storage tanks, 65 gal: 24 × 60 in.; 80 gal: 26 ×60 in.; 120 gal: 28 × 69 in.	Sunworks Selector solar energy collector, closed loop forced circulation	Solid state differential temperature control, sensors mechanically attached at the top of Solector and bottom of tank	1/25 hp circulator	Stonelined water heaters with 4.5 kw electric elements; 65, 80 and 120 gal.; 65 gal. has 15 ft^2 heat exchanger, 80 gal. has a 20 ft^2 coil, 120 gal. tank has a 40 ft^2 coil	Heat exchangers integral with tank
SUNWORKS DIVISION OF ENTHONE SOLECTOR PAK 3000	Systems can be adapted to any standard electric water heater of sufficient capacity	Sunworks Solector Solar Energy Collector	Solid state differential temperature control, sensors mechanically attached at bottom and top of Solectors and at bottom of tank	1/200 hp circulator	No tank supplied	Not required
THE WILCON CORP.	Liquid flat plate	Wilcon Solar Water Heating System 120-6	Solid state	Not specified	Not specified	Not specified
W.L. JACKSON MANUFACTURING CO.INC.	Closed-loop liquid collector	Water heating system, all aluminum panels	Differential thermostat	1/10 hp magnetically driven pump	Not specified	Not specified

Heat transfer fluid	Other components supplied with system	Design performance and capacity of system	Based on 300 Btu/hr/ft^2 incident On collector and total system efficiency, state dollar cost per Btu deliverable at 50°F ΔT	100°F ΔT	Applications	Limitations
Air	Air handling unit	See technical data sheet (included)	Dollar, per Btu, depending on geographic location and installed system cost of $22-$28 per ft^2, $7-15 per million Btu (assuming a 30-year loan)		Water heating for domestic and service use	n.a.
Closed loop fluid	panels, storage, tank and pump	Not specified	Not specified	Not specified	Domestic hot water	More efficient in south and southeast
None	Valves for complete installation	Capacity according to size	Not specified	Not specified	Residential and commercial hot water	Must have pitched lines, no scale-causing water
Propylene glycol	Valves and expansion tank necessary for complete installation	Capacity according to size	Not specified	Not specified	Residential and commercial hot water	Not specified
Sunsol 60 non-toxic freeze inhibitor with corrosion inhibitor	Installation manual, circulator differential controller, sensors, expansion tank, air purger, automatic air vents, auto T&P relief, tempering valve, vacuum relief, flow check	Should be sized for 20 gallons of storage per person per day on a year-round basis. Each collector will provide about 18 to 20 gallons of hot water that has been raised 65-70°F	System is under testing to determine system efficiency		Residential, commercial domestic hot water	None
Not required	Installation manual, circulator, differential controller, sensors, automatic air vent, vacuum relief, auto. T&P relief, tempering valve; flow check	System should be sized for 20 gallons of storage per person per day on a year-round basis. Each collector will provide 18 to 20 gallons of hot water that has been raised 65-75°F	System is under testing to determine system efficiency.		Residential and commercial domestic hot water	Can be effectively used in areas with 30 freezing day per year or less
Priority heat transfer fluid	Not specified	Not specified	Not specified	Not specified	Domestic hot water	Installed only by licensed dealers
Deionized water	Collector panels, water storage tank	Can supply up to 95 gallons of 150°F water per day; auxiliary heating element	Not specified	Not specified	Domestic hot water	Distilled water is changed every two years

183

Complete Systems: Swimming

	Type and dimensions	Collector trade name, design, and materials	Control specifications and attachment location for sensors	Components	Storage capacity if other than pool	Heat exchangers, if any
AQUASOLAR, INC. MODEL M2-20	Open-air pipe system, using 100 ft of pipe per 1,000 gals of pool water	Aquasolar pipe is black, made from Cycolac, has 3 internal fins, is in 20 ft. lengths. Pipes are fused together at job site, using a 180° fitting to make a serpentine design	70°F temperature differential controller on the fully automatic system. 1 sensor in manifold, other at the pool equipment	All component parts supplied by authorized distributors	n.a.	n.a.
FAFCO INC. SOLAR HEAT EXCHANGER, MODEL EP-1	Positive-pressure flat plate collector system. Dimensions: 51⅜ × 120 in. or 51⅜ × 96 in.	FAFCO Solar Heat Exchanger, parallel circular tubes with integral headers. Polypropylene coplymer with proprietary additive	FAFCO Solar Control. 12 variations available with valve relay output. Differential sensing and heat-or-cool logic	n.a.	n.a.	n.a.
H.C. PRODUCTS CO. ALCOA SOLAR HEATING SYSTEM, MODEL 26,36,45,46,48	Modular plastic panel: 45¾ × 45¾ × 1½ in.; effective absorber area 14.3 ft²	Alcoa Solar Heating System, unglazed, ABS plastic with Korad Film for protection from ultraviolet degradation; volumetric capacity at 1.82 gallons	Automatic venting and draindown; pressure relief valve. Solid state automatic temperature control optional	Collector modules	n.a.	n.a.
RAYPAK, INC. SOLAR-PAK, MODEL SK800 AND SK1000	Liquid flat plate, 44×96 in. (SK800); 44×120 in. (SK1000)	Solar-Pak: copper waterway with aluminum heat absorbing surface	Not specified	Not specified	n.a.	n.a.
SOLAR DEVELOPMENT, INC. SDI	Either high-flow or full-flow copper flat plate, 4×10 ft	Bare copper absorber plate or standard SD-5 collector. Bare plates use parallel flow; standard collector uses series or parallel flow	Bare copper system uses differential control, standard SD-5 collector has no controls (on in cool weather, off in warm season)	Bare copper absorbers or SD-5 standard collector, both supplied with complete roof mounting hardware	n.a.	None
SOLAR HOME SYSTEMS, INC. MODELS SPH-0001	Flat plate collector, liquid, unglazed: 50 ft²	Low temperature swimming pool heaters, Tedlar coated vinyl, 5×10 ft	Differential thermostat and control valve	Collector with integral manifold; controls	n.a.	n.a.

Pools

Design capacity	Projected longevity	Limitations
No restrictions on capacity of pool. Largest at this time is 180,000 gallons	15 to 20 years	Low temperature high-volume unit designed specifically for pool heating
2 to 8 gpm/collector, 4 gpm/collector optimum	In excess of 10 years	30 psi maximum operating pressure. Glazing of collectors voids warranty
Recommended flow rate 2-4 gpm per vertical panel row	Not specified	Operating pressure not to exceed 10 psig; minimum slope 15° for draining
300°F: test procedures similar to NBS guidelines used at factory test facility	Not specified	All components will withstand 300°F stagnant conditions
All pools can be heated if there is sufficient space to mount collectors. Factory will size systems at no charge	20 years	n.a.
110°F	Not specified	Fast flow low delta T is used. Since collector is unglazed temperature will not get much above ambient: 1 psi internal pressure

Complete Systems: Cooling

	Type and dimensions	Collector trade name, design and materials	Control specifications and attachment location for sensors	Components and materials	Heat transfer fluid	Storage type and capacity
ARKLA INDUSTRIES INC. SOLAIRE 36, MODEL WF-36	Absorption chiller	Solaire 36; 3 tons of cooling from 195°F solar heated water	Not specified	Not specified	Working fluid: lithium bromide as absorbent, water as refrigerant	n.a.
ARKLA INDUSTRIES INC. SOLAIRE 300, MODEL WFB-300	Absorption chiller	Solaire 300, 25 tons of cooling from 195°F solar heated water	Not specified	Not specified	Lithium bromide as absorbent, water as refrigerant	n.a.
YORK, DIV. OF BORG-WARNER	Absorption chiller	Available in 21 sizes; hot water temperatures as low as 150°F	Not specified	Not specified	Not specified	n.a.

Measurement and Data Collection

Company name/ product identification	Type of instrument	Sensitivity	Electrical or mechanical	Recording capability	Linearity	Temperature dependence
BAILEY INSTRUMENTS CHANNELYSER RECORDER MULTIPLEXER CHL-1, CHL-1R	Recorder multiplexer (allows up to 4 variables to be recorded on a single-pen strip-chart recorder or a digital printer)	n.a.	Electrical	None; device connects to a recorder	n.a.	n.a.
BARNES ENGINEERING, CO. THERMATRACE	Remote temperature measurement	Not specified	Electrical	A pushbutton on the top of the instrument holds a single reading	n.a.	Available in three temperature ranges: −10 to 60°C; 0 to 200°C; and 0 to 600°C
BARNES ENGINEERING CO. INSTATHERM	Thermal scanning instrument	0.5°C at 25°C	Electrical	Optional photographic recording system, using Polaroid SX-70 camera	n.a.	Five ranges: 10, 30, 100, 300, and 1,000°C

Heat exchangers	Design performance and capacity	Based on 300 Btu/hr/ft^2 incident on collector and the total system efficiency, dollar per Btu deliverable at a 120°F	Applications	Limitations
n.a.	COP: .61 with 175°F water, .67 with 180°F water, .71 with 185°F water, .73 with 190°F water, .72 with 195°F water, .72 with 200°F water	Not specified	Residential and small commercial cooling	At this time, we recommend the unit be installed indoors
n.a.	COP: .69 with 170°F water, .71 with 180°F water, .71 with 185°F water, .69 with 190°F water, .68 with 195°F water, .66 with 200°F water	Not specified	Commercial and industrial cooling	The machine is designed for indoor installation only
n.a.	Not specified	Not specified	Installation requiring 40 tons or more of full load cooling	Not specified

Response time	Mode of data storage	Input power requirements
n.a.	n.a	"C" batteries (Model CHL-1); floating batteries and built-in charger for ac and battery use (Model CHL-1R)
Instantaneous	n.a.	Standard replaceable batteries
Instantaneous	Photograph (optional)	Rechargeable batteries (batteries and charger supplied)

Continued

Company name/product identification	Type of instrument	Sensitivity	Electrical or mechanical	Recording capability	Linearity	Temperature dependence
ENTROPY LIMITED SUNPUMP SIM-100 AND SIM-200	Solar insolation meter	Solar radiation from 0.01 langleys/minute	Electrical	Electronic memory accumulates solar insolation	2% from 0.4 langleys/min. to 2.0 langleys/min	±2% accuracy at 20°C±25°C
THE EPPLEY LABORATORY, INC. MODEL NIP	Normal incidence pyrheliometer (measures solar radiation)	8 microvolts per watt/meter^2	Electrical	None. A recorder may be connected to the device.	±0.5% from 0 to 2,800 watts/m^2	±1% over ambient range of −20 to 40°C
THE EPPLEY LABORATORY, INC. LABORATORY, INC. MODEL ANG	Angstrom pyrheliometer (measures the intensity of solar radiation)	Very precise	Electrical	None. Instantaneous display on control unit (separate device)	n.a.	n.a.
THE EPPLEY LABORATORY, INC. MODEL KAR	Kendall absolute radiometer (measures thermal radiation)	Very precise	Electrical	None	n.a.	n.a.
THE EPPLEY LABORATORY, INC. MODEL PIR	Precision infrared radiometer (pyrgeometer). Measures undirectional terrestrial long-wave radiation (background radiation)	7 microvolts per watt/meter^2	Electrical	None. May be connected to a recorder	±1%, 0 to 700 watts/m^2	±1%, −20 to 40°C (nominal)
THE EPPLEY LABORATORY, INC. MODEL TUVR	Ultraviolet radiometer (photometer)	Approx. 150 microvolts per watt/meter^2	Electrical	None. May be connected to a recorder	±2% from 0 to 70 watts/meter^2	Less than 0.1% per degree from −40 to 40°C
THE EPPLEY LABORATORY, INC. MODEL PSP	Precision spectral pyranometer (measures sun and sky radiation)	9 microvolts per watt/meter^2	Electrical	None. May be connected to a recorder	±0.5% from 0 to 2,800 watts/m^2	±1% over ambient temp. range of −20 to 40°C
THE EPPLEY LABORATORY, INC. MODEL 8-48	Black and white pyranometer (measures global, i.e. total sun and sky, radiation)	11 microvolts per watt/meter^2	Electrical	None. May be connected to a recorder.	±1% from 0 to 1,400 watts/m^2	±1.5% constancy from −20 to 40°C

Response time	Mode of data storage	Input power requirements
100 milliseconds	Solid-state memory	Solar-operated, with battery-supplied energy for nighttime display of accumulated solar insolation
1 second	None	Connects to standard-type recording po-tentiometers (pyrheliometer has approx. 200 ohms impe-dance.) Not suitable for use with microam-meter-type recorders
Instantaneous	None	Must be used with conrol unit
Instantaneous	None	Must be used with control unit
2 seconds	None	None
milliseconds	None	None
1 second	None	None
3 to 4 seconds	None	None

Continued

Company name/product identification	Type of instrument	Sensitivity	Electrical or mechanical	Recording capability	Linearity	Temperature dependence
THE EPPLEY LABORATORY, INC MODEL ST1	Solar tracker (follows sun across sky; designed to accomodate a pyrheliometer or other instruments)	n.a.	Electrical	n.a.	n.a.	n.a.
THE EPPLEY LABORATORY, INC MODEL EQM EQUATORIAL MOUNT	Solar tracker (designed to accomodate as many as four pyrheliometers or other instruments)	n.a.	Electrical	n.a.	n.a.	n.a.
THE EPPLEY LABORATORY, INC MODELS 410, 411 AND 411-6140	Electronic integrator (readout device for solar instumentation)	n.a.	Model 410: mechanical counter; Model 411: electronic counter	Model 411-6140 only. Model 411 may be connected to a recorder	n.a.	0 to 50°C
THE EPPLEY LABORATORY, INC SPEEDMAX H AND W RECORDERS	Strip-chart recorders	n.a.	Electrical	Single-point, continuous line	n.a.	n.a.
RHO SIGMA, INC. RS 5080	Microprocessor for data acquisition	n.a.	Electrical	Yes. Device monitors 20 analog channels, 5 pulsed inputs and 8 discrete inputs	n.a.	n.a.
RHO SIGMA, INC. RS 1008	Photovoltaic pyranometer	±4% accuracy	Electrical	May be directly connected to a strip-chart recorder	Output direcly proportional to input	—2.2 at 0°C (with daum point at 30°C); +0.8% at 60°C

Response time	Mode of data storage	Input power requirements
n.a.	None	120 V, 60 Hz (or other speci-fied supply)
n.a.	None	120 V, 60 Hz (or other speci-fied supply)
n.a.	Digital printer	105/125 or 210/230 V ac, 50-60 Hz
1 second nominal	Paper tape	120 V, 60 Hz (50 Hz also avilable; 220 V also available)
n.a.	Printer output of all inputs plus several cal-culations and energy-related quantities	Not specified
Instantaneous	Strip-chart re-corder (optional)	None

Performance charts

These performance charts have been supplied by flat plate collector manufacturers. Not all the charts are comparable—data is not always on the same base, though we have tried to include only those charts that make it clear what the data base is—but in the interest of having a lot of information in one place, here they are.

COLLECTOR EFFICIENCY, SUNWAVE™ 420

Graph plotted BY MASSDESIGN from Acorn Structures, inc., data for 1976-1977

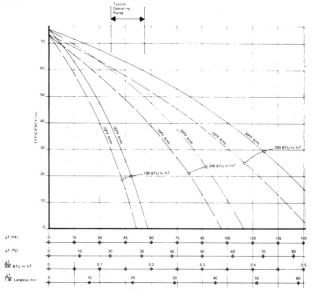

TRANSIENT EFFICIENCY, CONTEMPORARY SYSTEMS SERIES IV

Data represents useable thermal conversion efficiency at the output manifold. Unit installed in test bed fabricated to NBS test standards. Recommended input temperature for system is about 65°F; output, 105°F±25°F.

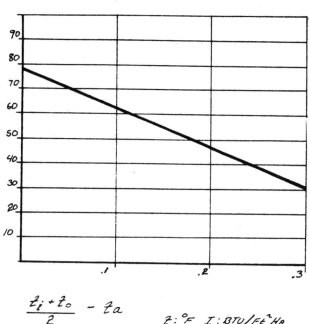

$$\frac{\frac{t_i + t_o}{2} - t_a}{I}$$

$$t : °F, \quad I : BTU/FT^2 HR.$$

INSTANTANEOUS EFFICIENCY, CHAMBERLAIN

Efficiency curves with one and two covers and black-painted absorber plate.

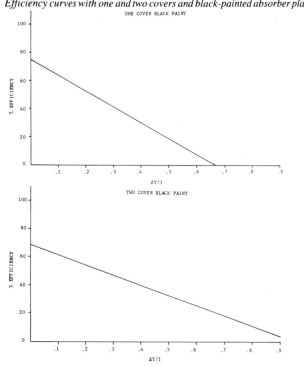

Efficiency curves with one and two covers and black chrome coated absorber plate.

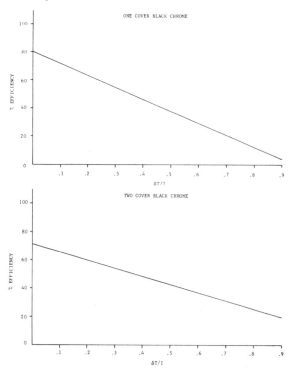

THERMAL EFFICIENCY CURVE, AMERICAN SOLARIZE HONEYCOMB

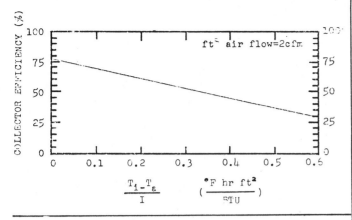

EFFICIENCY CURVE, BURTON FLAT PLATE COLLECTOR

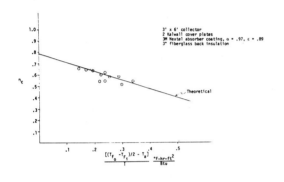

NBS EFFICIENCY, REYNOLDS

JANUARY 15 — FEBRUARY 9, 1976
COLLECTOR TEST RESULTS
VIRGINIA POLYTECHNIC INSTITUTE &
STATE UNIVERSITY

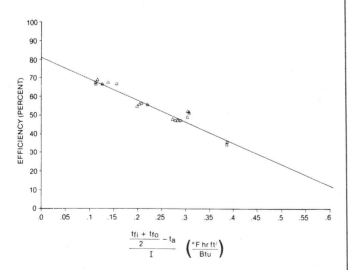

EFFICIENCY, ENERGY APPLICATIONS INC. MODEL 1000

Insolation figures taken from pyrheliometer data recorded January 15, 1975, at Kennedy Space Center. Normal radiation displayed on upper curve was used for calculating efficencies.

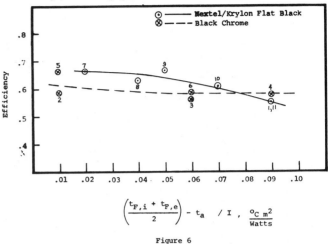

$$\left(\frac{t_{F,i} + t_{P,e}}{2}\right) - t_a \Bigg/ I \ , \ \frac{^{\circ}C \ m^2}{Watts}$$

Figure 6

EFFICIENCY, DELTA T MODEL 15 ABSORBER PLATE

Comparison of Model 15 absorber with typical flat plate collector with double glass cover.

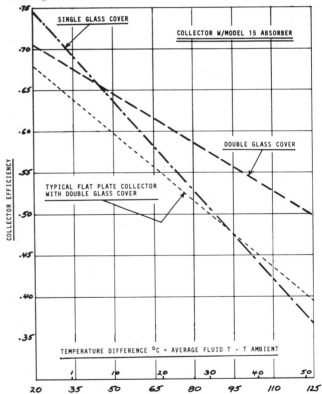

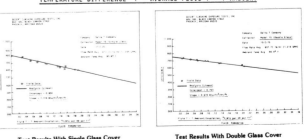

Test Results With Single Glass Cover   Test Results With Double Glass Cover

193

PERFORMANCE CURVE, GULF THERMAL CORP.

Performance measured by Energy Design Associates, Gainsville, according to NBS criteria developed for HUD.

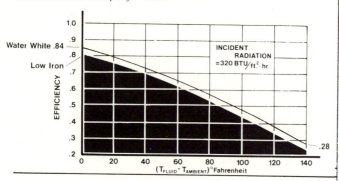

RECOVERABLE ENERGY, JACKSON'S WATER HEATING SYSTEM

Nomograph is based on experimental data by Energy Converters, Inc. and W.L. Jackson Mfg. Co.

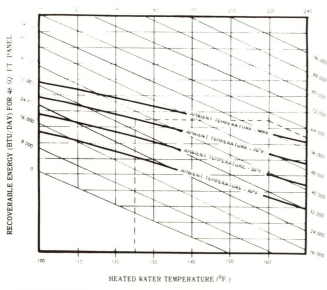

EFFICIENCY, LOF SUNPANEL

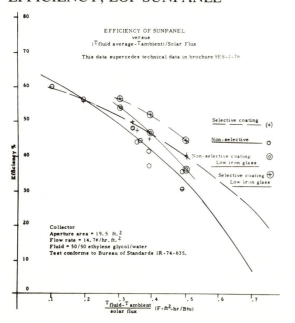

EFFICIENCY, PAC (PEOPLE/SPACE CO.)

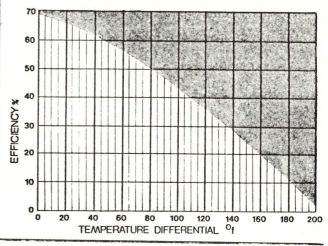

EFFICIENCY, RAYPAK, INC.

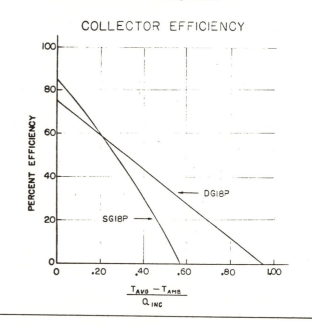

EFFICIENCY, RAYPAK SOLAR-PAK POOL PANEL

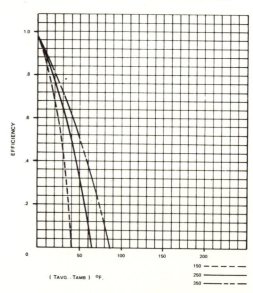

194

EFFICIENCY, SOLAR CORP. OF AMERICA MARK III-A

Copper aluminum absorber, black chrome coating, single glazing.

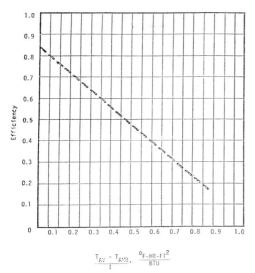

$$\frac{T_{AV} - T_{AMB}}{I}, \quad \frac{^{\circ}F-HR-FT^2}{BTU}$$

EFFICIENCY, SOLARON CORP. SERIES 2000

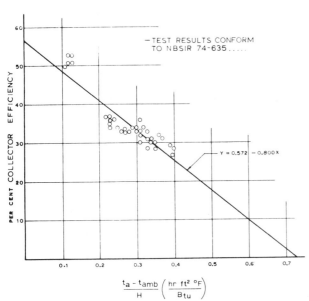

—TEST RESULTS CONFORM TO NBSIR 74-635.....

$Y = 0.572 - 0.800 X$

$$\frac{t_a - t_{amb}}{H} \left(\frac{hr\ ft^2\ {}^{\circ}F}{Btu} \right)$$

FLOW RATE = 2 CFM/FT² COLLECTOR

$t_a = \frac{1}{2}(t_i + t_o)$

H = INSOLATION STRIKING GROSS COLLECTOR AREA

EFFICIENCY, SUNSPOT MODEL 3200

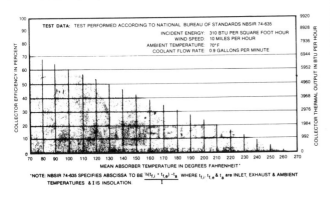

EFFICIENCY, SUNEARTH, INC. MODEL 3296-1

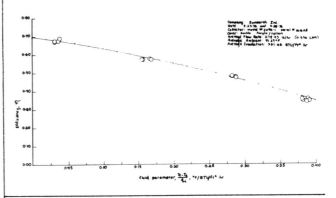

EFFICIENCY, THERMOTUBE MODEL 45 BG

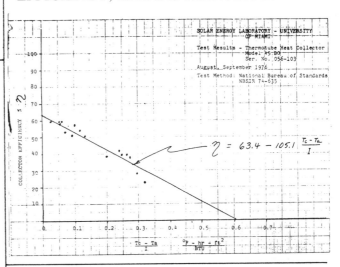

$$\eta = 63.4 - 105.1 \frac{T_c - T_a}{I}$$

EFFICIENCY, FUTURE SYSTEMS' SUN*TRAC

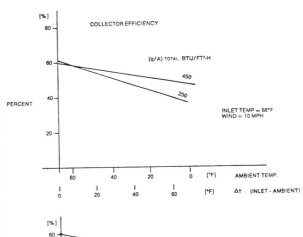

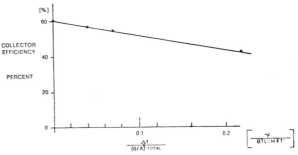

195

Manufacturer's DIRECTORY

A.C.M. INDUSTRIES, INC.
Box 185
Clifton Park, N.Y. 12065
Phone: (518)371-2140

ACORN STRUCTURES, INC.
P.O. Box 250
Concord, Mass. 01742
Phone: (617)369-4111

ACUREX AEROTHERM
485 Clyde Ave.
Mountain View, Cal. 94042
Phone: (415)964-3200

AERODESIGN CO.
P.O. Box 246
Alburtis, Penn. 18011
Phone: (215)967-5420

ALL SUNPOWER, INC.
10400 S.W. 187th St.
Miami, Fla. 33157

ALPHA SOLARCO
Suite 2230, 1014 Vine St.
Cincinnati, Ohio 45202
Phone: (513)621-1243

ALTEN CORP.
2594 Leghorn St.
Mountain View, Cal. 94043
Phone: (415) 969-6474

AMERICAN SOLARIZE INC.
19 Vandeventer Ave.
Princeton, N.J. 08540
Phone: (609) 924-5645

AMETEK, INC.
Power Systems Group
One Spring Ave.
Hatfield, Pa. 1944
Phone: (215) 822-2971

ANACONDA COMPANY, Brass Div.
414 Meadow St.
Waterbury, Conn. 06720
(203) 574-8500

A.O. SMITH CORP.
P.O. Box 28
Kanakee, Ill. 60901
Phone: (815) 933-8241

AQUEDUCT COMPONENT GROUP
1537 Pontius Ave.
Los Angelos, Cal. 90025
Phone: (213) 479-3911

AQUASOLAR INC.
1232 Zachini Ave.
Sarasota, Fla. 33577
Phone: (813) 366-7080

ARKLA INDUSTRIES INC.
P.O. Box 534
Evansville, Indiana 47704
Phone: (812) 424-3331

BAILEY INSTRUMENTS, INC.
P.O. Box 5068
Saddle Brook, N.J. 07662
Phone: (201) 845-7252

BARD MANUFACTURING CO.
P.O. Box 607
Bryan, Ohio 43506
Phone: (419) 443-2993

BARKER BROTHERS
207 Cortez Ave.
Davis, Cal. 95616
Phone: (916) 756-4558

BARNES ENGINEERING CO.
30 Commerce Road
Stamford, Conn. 06904

BERRY SOLAR PRODUCTS
Woodbridge at Main
P.O. Box 327
Edison, N.J. 08817
Phone: (201) 549-3800

BURKE INDUSTRIES, INC.
2250 So. Tenth St.
San Jose, Cal. 95112
Phone: (408) 297-3500

BURTON INDUSTRIES, INC.
243 Wyandanch Ave.
North Babylon, N.Y. 11704
Phone: (516) 643-6660

CY/RO INDUSTRIES
Wayne, N.J. 07470
Phone: (201) 839-4800

CALMAC MANUFACTURING CORP.
150 S. Van Brunt St.
Englewood, N.J. 07631
Phone: (201) 569-0420

CHAMBERLAIN MANUFACTURING CORP.
R & D Division
P.O. Box 2545
Waterloo, Iowa 50705
Phone: (319) 232-6541

CHAMPION HOME BUILDERS CO.
Solar Division
5573 E. North St.
Dryden, Mich. 48428

CHASE-WALTON ELASTOMERS, INC.
27-S Apsley St.
Hudson, Mass. 01749
Phone: (617) 485-5600

COLE SOLAR SYSTEMS
440A East St. Elmo Rd.
Austin, Texas 78745
(512) 444-2565

COLUMBIA CHASE SOLAR ENERGY DIV.
55 High St.
Holbrook, Mass. 02343
Phone: (617) 767-0513

CONTEMPORARY SYSTEMS, INC.
68 Charlonne St.
Jaffrey, N.H. 03452
Phone: (603)532-7972

**CUSTOM SOLAR
HEATING SYSTEMS**
P.O. Box 375
Albany, N.Y. 12201
(518)438-7358

DAYSTAR CORP.
90 Cambridge St.
Burlington, Mass. 01803
Phone: (617)272-8460

DEL SOL CONTROL CORP.
11914 U.S. #1
Juno, Fla. 33408
Phone: (305)626-6116

DELTA T CO.
2522 West Holly St.
Phoenix, Ariz. 85009
Phone: (602)272-6551

DESOTO, INC.
1700 South Mt. Prospect Rd.
Des Plaines, Ill. 60018
Phone: (312)269-6611

DOW CORNING CORP.
Solar Energy Div.
Midland, Mich. 48640
Phone: (517)496-4000

DUPONT CO.
Plastics and Resins Dept.
Wilmington, Del. 19898
Phone: (302)999-3456

DUTCHER INDUSTRIES, INC.
solar Energy Div.
7617 Convoy Court
San Diego, Cal. 92111
Phone: (714)279-7570

EL CAMINO SOLAR SYSTEMS
5330 Debbie Lane
Santa Barbara, Cal. 93111

ENERGY APPLICATIONS, INC.
830 Margie Dr.
Titusville, Fla. 32780
Phone: (305) 269-4893

ENERGY CONVERTERS, INC.
2501 N. Orchard Knob Ave.
Chattanooga, Tenn. 37406
Phone: (615) 624-2608

ENTROPY LIMITED
5735 Arapahoe Ave.
Boulder, Colo. 80303
Phone: (303) 443-5103

THE EPPLEY LABORATORY, INC.
12 Sheffield Ave.
Newport, R.I. 02840
Phone: (401) 847-1020

FAFCO NC.
138 Jefferson Drive
Menlo Park, Cal. 94025
Phone: (415) 321-6311

FALBEL ENERGY SYSTEMS CORP.
472 Westover Road
Stamford, Conn. 06902
Phone: (203) 357-0626

FORD PRODUCTS CORP.
Ford Products Road
Valley Cottage, N.Y. 10989
Phone: (914) 358-8282

FUTURE SYSTEMS, INC.
12500 West Cedar Drive
Lakewood, Colo. 80228
Phone: (303) 989-0431

GARDEN WAY PUBLISHING CO.
Dept. 130ZZ
Charlotte, Vt. 05445
(802) 425-2147

GENERAL ELECTRIC CO.
P.O. Box 13601
Philadelphia, Pa. 19101
Phone: (215) 962-2112

**THE GLASS-LINED WATER
HEATER CO.**
13000 Athens Ave.
Cleveland, Ohio 44107
Phone: (516) 521-1377

GOULD LABORATORIES
540 East 105th St.
Cleveland, Ohio 44108
Phone: (216) 851-5500

GRUNDFOS PUMPS CORP.
2555 Clovis Ave.
Clovis, Cal. 93612
Phone: (209) 299-9741

GULF THERMAL CORP.
P.O. Box 13124, Airgate Branch
Sarasota, Fla. 33578
Phone: (813) 355-9783

HALSTEAD AND MITCHELL
P.O. Box 1110
Scotsboro, Ala. 35768

HAWTHORNE INDUSTRIES, INC.
1501 S. Dixie Highway
West Palm Beach, Fla. 33401
Phone: (305) 659-5400

H.C. PRODUCTS CO.
Div. of Aluminum Co. of America
P.O. Box 68
Princeville, Ill. 61559
Phone: (412) 553-3185

HEILMANN ELECTRIC
127 Mountainview Rd.
Warren, N.J. 07060
Phone: (201) 757-4507

HELIOTROPE GENERAL
3733 Kenora Drive
Spring Valley, Cal. 92077
Phone: (714) 460-3930

HEXCEL
11711 Dublin Blvd.
Dublin, Cal. 94566
Phone: (415) 828-4200

HIGH PERFORMANCE PRODUCTS, INC.
25 Industrial Park Rd.
Hingham, Mass. 02043

HONEYWELL INC.
1885 Douglas Drive North
Minneapolis, Minn. 55422
Phone: (612) 542-7500

ILSE ENGINEERING, INC.
7177 Arrowhead Rd.
Duluth, Minn. 55811
Phone: (218) 729-6858

INDEPENDENT LIVING, INC.
5715 Buford Highway, N.E.
Suite 205
Doraville, Ga. 30340

INTERNATIONAL RECTIFIER, SEMICONDUCTOR DIV.
233 Kansas St.
El Segundo, Cal. 90245
Phone: (213) 322-3331

KALWALL CORP.
1111 Candia Rd.
Manchester, N.H. 03103
Phone: (603) 668-8186

KOLDWAVE DIV. OF HEAT EX-CHANGERS, INC.
8100 N. Monticello
Skokie, Ill. 60076
Phone: (312) 267-8282

KTA CORP.
12300 Washington Ave.
Rockville, Md. 20852
Phone: (301) 468-2066

LENNOX INDUSTRIES INC.
P.O. Box 250
200 South 12th Ave.
Marshalltown, Iowa
Phone: (515) 754-4011

LIBBEY-OWENS-FORD GLASS CO.
1701 E. Broadway St.
Toledo, Ohio 43605
Phone: (419) 247-4350

L.M. DEARING ASSOCIATES, INC.
12324 Ventura Bvd.
Studio City, Cal. 91604
Phone: (213) 796-2521

3M COMPANY
3M Center, Bldg. 223-2
St. Paul, Minn. 55101
Phone: (612) 733-2184

MANN-RUSSELL ELECTRONICS, INC.
1401 Thorne Rd.
Tacoma, Wash. 98421
Phone: (206)383-1591

MARK M MANUFACTURING
R.D. #2, Box 250
Rexford, N.Y. 12148
Phone: (518) 371-9596

MARTIN PROCESSING, INC.
P.O. Box 5068
Martinsville, Va. 24112
Phone: (703) 629-1711

MCARTHURS, INC.
P.O. Box 236
Forest City, N.C. 28043
(704) 245-7223

MND, INC.
P.O. Box 15534
Atlanta, Ga. 30333

NATURAL POWER INC.
New Boston, N.H. 03070
Phone: (603) 487-2426

NEW JERSEY ALUMINUM
P.O. Box 73
North Brunswick, N.J. 08902
Phone: (toll free) (800) 631-5856

NORTHRUP, INC.
302 Nichols Dr.
Hutchins, Texas 75141
Phone: (214) 225-4291

NUCLEAR TECHNOLOGY CORP.
P.O. Box 1
Amston, Conn. 06231
Phone: (203) 537-2387

O.E.M. PRODUCTS, INC.
220 W. Brandon Blvd.
Brandon, Fla. 33511
Phone: (813) 689-1182

OLIN BRASS, ROLL-BOND PRODUCTS
East Alton, Ill. 62024
Phone: (618) 258-2443

OWENS-CORNING FIBERGLAS CORP.
Tank Marketing Div.
Fiberglas Tower
Toledo, Ohio 43659
Phone: (419) 248-8063

PAYNE, INC.
1910 Forest Drive
Annapolis, Md. 21401
Phone: (301) 268-6150

PEOPLE/SPACE CO., PAC DIV.
49 Garden St.
Boston, Mass. 02114
Phone: (617) 742-8652

PPG INDUSTRIES INC.
One Gateway Center
Pittsburg, Pa. 15222
Phone: (412) 434-3555

RAYPAK, INC.
31111 Agoura Rd.
Westlake Village, Cal. 91361
Phone: (213) 889-1500

REYNOLDS METALS CO.
6601 W. Broad St.
Richmond, Va. 23261
Phone: (804) 281-4734

RESOURCE TECHNOLOGY CORP.
151 John Downey Dr.
New Britain, Conn. 06051
Phone: (203) 224-8155

REVERE COPPER AND BRASS INC.
Solar Energy Dept.
P.O. Box 151
Rome, N.Y. 13440
Phone: (315) 338-2401

RHO SIGMA, INC.
15150 Raymer St.
Van Nuys, Cal. 91405
Phone: (213) 342-4376

RICHDEL, INC.
1851 Oregon St.
Carson City, Nev. 89701
Phone: (702) 832-6786

ROBERTSHAW CONTROLS CORP.
100 W. Victoria
Long Beach, Cal. 90805
Phone: (213) 638-6111

ROHM AND HAAS CO.
Independence Hall West
Philadelphia, Pa. 19105
Phone: (215) 592-3000

SENSOR TECHNOLOGY INC.
21012 Lassen St.
Chatsworth, Cal. 91311
Phone: (213) 882-4100

SHELDAHL, ADVANCED PRODUCTS DIV.
Northfield, Minn. 55057
Phone: (507) 645-5633

SIMONS SOLAR ENVIRONMENTAL SYSTEMS
24 Carlisle Pike
Mechanicsburg, Pa. 17055
Phone: (717) 697-2778

SKYTHERM PROCESSES AND ENGINEERING
2424 Wilshire Blvd.
Los Angeles, Cal. 90057
Phone: (213) 389-2300

SOLAR COMFORT SYSTEMS, MFG.
4853 Cordell Ave. Suite 606
Bethesda, Md. 20014
Phone: (301) 652-8941

SOLAR CONTROL CORP.
5595 Arapahoe Rd.
Boulder, Colo. 80302
Phone: (303) 449-9180

SOLAR CORP. OF AMERICA
100 Main St.
Warrenton, Va. 22186
Phone: (703) 347-7900

SOLAR DEVELOPMENT, INC.
4180 Westroads Dr.
West Palm Beach, Fla. 33407
Phone: (305) 842-8935

SOLAR ENERGY CO.
Drawer 649
Gloucester Point, Va. 23062

SOLAR ENERGY PRODUCTS, INC.
1208 N.W. 8th Ave.
Gainesville, Fla. 32601

SOLAR ENERGY RESEARCH CORP.
701B South Man St.
Longmont, Colo. 80501
Phone: (303) 772-8406

SOLARGENICS, INC.
9713 Lurline Ave.
Chatsworth, Cal. 91311
Phone: (213) 998-0806

SOLARGIZER CORP.
220 Mulberry St.
Stillwater, Minn. 55082
Phone: (612) 439-5734

SPIRAL TUBING CORP.
533 John Downey Dr.
New Britain, Conn. 06051
Phone: (203) 224-2409

STATE INDUSTRIES, INC.
Ashland City, Tenn. 37015

STURGES HEAT RECOVERY, INC.
P.O. Box 397
Stone Ridge, N.Y. 12481
Phone: (914) 687-0281

SOLAR KINETICS, INC.
147 Parkhouse St.
Dallas, Texas 75207

SOLARKIT OF FLORIDA, INC.
1102 139th Ave.
Tampa, Fla. 33612
Phone: (813) 971-3934

SOLARON CORP.
300 Galleria Tower
720 S. Colorado Blvd.
Denver, Colo. 80222

SOLAR PHYSICS CORP.
P.O. Box 357
Lakeside, Cal. 92040
Phone: (714) 561-4531

SOLAR POWER CORP.
5 Executive Park Dr.
North Billerica, Mass. 01862
Phone: (617) 667-8376

SOLAR TECHNOLOGY CORP.
2160 Clay Street
Denver, Colo. 80211
Phone: (303) 455-3309

THE SOLARAY CORP.
2414 Makiki Hgts. Dr.
Honolulu, Hawaii 96822
Phone: (808) 533-6464

SUNEARTH SOLAR PRODUCTS CORP.
R.D. 1, Box 337
Green Lane, Pa. 18054
Phone: (215) 699-7892

SUNPOWER INDUSTRIES, INC.
10837 S.E. 200th
Kent, Wash. 98031
Phone: (206) 854-0670

SUNSHINE MFG. CO.
4870 SW Main #4
Beaverton, Ore. 97005
Phone: (503) 643-6172

SUNSTREAM,
A Div. of Grumman Houston Corp.
P.O. Box 365
Bethpage, N.Y. 11714
(516) 575-0574

SUNWORKS, DIV. OF ENTHONE
P.O. Box 1004
New Haven, Conn. 06508
Phone: (203) 934-6301

SOUTHERN LIGHTING MFG.
501 Elwell Ave.
Orlando, Fla. 32803

TECHNOLOGY APPLICATIONS LABORATORY
1670 Highway A-1-A
Satellite Beach, Fla. 32937
(305) 777-1400

TRANTER, INC.
735 E. Haxel St.
Lansing, Mich. 48912
Phone: (213) 882-4100

THOMASON SOLAR HOMES, INC.
609 Cedar Ave., Oxon Hill
Fort Washington, Md.
Washington, D.C. 20022

UNIT ELECTRIC CONTROL, INC.
Sol-Ray Div.
130 Atlantic Cr.
Maitland, Fla. 32751
Phone: (305) 831-1900

URETHANE MOLDING, INC.
RFD #3, Rt. 11
Laconia, N.H. 03246
Phone: (603) 524-7577

WESTINGHOUSE ELECTRIC CORP.
Comm-Indust. Air Cond. Div.
P.O. Box 2510
Staunton, Va. 24401
Phone: (703) 886-0711

WEST WIND CO.
Box 1465
Farmington, N.M. 87401
(505) 325-4949

THE WILCON CORP.
3310 S.W. 7th St.
Ocala, Fla. 32670
Phone: (904) 732-2550

W.L. JACKSON MFG. CO. INC.
P.O. Box 11168
Chattanooga, kTenn. 37401
Phone: (615) 867-4700

YORK, DIV. OF BORG-WARNER
P.O. Box 1592
York, Pa. 17405

ZOMEWORKS CORP.
P.O. Box 712
Albuquerque, N.M. 87103
Phone: (505) 242-5354

SOLAR VIDEO

INTRODUCING a new series of video-tape programs covering many of the prominent issues and individuals now moving the solar community—geared for students, professionals, and laymen who wish to explore the concepts and implications of solar energy application.

The Solar Age

A broad spectrum, hour-long documentary that looks at the diverse forms of solar energy use. The most extensive program of the series, it deals with the dominant issues within the solar community and serves as an overview with leading solar architects, builders, engineers and historians: the people who are making it happen!

A Community Approach

How an inner-city community approached solar retrofitting and partial self-sufficiency. Take a look at a gutted out apartment building on New York City's lower east side that was refurbished by tenants who now own it through "sweat equity".

Moving Towards Mass Production

This tape shows how industry has adapted to the coming of solar. Specifically the relationship between the Solaron Corporation and the Choice Vend Corporation, large vending machine manufacturer. Featuring long time solar pioneer George O. G. Lof.

Solar Reality: A Dialogue with John I. Yellott

The encyclopedic solar pioneer talks about the past, present, and future of solar.

The Karen Terry House
Architecture, Building and the Sun:

A documentary account of the conception, construction, and performance of one of the most sophisticated "passive" projects thus far, featuring Karen Terry, builder; David Wright, architect; and B. T. (Buck) Rogers.

Solar Law

William Thomas, a research attorney for the American Bar Foundation, talks on the many legal implications and problems facing the solar industry, and discusses solar legal research done under a grant from the National Science Foundation.

ALL TAPES AVAILABLE IN HALF-INCH OR THREE-QUARTER CASSETTE FOR MORE INFORMATION WRITE:

SOLARVISION INC.,
BOX A, HURLEY, N. Y. 12443

Resources and Services

This directory includes architects and engineers, designers and consultants, builders, retailers, people who offer information and resources. Listings are state-by-state to enable you to find the ones nearest you, but many offer regional or national services—indicated under the heading "Geographical range."

Alaska

THE INSTITUTE OF WATER RESOURCES, *University of Alaska, Fairbanks, Alaska 99701.* Bibliographic materials on solar applications; work with solar water heating underway; willing to dispense local advice and especially interested in coupled heat pump-ground storage of solar heat, and cold climate hot water heaters. **Geographical range:** Alaska. **Contact:** Richard Seifert.

Arizona

GOETTL BROS. METAL PRODUCTS, INC., *2005 East Indian School Road, Phoenix, Ariz. 85016.* Currently involved in the research, development, and testing of solar systems. Will manufacture and install solar systems in 1977. **Contact:** Cliff Conn; phone: (602) 957-9800.

J & B SALES CO., *3441 N. 29th Ave., Phoenix, Ariz. 85017.* A diversified distributor agent for companies that have equipment for use on solar energy systems. **Contact:** William E. Seginski; phone: (602) 258-1545.

California

F.E.ADAMS, P.E., *18154 Bancroft Road, Monte Sereno, Cal. 95030.* Solar systems specialist. **Contact:** Fran Adams; phone: (408) 354-4974.

APPROVED ENGINEERING TEST LABORATORIES, *15720 Ventura Blvd., Suite 420, Encino, Cal. 91316.* Full service high technology testing laboratory. Presently participating in the first round evaluation of the Dept. of Commerce, NBS, National Voluntary Laboratory Accreditation program for Solar Laboratories to ASHRAE 93P. **Geographical range:** national. **Contact:** Robert A. Finch; phone: (213) 783—5985.

ALTAS CORP., *2060 Walsh Ave., San Tomas Business Park, Santa Clara, Cal. 95050.* Research and development emphasis on the use of solar heat in residential, commercial, and industrial applications. Special interest in the heat transfer aspects of solar energy equipment, including such associated fields as fluid dynamics and thermodynamics in the design and analysis of cost-optimized systems. Development of new design criteria for solar systems. **Contact:** Francis de Winter; phone: (408) 246-9664.

E.R. BAUGH, *2690 Wallingford, San Marino, Cal. 91108.* Engineering consultation for energy conservation, retrofitting, and sewage water recycle systems in buildings. **Geographical range:** southern California. **Contact:** E.R. Baugh.

BERKELEY SOLAR GROUP, *3026 Shattuck Ave., Berkeley, Cal. 97405.* Solar system design; computer modeling of buildings and solar energy systems; passive heating and cooling. **Contact:** Chip Barnaby; phone: (415) 843-7600.

BLUE SKIES RADIANT HOMES, *40819 Park Ave., Hemet, Cal. 92343.* Developers of full solar homes and design consultants. Thirty-three units completed. **Contact:** Warren Buckmaster; phone:(714) 658-2070.

BUFFALOW'S, INC., *1245 Space Park Way, Mountain View, Cal. 94043.* Design, fabrication, and installation capabilities for residential, commercial, and industrial applications of solar, air conditioning and heating, ventilation and exhaust, clean rooms, refrigeration, etc. Three years experience with solar. **Geographical range:** generally northern California. **Contact:** Henry Buffalow Jr.; phone: (415) 961-7750.

CHAPMAN ENERGY SYSTEMS, *20743 Strathern St., Canoga Park, Cal. 91306.* Design and contracting of solar and wind energy systems, residential and commercial applications. Mechanical, electrical and prototype industrial design services. Four years in field. **Geographical range:** mainly western United States, but can work any area. **Contact:** Alan Arthur Tratner, Howard Chapman.

FEDERICO GRABIEL, *806 North Lillian Way, Los Angelos, Cal. 90038.* Consulting service for on-site use of solar energy; teaching and lecturing. Author and inventor. **Geographical range:** not limited. **Contact:** Federico Grabiel; phone: (213) 465-2958.

GRUNDFOS PUMPS CORP., *2555 Clovis Ave., Clovis, Cal. 93612.* Sizing and selection of alternative types of water pumps for solar related applications. Manufacturing pumps for water heating systems for thirty years. **Geographical range:** all countries in the world. **Contact:** Dave J. Reindl, solar applications engineer; phone: (209) 299-9741.

HARDISON AND KOMATSU ASSOCIATES, *522 Washington St., San Francisco, Cal. 94111.* Architects, consultants on solar space-heating installations, ground water heat pump installation, drafting of energy-related legislation. **Geographical range:** California and Nevada. **Contact:** William S. Taber, Jr., phone: (415) 981-2025.

HAYAKAWA ASSOCIATES, *1180 South Beverly Drive, Los Angeles, Cal. 90-035.* Consulting engineers. **Contact:** Kathleen Thornton; phone: (213) 879-4477.

INTERACTIVE RESOURCES, INC., *117 Park PLace, Point Richmond, Cal. 94801.* A multi-disciplinary consulting service in the areas of energy management, design and implementation, energy conservation. Consultants to various solar projects. **Geographical range:** western U.S. **Contact:** Dale Sartor; (415) 236-7435.

NATURAL HEATING SYSTEMS, *207 Cortez Ave., Davis, Cal. 95616.* Complete

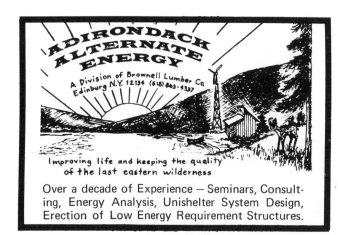

design and specification of systems, site and feasibility analyses, computer simulation to determine useful collector heat output, and consulting on passive heating design. Three years experience. **Geographical range:** northern California and greater area. **Contact:** David Springer; phone: (916) 756-4558.

JAMES L. RUHLE & ASSOCIATES, *P.O. Box 4301, Fullerton, Cal. 92631.* Audio-visual materials—slide programs and filmstrips—covering energy-related matters. **Contact:** James L. Ruhle; phone: (714) 526-6120.

SHARPE SOLAR SYSTEMS, *2114 Woolard Drive, Bakersfield, Cal. 93304.* Full-range solar service: books, do-it-yourself supplies, solar items, pool heaters, energy-efficient housing, architectural consulting and design, domestic water, space heating, agricultural services, etc. **Geographical range:** Kern and Tulare Counties. **Contact:** Charles Sharpe: phone: (805) 832-7259.

SIERRA SOLAR SYSTEMS, *P.O. Box 310, Nevada City, Cal. 95959.* Consulting and systems design; market solar equipment, including "build your own" solar collector kit. **Geographical range:** northern California. **Contact:** Karl R. Stewart; phone: (916) 272-3444.

SOLAR ACCESS, *320-1 Cedar St., Santa Cruz, Cal. 95060.* Alternative energy department store carrying total energy systems, energy conservation products, water conservation products, books, plans, materials, tools, wood stoves, etc. **Geographical range:** central coastal California. **Contact:** James Bukey; phone: (408) 426-3100.

SOLAR ENERGY DIGEST, *P.O. Box 17776, San Diego, Cal. 92117.* A monthly twelve-page newsletter covering the frontiers of solar. Books and plans are also available. $28.50 per year. **Contact:** William B. Edmondson, publisher; phone: (714) 277-2980.

SOLAR SYSTEMS, INC., *P.O. Box 931, Livermore, Cal. 94550.* Solar heating system and solar instrumentation design. Presently developing on-line computer-based solar installation sizing service. Developed instrumentation system for measurement of insolation at Lawrence Livermore Laboratory. **Geographical range:** San Francisco Bay Area. **Contact:** George Bush; phone: (415) 447-9286.

SOLAR SYSTEM SALES, *180 Country Club Drive, Novato, Cal. 94947.* Consulting and sales for solar space, water, and pool-heating equipment. **Geographical range:** Marin and Sonoma counties. **Contact:** George Walters; phone: (415) 883-7040.

SOLAR UTILITIES COMPANY, *406 North Cedros, Solana Beach, Cal. 92075.* Complete solar system design and engineering; performance evaluations. Over fifty installations, including solar stills, to date. **Geographical range:** San Diego region. **Contact:** Jack Schultz.

SOLARC, SOLAR ENERGY IN ARCHITECTURE, *P.O. Box 4233, Whittier, Cal. 90607.* Solar energy design and consultation. Both principals are architects and are co-authors of *Solar Primer One.* **Geographical range:** southern California. **Contact:** James F. Carlson; phone: (213) 693-7248.

SOLPUB CO., *1831 Weston Cir., Camarillo, Cal. 93010.* Consultants in solar space and domestic hot water heating to architects and engineers. Three years experience as manager of Navy solar and alternative energy R&D program, author of *Navy Design Manual for Solar Heating of Buildings and Domestic Hot Water.* Geographical range: southern California. **Contact:** Dr. Richard L. Field; phone: (805) 482-6651.

SOUTHWEST ENERGY MANAGEMENT, INC.. *8290 Vickers St., Suite B, San Diego, Cal. 92111.* Design, construc-tion, and component supply group; also perform energy management reports for industry (Beatrice Foods; Ralston Purina). Have designed and built solar installations in the San Diego area. **Geographical range:** San Diego-Los Angeles. **Contact:** Brian E. Langston; phone: (714) 292-5185.

WARREN STEELE AND ASSOCIATES, *3518 Cahuenga Boulevard West, Los Angeles, Cal. 90068.* Consulting mechanical and solar engineers. Among their projects: Torrance, California,High School gyms—solar systems, 167 collectors. **Geographical range:** western states and Pacific. **Contact:** R. Warren Steele; phone: (213) 851-4175.

STERN RESEARCH CORP., *Route 3, Box 274, Orcutt Rd., San Luis Obispo, Cal. 93401.* Financial and/or technical support for energy development. Several years of interest and technical experience in geothermal, solar, and wind energy. **Geographical range:** California and Illinois. **Contact:** Robert L. Stern; phone: (805) 543-4444.

SUNSAVER, DIV. OF NORTHERM INC., *Box 686, 4645 Belvedere Dr., Julian, Cal. 92036.* Solar design service for buildings and products. Offers sixteen page solar design manual to builders, architects, engineers and homeowners for $3. Three years in solar research. **Geographical range:** U.S., Europe, and the Middle East. **Contact:** C.R. Braathen; phone: (714) 765-1793.

SUNTEK RESEARCH ASSOCIATES, *506 Tamal Plaza, Corte Madera, Cal. 94925.* Materials R & D on phase change heat storage, heat mirrors, thin films, plastic films, transparent insulation, and testing of optical, accelerated aging, and insulation. Five years of similar materials R & D work culminating in the development of transparent insulation, Thermocrete, and variable transmission materials. **Geographical range:** anywhere. **Contact:** Day Chahroudi/John Brooks/Blair Hamilton; phone: (415) 924-6887.

SUNWAY, *1301 Berkeley Way, Berkeley, Cal. 94702.* Solar energy store carrying a variety of solar products and a large selection of books pertaining to solar. **Phone:** (415) 843-4019.

THACHER & THOMPSON, *215 Oregon St., Santa Cruz, Cal. 95060.* General contractors will design and build solar homes. **Geographical range:** Santa Cruz area. **Contact:** Richard R. Rahders; phone: (408) 426-4683.

UNIVERSAL HERITAGE INVESTMENTS CORP., *Wells Fargo Plaza, 21535 Hawthorne Blvd., Del Amo Financial Center, Torrance, Cal. 90503.* Brokerage firm; stock information on companies that are publicly traded and heavily involved in solar energy. **Contact:** Stuart Tsujimoto; phone: (213) 370-8531.

VANTECH CORPORATION, *5555 Magnatron Blvd., Unit J, San Diego, Cal. 92111.* Research, design, and manufacture of solar and alternate energy systems. Operating as a company dedicated to alternate energy for six years. **Geographical range:** national. **Contact:** John van Geldern; phone: (714) 292-9614.

Colorado

CROWTHER SOLAR GROUP, *310 Steele Street, Denver, Colo. 80206.* Architectural design firm specializing in energy conservation and natural energy use, including solar; analysis of building energy requirements and utilities consumption; passive and active systems design. Have designed numerous solar and energy conserving builings. **Contact:** Richard L. Crowther AIA; phone: (303) 355-2301.

ECS, INC.
P.O. Box 1063
Boulder, Co. 80306
Systems design, training courses, computer modeling; will work with or train designers, builders, contractors, academicians, engineering firms. **Contact:** Dr. Jan F. Kreider, president; phone: (303) 447-2218.

ENERGY EQUIPMENT CO., *327 W. Vermijo Ave., Colorado Springs, Colo. 80903.* Sales of solar energy conserving equipment and services. **Geographical range:** Rocky Mountain States. **Contact:** Peter Wood; phone: (303) 475-0332.

FEDERAL ENERGY CORP., *909 17th St., Suite 308, Denver, Colo. 80202.* Engineering and design firm for solar

heating units, both water and air products. Two years R & D on solar products. **Geographical range:** U.S. and Canada. **Contact:** Grant Dissette; phone: (303)623-1097.

FUTURE SYSTEMS, INC., *12500 West Cedar Drive, Lakewood, Colo. 80228.* Testing of solar collectors in accordance with proposed standards in NBS Technical Note #899. **Geographical range:** Rocky Mountains and Western Plains regions. **Contact:** William V. Thompson; phone: (303) 989-0431.

KINETICS CORP., *608 Pearl Street, Boulder, Colo. 80302.* Environmental planners, designers, and builders of energy-conserving solar buildings. Completed two building, eight-unit solar condominium in Boulder *(Solar Age, Jan., '77).* **Contact:** Robert White; phone: (303) 449-2979.

DR. JAN F. KREIDER, P.E., *1929 Walnut Street, Boulder, Co. 80302.* Solar system mechanical design, solar-mechanical systems consultant, solar economic feasibility studies, training courses for architects, engineers, and equipment installers. Author (with Dr. Frank Kreith) of two solar energy books published by McGraw-Hill. Geographical range: national. Designer, with Swanson-Rink Engineers of Denver, of 40,000 square foot solar collector. **Contact:** Dr. Kreider; phone: (303) 447-2218.

DR. FRANK KREITH, *1485 Sierra Dr., Boulder, Colo. 80302.* Consulting service for design, optimization, and economic analysis of solar thermal systems. Consultant to state and local government agencies in the U.S., Europe and the Middle East. **Geographical range:** anywhere in the world. **Contact:** Dr. Frank Kreith, P.E.

PROFESSIONAL DESIGN BUILDERS, INC., *P.O. Box 275, Loveland, Colo. 80537.* Consulting engineers for solar design both passive and active; analyze heat loss, thermal gain, thermal storage characteristics, structure strength, site orientation. **Geographical range:** Front range Colorado and Wyoming. **Contact:** Ivar W. Larson; phone: (303) 669-2650.

SOLAR ENERGY RESEARCH CORP., *701B South Main Street, Longmont, Colo. 80501.* Consultant for residential, industrial, commercial solar heating system designs. Computer analysis of solar heating system performance. Preparation of reports, plans and specifications for solar equipment. Product evaluation, development, testing, and business consulting. Five years in solar field; holds patent on Thermo-Spray™ system. **Geographical range:** will travel. **Contact:** James B. Wiegand, president; phone: (303) 772-8406.

SOLAR SCIENCES, *800 Grant, Suite 410, Denver, Colo. 80203.* Researching and developing double-axis tracking concentrator for domestic and commercial (car washer, laundromats) hot water and space heat. Company in existence since February 1976. **Geographical range:** product will be designed to operate nationwide. **Contact:** George Robertson; phone: (303) 837-1268.

SOLARON CORPORATION, *300 Galleria Tower, 720 S. Colorado Blvd., Denver, Colo. 80222.* System design and engineering assistance; design manual covering solar system engineering, architectural requirements, and economics; designed and installed over 135 complete systems. **Geographical range:** continental U.S. and Canada. **Contact:** Jim Junk, manager, systems engineering.

SUN MOUNTAIN DESIGN, *229 South St., P.O. Box 797, Breckenridge, Colo. 80424.* Solar, cold weather, and environmental design. Two years experience in solar design. **Geographical range:** Rocky Mountain region. **Contact:** Peter P. Brower; phone: (303) 453-6130.

WALTON-ABEYTA AND ASSOCIATES, INC., *1221 S. Clarkson Street, Suite 222, Denver, Colo. 80210.* Mechanical-electrical-environmental consulting engineers; design, specifications, construction documents, economic analyses, feasibility studies. Among their commercial and residential solar projects: Solar Centre Office Building, Denver, and Robert Redford residence, Sundance, Utah. **Geographical range:** registered engineers in thirty states. **Contact:** Monty E. Abeyta; phone: (303) 777-3068.

PETER WOOD, *327 Vermijo Ave., Colorado Springs, Colo. 80903.* Design of solar systems and component selection; design of energy conserving structures. **Geographical range:** Illinois to California. **Contact:** Peter Wood; phone: (303) 475-0332.

Connecticut

CONSOLAR, *Litchfield, Conn. 06759.* Consultation in building design for solar energy in new construction or retrofitting. **Geographical range:** New England, New York. **Contact:** Walter J. Gress, Jr.

DESIGN ASSOCIATES, *Nettleton Hollow, Washington, Conn. 06793.* Design and/or construction of passive solar heated, energy conserving buildings. Two completed and operating residences; presently planning small passive solar community. **Geographical range:** western and central Connecticut. **Contact:** Forbes Morse; phone: (203) 868-2900 or Stephen Lasar; phone: (203) 354-0855.

ENERGY RESOURCE CORP., *88 Main St., New Canaan, Conn. 06840.* Architectural and engineering design of solar heating for houses and commercial applications for both space and hot water; marketing and installation of energy conservation devices and materials such as transient voltage suppressors, insulation, storm windows, weatherstripping, and solar hot water heater package. **Geographical range:** Fairfield County, Conn.; Westchester and Metropolitan New York. **Contact:** Alfred Alk; phone: (203) 972-0150.

KAMAN AEROSPACE CORP., *Old Windsor Rd., Bloomfield, Conn. 06002.* Consulting in overall solar concepts, thermal analysis, system design, controls, data acquisition and economics. R & D capability to investigate and demonstrate systems and components, particularly high technology. Solar research and evaluation system available to test and develop collectors or other components of outside organizations or individuals. Consultant to state of Connecticut on Hamden project to provide solar-heated housing for the elderly. Anticipate manufacture of solar products in the near future. **Geographical range:** northeast U.S. **Contact:** Dr. Edward Kush, research group; phone: (203) 242-4461, Ext. 336.

KELLY ROOFING-RESTORATION INC., *50 Randolf Ave., Waterbury, Conn. 06710.* Manufacture and sell fully insulated modular roofing segments that may be modified to flat plate solar collectors. **Contact:** Thomas L. Kelly; phone: (203) 754-7950.

MOORE GROVER HARPER
PC, Essex. Conn. 06426.
Architects and planners of more than twenty-five solar heated and energy conserving buildings and complexes including the Norwich, Connecticut Armory, solar heated, to be completed in 1977. Geographical range: worldwide. **Contact:** William H. Grover; phone (203) 767-0101.

NATIONAL SOLAR CORP., *Novelty Lane, Essex, Conn. 06426.* Solar systems engineers and manufacturers. In operation since 1975. **Geographical range:** emphasis on Northeast. **Contact:** Anthony Easton; phone: (203) 767-1644.

PINE ASSOCIATES INC., *Spring Lane, P.O. Box 305, Farmington, Conn. 06032.* Research, design, consulting, and construction of residential and light commercial buildings. Currently working on solar projects. **Geographical range:** Connecticut—elsewhere for design/consulting only. **Contact:** Thomas E. Anderson; phone: (203) 677-9330.

JAMES ADDISON POTTER, *12 Green House Boulevard, West Hartford, Conn. 06110.* Professional engineer marketing solar and wind-power components and structural elements for self-sufficient residences. **Geographical range:** United States on purchase order basis; worldwide via letters of credit. **Contact:** James A. Potter; phone: (203) 628-0911.

DR. K. RAMAN, *28-A Ambassador Drive, Manchester, Conn. 06040.* Consultant in solar energy design and energy conservation. **Geographical range:** not given. **Contact:** K. Raman; phone: (203) 649-9122.

SUNWORKS, DIVISION OF ENTHONE, INC., *P.O. Box 1004, New Haven, Conn. 06508.* Architectural and engineering staff will consult with qualified architects, engineers, and contractors involved in solar system design. Computer program for determining collector performance is available. Founded in 1973. **Geographical range:** fifty states, Canada, Mexico, Bahamas. **Contact:** Floyd Perry, product manager or Ryc Loope, technical and design services; phone: (203) 934-6301.

Delaware

COOPERSON & BRECK ACCOCIATES, *4000 Thompsons Bridge Rd., P.O. Box, Montchain, Del. 10710.* Architectural and engineering design services, particularly life cycle costing of building energy systems; design of solar systems deemed economical in life cycle evaluation; presently designing solar system for Wilmington Swim Club, Wilmington, Del. **Geographical range:** middle Atlantic states. **Contact:** Jay Cooperson; phone: (302) 478-1815.

SOLAR ENERGETICS, INC., *301 South West Street, Wilmington, Del. 19899.* Distribution of products and technology for solar energy systems. Have engineered solar residences and water heaters. **Geographical range:** Middle Atlantic States. **Contact:** William Anderson; phone: (302) 654-3252.

District of Columbia

CENTER FOR SCIENCE IN THE PUBLIC INTEREST, *1757 S Street, N.W., Washington, D.C. 20009* A nonprofit, tax-exempt research organization working in energy policy, environmental protection, and consumer concerns. Among their reports; "Solar Energy: One Way to Citizen Control." Monthly newsletter: "People & Energy." **Contact:** Ken Bossong or Alan Okagaki; phone: (202) 332-4250.

ARTHUR COTTON MOORE AND ASSOCIATES
1214 28th St.
Washington, D.C. 20007
Designers of systems and buildings. Among their jobs: the science building at the Madeira School, Greenway, Va. **Contact:** Patricia Moore; phone: (202) 337-9083.

SOLAR ENERGY INSTITUTE OF AMERICA (SEINAM), *P.O. Box 9352, Washington, D.C. 20005.* SEINAM ($15 per year individuals and $45 per year firms) keeps members apprised of activities and events in the field, and entitles them to a 600-page catalog of products and services, and a newsletter. In broadscale operation since early 1976. **Geographical range:** worldwide, members in forty-eight states and five countries. **Contact:** Luana Moore, vice-president; phone: (301) 853-2335.

THE SKY IS FALLING, *1200 9th Street, NW, Washington, D.C. 20001.* Energy store with display and sales, retail or wholesale, of a wide range of solar heating equipment and components, representing a wide range of manufacturers; collectors, storage tanks, controls, and other components. Installation. **Contact:** Drew Stamps; phone: (202) 387-0200.

Florida

ARIES CONSULTING ENGINEERS, INC., *115 N.E. Seventh Ave., Gainesville, Fla. 32601.* Consulting and systems design for solar and hybrid solar/conventional heating, cooling and large scale hot water systems. Twelve residential, two commercial projects completed. **Geographical range:** worldwide. **Contact:** William J. Fielder; phone: (904) 372-6687.

GRAY PRODUCTS CORP., *Solar Division, 3800 N.W. 35th Ave., Miami, Fla. 33142.* Design, install and sell wholesale and retail components for solar systems. Mechanical contractors and engineers since 1941, specializing in refrigeration, steam, and air conditioning. **Geographical range:** Dade, Broward and Monroe counties in Florida. **Contact:** Robert Beck; phone: (305) 633-7561.

SOLAR ENERGY PRODUCTS, INC., *1208 N.W. 8th Avenue, Gainesville, Fla. 32601.* Designer, manufacturer, marketing, distributor, dealer, installer. In business four years. **Geographical range:** international. **Phone:** (904) 377-6527

UNIVERSAL SOLAR AIDS, INC., *P.O. Box 731, Fern Park, Fla. 32730.* Solar energy equipment for six counties in central Florida: Orange, Seminole, Volusia, Lake, Osceola, and Brevard. **Contact:** Russ Flanigan; phone: (305) 831-9721.

Georgia

ELLISON DEVELOPMENT CO., INC., *Solar Products Branch, 6809 Forrest Road, Columbus, Ga. 31907.* A multi-service firm for consulting, design, and fabrication of active and passive solar systems. Have designed and installed numerous solar systems. **Geographical range:** within a 100-mile range of Columbus, Georgia. **Contact:** William L. Wren; phone: (404) 561-9117.

WHELCHEL SOLAR ENTERPRISES, INC., *2050 Carroll Ave., Atlanta, Ga. 30341.* Solar system consultation, engineering, and installation for industrial, commercial, and residential use. **Contact:** Henry C. Whelchel; phone: (404) 458-2311.

Illinois

SCHMIDT, GARDEN & ERIKSON, *104 S. Michigan Ave., Chicago, Ill. 60603.* Architects and engineers providing building and solar systems design. Eighty years experience in planning and design of medical, research, corporate, and educational facilities. **Geographical range:** U.S.A. **Contact:** Louis A. Michelsen; phone: (312) 332-5070.

SHEAFFER & ROLAND, INC., *20 North Wacker Dr., Chicago, Ill. 60606.* Consulting, architectural, and engineering services for energy conservation and solar energy harvesting. Specialization in space and domestic water heating and in designing and building systems for waste management, including anaerobic digesters. **Contact:** John H. Martin; phone: (312) 236-9106.

WILLIAM Y. ZAKROFF, *2233 Grey Ave., Evanston, Ill. 60201.* Management consultation for marketing solar products, including new product introduction, application of marketing audit concept, setting up distribution channels, solar industry evaluations, product evaluations, and general management. Presently writing a marketing audit on collector panels manufactured by Chamberlain Mfg. Co. **Geographical range:** national. **Contact:** W.Y. Zakroff.

Iowa

BROOKS BORG AND SKILES, *Hubbell Building, Des Moines, Iowa 50309.* Architectural and engineering firm with experience in solar design. Designers of Raccoon Valley State Bank in Adel, Iowa for solar heating and cooling with a 25-ton Arkla chiller and 3,000 square feet of PPG panels. **Geographical range:** national. **Contact:** Thomas J. Van Horn.

THE SOLARWAY, INC., *6412 Washington Ave., Des Moines, Iowa 50322.* Solar home builder. **Contact:** G.C. Corrigan; phone: (515) 277-7760.

Kansas

JOHN C. BYRAM, JR., *Suite 113, Corinth Plaza Building, 8340 Mission Rd., Shawnee Mission, Kan. 66206.* Feasibility analyses for solar heating and cooling equipment—consulting; real estate appraiser and analyst. **Geographical range:** midwest. **Contact:** John Byram; phone: (913) 381-7111.

Kentucky

KENTUCKY SOLAR SYSTEMS, *1829 Cargo Court, Bluegrass Research and Industrial Park, Louisville, Ky. 40299.* Supplier, servicer, and installer of solar heating systems for hot water, space, and pool heating. Design and consultation services for solar applications. One year in the field. **Geographical range:** Kentucky and southern Indiana. **Contact:** R.D. Sanders; phone: (606) 491-7500.

Maine

MAINE AUDUBON SOCIETY, *Gilsland Farm, 118 Old Route One, Falmouth, Maine 04105.* Will provide general information to consumers contemplating installation of some form of solar heating; reference library, guided tours of completed solar/wood heated installation. **Geographical range:** Maine, New England. **Contact:** Erika Morgan; phone: (207) 781-2330.

Maryland

MARTIN BECK ASSOCIATES, INC., *762 Fairmont Ave., Towson, Md. 21204.* Full architectural design services in conjunction with the installation of solar equipment, including feasibility studies for incorporation of solar systems in existing and planned facilities. Have reviewed solar energy grant applications for HUD. **Geographical range:** Maryland, Washington,

D.C., northern Virginia. **Contact:** Peter C. Powell; phone: (301) 828-6000.

INTERNATIONAL COMPENDIUM, *10762 Tucker Street, Beltsville, Md. 20705.* Information company designed to serve solar industry and consumers; large solar energy research library; research reports; bibliographies; literature searches; architectural and engineering. Five years involvement in solar field. **Geographical range:** international. **Contact:** Carlo La Porta; phone: (301) 937-0040.

LAPICKI/SMITH ASSOCIATES, *617 Park Ave., Baltimore, Md. 21201.* Design services for residential and commercial solar energy projects, new buildings or retrofitting. **Geographical range:** Mid-Atlantic. **Contact:** Shannon P. Kennedy; phone: (301) 685-4900.

MUELLER ASSOCIATES, INC., *1900 Sulphur Spring Rd., Baltimore, Md. 21227.* Consulting engineers providing design for space heating and cooling, domestic hot water, process heat, cost analysis, and evaluation of solar system performance. **Geographical range:** Baltimore-Washington D.C. area. **Contact:** Andrew J. Parker; (301) 247-5666.

NATIONAL SOLAR HEATING AND COOLING INFORMATION CENTER, *P.O. Box 1607, Rockville, Md. 20850.* Established by Department of Housing and Urban Development and ERDA to help people become aware of practical applications of solar energy and to encourage public and industry use of the same. In response to letters and phone calls the Center will send appropriate responses to questions. **Geographical range:** primarily national, some international. **Contact:** Toll-Free phones: (800) 523-2929 or (in Pennsylvanis) (800) 462-4983.

PAYNE, INC., *1910 Forest Drive, Annapolis, Md. 21401.* Solar system design and optimization; consultants; flat plate solar panels designed and built and installed in swimming pool applications. **Geographical range:** continental U.S. **Contact:** Peter R. Payne; phone: (301) 268-6150.

WILLIAM D. POTTS, AIA, *Providence Church, Glenelg, Md. 21737.* Architect working on design of solar-assisted heat pump system for space heating and hot water for 17,000-square-foot multi-purpose center for Baltimore, Md. Solar greenhouses; passive systems. **Geographical range:** Baltimore/Washington D.C. area. **Contact:** William Potts; phone: (301) 465-8285.

SOLIFE, *111 Old Quarry Rd., Guilford, Conn. 06437 and 9242 Hobnail Court, Columbia, Md. 21043.* Energy conscious and solar design and construction; architects and engineers. In existence for three years. **Geographical range:** New England and the Mid-Atlantic states. **Contact:** in Connecticut—Carleton Granbery; phone: (203) 453-2449; in Maryland—Greg Mitchell; phone: (301) 997-0955.

Massachusetts

CENTER FOR ECOLOGICAL TECHNOLOGY, *P.O. Box 427, Pittsfield, Mass. 01201.* A non-profit organization funded to research, develop, and demonstrate ecologically feasible technologies. Will do consultation and design for passive and active solar projects. **Contact:** J. David Nisson; phone: (413) 447-9109.

EDWARD C. COLLINS ASSOCIATES, *Box 284, Lincoln, Mass. 01773.* Designers of energy conserving and producing buildings. Have completed two solar-assisted HVAC system condominiums. **Geographical range:** New England, New York and California, to date. **Contact:** Keith B. Gross; phone: (617) 259-0420.

ECO-ENERGETICS, *P.O. Box 7, Easthampton, Mass. 01027.* Complete engineering of solar oriented heating systems—active and passive—and domestic hot water systems. Engineering back-up to those installing solar systems on their own (DIY). Developed "Ecolector I" solar absorber kit—a do-it-yourself absorber panel. **Geographical range:** western Massachusetts and northwestern Connecticut. **Contact:** R.G. Bettini, PE or R.E. Pinkos, BSME.

ENERGY ALTERNATIVES, INC., *69 Amherst Road, Leverett, Mass. 01054.* Sellers of solar, wind, and wood energy equipment. **Contact:** Klaus Kroner; phone: (413) 549-3635.

ENERGY MARKETING ASSOCIATES, INC., *P.O. Box 488 Kenmore Station, Boston, Mass. 02215.* Marketing specialists for solar and other energy products; publishers of *The Buyer's Book of Solar Water Heaters;* marketing and merchandising services; energy shows and expositions. **Geographical range:** New England; the northeast. **Contact:** Michael Silverstein; phone: (617) 661-8118.

HOAGLAND, MACLACHLAN & CO., *8 Grove Street, Wellesley, Mass. 02181.* Management consulting firm doing industry studies, economic analyses, and forecasts of the market for solar energy products. Prepared Frost & Sullivan study on the market for solar energy products in 1975. **Geographical range:** worldwide. **Contact:** Robert H. MacLachlan, president; phone: (617) 237-5777.

MASSDESIGN ARCHITECTS AND PLANNERS, INC., *18 Brattle St., Cambridge, Mass. 02138.* Architectural, planning, and solar design services; energy conservation research and design; solar and energy conservation feasibility studies. Designed Massachusetts Audubon Society headquarters. **Geographical range:** north temperate and cold climates—U.S. and Canada; east coast from Virginia north. **Contact:** Gordon F. Tully; (617) 491-0961.

THE PEOPLE/SPACE COMPANY *49 Garden St., Boston, Mass. 92114,* Architectural design of new and retrofit solar buildings including Grassy Brook Village, Brookline, Vermont. Progressive Architectural Design Award for solar residence in 1973. **Geographical range:** USA. **Contact:** Robert Shannon or Michael Epp; phone: (617) 742-8652.

REINHARDT ASSOCIATES, INC., *1500 Main Street, Suite 2400, Springfield, Mass. 01115, and 30 Lafayette Square, Vernon Conn. 06066.* Architects, engineers, and planners. Complete building design services for new construction or renovation, including programming, siting, site and civil engineering and landscape architecture, architecture, structural, mechanical, and electrical engineering, estimating analysis of energy conservation, solar and alternative energy sources. HUD grant (with Massdesign) for retrofit in Northampton, Mass. **Geographical range:** primarily New England and New York. **Contact:** Robert P. Foley or Douglas C. Goodman; phone: (413) 781-8080.

SIPPICAN SOLAR SYSTEMS, *14 Ichabod Lane, Marion, Mass. 02738.* Solar contractors and engineers; full service work from design stage through completion of installation and start-up, including annual service contracts. Installed 2,500 square feet of collectors to date. **Geographical range:** S.E. Mass., Cape Cod, Rhode Island. **Contact:** Warren J. Mackensen; phone: (617) 748-2870.

ROBERT O. SMITH & ASSOCIATES, *55 Chester St., Newton, Mass. 02161.* Consulting engineers for the design of solar heating systems for building and water supplies. **Contact:** Robert Smith; phone: (617) 965-5428.

SOLAR DESIGN ASSOCIATES, *P.O. Box 153, Sharon, Mass. 02067.* Design services for solar architecture and energy systems engineering. Several solar residences completed. **Geographical range:** New England. **Contact:** Steven J. Strong; phone: (617) 828-7115.

SOLAR HEAT CORPORATION, *108 Summer St., Arlington, Mass. 92174.* Solar system design and engineering. Have completed five solar installations. **Contact:** Mark Hyman; phone: (617) 646-5763.

SOLAR HEATING CO. OF BOSTON, *70 Lansdowne Rd., Arlington, Mass. 02174.* Installation, sales, and consultation for solar heating systems. **Contact:** William Ravenscroft; phone: (617) 646-5127.

SOLAR RAY COMPANY, *P.O. Box 444, Fitchburg, Mass. 01420.* Solar water heating sales and manufacture. **Contact:** Joseph A. Lanciault.

SOLAR SOLUTIONS, INC., *Mill Village. Route 20, Sudbury, Mass. 01776.* Design and installation of solar climate control systems; energy efficiency analysis and heat-scanning of residences and commercial sites; sales and installation of energy-efficient devices. **Geographical range:** southern New England. **Contact:** Lew Boyd; phone: (617) 275-0111.

SOLSEARCH ARCHITECTS, *15½ Roberts Rd., Cambridge, Mass. 02138.* Energy-conscious architecture firm ready to apply energy conservative and alternative energy; design, consultation, and construction management. Designed the Prince Edward Island Ark for the New Alchemy Institute. **Geographical range:** East coast, eastern Canada. **Contact:** in U.S.: Ole Hammarlund; phone: (617) 332-5094; in Canada: David Bergmark, The Ark, Souris R.R. 4, P.E.I. Canada; phone: (902) Cardigan 181.

STEVENS ENGINEERING, *271 Washington Street, Canton, Mass. 02021.* Solar architecture and solar energy systems engineering for residential and commercial projects. Several solar residences designed. **Geographical range:** New England. **Contact:** Steven J. Strong; phone: (617) 828-7115.

WARMRAYS, *P.O. Box 896, Pittsfield, Mass. 01201.* Solar and environmental housing design and contracting; installation of hot water systems. Research in the environmental housing field. **Geographical range:** Berkshire and Hampshire counties in western Massachusetts. **Contact:** (413) 445-4858.

Michigan

FAIRBROTHER AND GUNTHER, INC., *325 Fuller Ave., N.E., Grand Rapids, Mich. 49503.* Consulting mechanical and electrical engineers doing feasibility and design for solar systems for space heating and process water, for commercial, institutional, and industrial buildings. Four years solar involvement. **Geographical range:** Michigan, Indiana and Ohio. **Contact:** Kenneth E. Gunther; phone: (616) 451-8476.

PRESTON SOLAR DESIGNS, *760 Gibbs Drive, Ithaca, Mich. 48847.* Specialize in do-it-yourself solar installations for residences. Have built solar house. **Geographical range:** Michigan. **Contact:** Lawrence P. Brown; phone: (517) 875-4008.

Minnesota

ALTERNATIVE SOURCES OF EN—ERGY, INC., *Route 2, Milaca, Minn. 56353.* An energy information firm and publishers of *Alternative Sources of Energy Magazine,* a bimonthly focusing on discussion of new developments in the field of appropriate technology and energy conservation; has been published for six years. **Geographical range:** United States and 45 foreign countries. **Contact:** Donald L. Marier; phone: (612) 983-6892.

I E ASSOCIATES, *5 Woodcrest Dr., Burnsville, Minn. 55337.* Consulting firm specializing in solar heating/cooling, bioconversion technologies, and energy management and conservation strategies. Past development work has included greenhouse and hydroponic designs that have included solar. **Contact:** Tom P. Abeles; phone: (612) 890-3480.

NORTHERN SOLAR POWER CO., *311 S. Elm St., Moorhead, Minn. 56560.* Consulting engineers for both active and passive solar systems for new or existing buildings. Constructed a low-cost forced-air retrofit solar system now being tested. **Geographical range:** North Dakota-Minnesota border area. **Contact:** Bruce Hilde; phone: (218) 233-2515.

SCIENTIFIC SUNSOURCE SYSTEMS *5 Woodcrest Drive, Burnsville, Minn. 55337.* Architectural and engineering services in the area of solar and energy conservation. Hold and license patents in the area of "black liquid" solar collectors and solar electrogenerative systems. **Geographical range:** U.S. and Canada. **Contact:** Tom P. Abeles; phone: (612) 890-3480.

Missouri

HEBENSTREIT CO., *P.O. Box 10148, Kansas City, Mo. 64111.* Agents for manufacturers of solar equipment. Water, pool, and space heating equipment. Thirty years in water heater sales and service. **Geographical range:** Kansas and Western Missouri. **Contact:** Bill Hebenstreit; phone: (816) 561-4661.

ARTHUR HALL PEDERSON, *532 Lake Ave., Webster Groves, Mo. 63119.* Consulting engineer in solar system design. **Contact:** Arthur Pederson; phone: (314) 962-4176.

Mississippi

READY SOLARISTICS, *4566 Office Park Dr., Jackson, Miss. 39206.* Full service solar center; system design, architectural lanning and design, project analysis for central solar power systems. **Geographical range:** Mississippi. **Phone:** (601) 982-4258.

Montana

L. CLARK MACDONALD, *Bootlegger Trail, Great Falls, Mont. 59401.* Dealer and installer of heat pumps. Currently working under state grant for solar domestic hot water. **Geographical range:** north-central Montana. **Contact:** L. Clark MacDonald; phone: (406) 452-5967.

Nebraska

THE CLARK ENERSEN PARTNERS, *1515 Sharp Building, Lincoln, Neb.68508.* Architects and engineers will design solar heating, domestic hot water and cooling systems. Have designed ten such systems to date. **Geographical range:** mainly Nebraska, but will accept projects elsewhere. **Contact:** Charles L. Thomsen; phone: (402) 477-9291.

GOLLEHON, SCHEMMER & ASSO-CIATES, *12100 West Center Rd., Suite 520, Omaha, Neb. 68144.* Architects, engineers, and planners in the solar energy field. Currently working on a solar-assisted heat pump system for the Omaha Housing Authority Child Care/Recreation Center. **Contact:** Roger Wozny; phone: (402) 333-4800.

IONIC SOLAR, *4630 Dodge St., Omaha, Neb. 68132.* Solar system design and installation; hardware supply; lectures and feasibility studies. **Geographical range:** Nebraska, Iowa, eastern Kansas, western Missouri. **Contact:** Garry D. Harley, AIA; phone: (402) 551-1441.

New Hampshire

CLEAN ENERGY SYSTEMS, *63 Maple Avenue, Keene, N.H. 03431.* Registered mechanical engineer will provide advice on applications and installation of solar systems. **Geographical range:** New Hampshire. **Contact:** H.Hamilton Chase; phone: (603) 352-0083.

COMMUNITY BUILDERS, *Canterbury, N.H. 03224.* Builders contracters with four solar houses completed in 1975-6. **Geographical range:** central New Hampshire. **Contact:** Don Booth; phone: (603) 783-4743

DONAVAN & BLISS, *Chocorua, N.H. 03817.* Professional engineering design firm offering services for the design of buildings and associated heating (or cooling) systems, where purchased energy requirements for heating or cooling are to be kept unusually low. **Contact:** Raymond W. Bliss.

EVOG CONSTRUCTION CO., INC., *P.O. Box 36, Hebron, N.H. 03241.* Builder and designer of solar homes and solar systems. **Geographical range:** central New Hampshire. **Contact:** Richard Holt; ne: (603) 744-8918.

FLETCHER/MYERS CONTRACTING INC., *P.O. Box 107, East Derry, N.H. 03041.* Designers and builders of solar houses and other small industrial buildings.

HUD grant recipient. **Geographical range:** southern New Hampshire, northern Massachusetts. **Contact:** Charles Myers; phone: (603) 434-0078.

ROSETTA STONE ASSOCIATES, INC., *142 Main St., Nashua, N.H. 03060.* Scientific and technical translations in all major languages of the world. **Geographical range:** unlimited. **Contact:** John Furey; phone: (603) 882-1760.

RUSSELL G. PAGE, *1 Ridgewood Drive, Concord, N.H. 03301.* HVAC and solar designer. **Geographical range:** Maine, New Hampshire, and Vermont. **Contact:** Russell Page; phone: (603) 228-0111.

TOTAL ENVIRONMENTAL ACTION, INC., (TEA), *Church Hill, Harrisville, N.H. 03450.* Passive and active solar and energy-conserving building design; wind systems; monthly workshops on alternative energy and living; special architectural seminars. Many solar designs and studies completed. **Geographical range:** worldwide. **Contact:** Hilda Wetherbee; phone: (603) 827-3374.

New Jersey

AMERICAN SOLARIZE INC., *19 Vandeventer Ave., Princeton, N.J. 08540.* Consultants and designers for air systems for residential, commercial, and industrial application. Have designed and built solar water stills in Petersburg, South Africa. **Geographical range:** open, depending on type and size of project. **Contact:** Joe Beudis; phone: (609) 924-5645.

CHARLES B. ANTHONY, INC., *97 Palisades Ave., Bogota, N.J. 07603.* General contractors and builders of solar homes and solar retrofitting. **Contact:** Gerald Anthony; phone: (201) 489-5199.

DOUG KELBAUGH, *70 Pine Street, Princeton, N.J. 08540.* Architect and design consultant to solar projects. Several solar houses, industrial and institutional buildings designed; special experience in passive systems. **Geographical range:** Northeast U.S. **Contact:** Doug Kelbaugh; phone: (609) 924-2703.

MEASE ENGINEERING ASSOCIATES, *Box 322, Netcong, N.J. 07857.* Engineering company working in alternative energy: wind, solar, methane, and water power; energy-efficient house design; presently working with several clients designing passive solar houses. **Geographical range:** unlimited. **Contact:** Michael J. Mease; phone: (201) 347-3769.

PRINCETON ENERGY GROUP, *245 Nassau St., Princeton, N.J. 08540.* Architectural and engineering consultation for all

kinds of alternative energy. **Geographical range:** New Jersey, Philadelphia to N.Y.C. **Contact:** Lawrence L. Lindsay; phone: (609) 924-7639.

SIGMA ENGINEERING, *P.O. Box 6, Oldwick, N.J. 08858.* Engineering and design of mechanical and electrical systems for buildings; founded in 1971. **Geographical range:** New Jersey, eastern Pennsylvania, New York. **Contact:** John R. Brady; phone: (201) 439-2979.

SOLAR-EN, *P.O. Box 64, Denville, N.J. 07834.* Sales and installation for swimming pool and hot water systems. **Geographical range:** New Jersey. **Contact:** Martin Spiegel; phone: (201) 361-2300.

SOLENCO CORP., *175 River Road, Flanders, N.J. 07836.* Engineers the design of solar collectors and systems for residential heating. Consulting to contractors and individuals. Pool heating installations. **Geographical range:** northwest New Jersey. **Contact:** Henry A. Pontious; phone: (201) 584-5055.

SOLAR ENERGY SYSTEMS OF NEW JERSEY, *134 Ocean Ave., Monmouth Beach, N.J. 07750.* Installers of domestic hot water systems and swimming pool heaters. Have developed and will build custom solar greenhouses. **Geographical range:** not given. **Contact:** Carl J. Reinertsen.

New Mexico

NEW MEXICO SOLAR ENERGY ASSOCIATION, *P.O. Box 2004, Santa Fe, N.M. 87501.* An information dissemination center: publishes a monthly bulletin and quarterly journal; holds workshops, conferences and seminars;library and bookstore; evaluates plans and sketches; office staff maintained since early 1976. **Geographical range:** 1,100 members in forty-two states and nine foreign countries. **Contact:** Keith Haggard; phone: (505) 983-2861.

SIGMA ENERGY PRODUCTS, *4100 Edith Bvld. N.E., Albuquerque, N.M. 87107.* Designers and installers of domestic hot water, swimming pool, and space heating systems. **Contact:** Don Davis, marketing manager; phone: (505) 345-8561.

S.W. ENERGY OPTIONS, *Rt. 8, Box 30-H, Silver City, N.M. 88061.* General consultation and installation, design and assistance for solar heating, wind conversion, and methane generation. **Geographical range:** Southwest, New Mexico. **Contact:** Bill Carlis; phone: (505) 538-9598.

SUN SYSTEMS, DIV. OF YUCCA BUILDERS OF SANTE FE, *528 Franklin Ave., Santa Fe, N.M. 87501.* Engineering, installation, and sales of solar systems. Background in general electrical and mechanical engineering. **Geographical range** northern New Mexico to Colorado. **Contact:** Don Gallard; phone: (505) 988-1844.

TECHNOLOGY APPLICATIONS CENTER (TAC), *University of New Mexico, 2500 Central Ave. S.E., Albuquerque, N.M. 87131.* 1. Custom tailored information searching and retrieval, 2. publication of energy bibliographies, 3. dissemination of NASA technology, 4. remote sensing. Have access to ninety computerized information bases to compile published energy bibliographies (solar, hydrogen, heat pipe, remote sensing) and to meet individual information needs for organizations and individuals. Ten years in the field. **Geographical range:** national. **Contact:** Jay Ven Eman or Marjorie Hlava; phone: (505) 277-3622.

WEST WIND CO., *Box 1465, Farmington, N.M. 87401.* Design, consultants, and manufacturers of electronic equipment, for solar and wind energy applications. Active and passive solar design consultants. **Geographical range:** unlimited. **Contact:** Geoffrey Gerhard; phone: (505) 325-4949.

New York

ABC SOLAR SYSTEMS, INC., *329 Central Ave., Albany, N.Y. 12206.* Consulting services in solar space, hot water, and swimming pool heating. Insulation for energy conserving buildings. **Geographical range:** Northeast and Central Atlantic States. **Contact:** Paul Klinger; phone: (518)465-5011.

AB.P SOLARTECH, *330 West 45th St., New York, N.Y. 10036.* Solar consultant for stages from planning through installation. In business three years; several small installations done. **Geographical range:** Quebec to Key West. **Contact:** Bruce Polak; phone: (212)JU2-4240.

A.C.M. INDUSTRIES, INC., *Box 185, Clifton Park, N.Y. 12065.* Design and manufacture components to erect energy-efficient residential, commercial, institutional and industrial buildings. **Geographical range:** northeast, including Canada, Ohio and Indiana. **Contact:** George Keleshian; phone: (518)371-2140

ADIRONDACK ALTERNATE ENERGY, *Div. of Brownell Lumber Co., Edinburg, N.Y. 12134.* Builders and designers of a low energy requirement house. **Contact:** Bruce Brownell; phone: (518) 863-4338.

ASTRAL SOLAR CORP., *448 East 87th Street, New York, N.Y. 10028.* Professional engineers doing design, consultation, research and development. Four years in the solar field. **Geographical range:** global. **Contact:** Tom Grayson; phone: (212) 722-3746.

CENTER FOR ENERGY POLICY AND RESEARCH OF THE NEW YORK INSTITUTE OF TECHNOLOGY, *Old Westbury, New York 11568.* Center provides energy hot line, energy information and referral service, energy management seminar series, and color television program, *Solar Energy Today.* **Geographical range:** adjacent areas of New York, New Jersey, and Connecticut. **Contact:** NYIT Hot LIne; phone: (516) 686-7744.

CUSTOM SOLAR HEATING SYSTEMS, *P.O. Box 375, Albany, N.Y. 12201.* Design and installation of solar systems for residential and commercial buildings; design of heat storage systems; specialize in use of air systems. **Geographical range:** 200-mile radius of Albany, N.Y. **Contact:** Donald Porter; phone: (518) 438-7358.

DAS/SOLAR SYSTEMS, *201 Sixth Ave., Brooklyn, N.Y. 11217.* Design, engineering, and installation of solar systems for domestic hot water, heating, and cooling. **Geographical range:** northeastern United States. **Contact:** William Ross; phone: (212) 636-1471.

G. RAYMOND DE RIS, *Hawthorne Valley, Ghent, N.Y. 12075.* Design and planning for passive solar residences; completed design of own house. **Geographical range:** New York, Massachusetts, Vermont. **Contact:** G. Raymond de Ris.

HAROLD A. DENKERS, *102 South Allen St., Albany, N.Y. 12208.* Engineering, architecture, and solar design for new and existing residential and commercial structures. Have designed solar retrofit office building. **Geographical range:** New York and New England. **Contact:** Harold A. Denkers; phone: (518) 438-8183.

ENERGY PLANNING INC., *P.O Box 387, 100 Bluff Drive, East Rochester, N.Y. 14445.* Designers and construction managers for industrial energy management systems, energy conservation programming, air conditioning, industrial ventilation. "Many years experience in field." **Geographical range:** greater Rochester area. **Contact:** Fred J. Belluscio.P.E., vice president; phone: (716) 385-4280.

FLACK AND KURTZ, *29 West 38th St., New York, N.Y. 10018.* Consulting engineers and planners. Energy resource analysis and planning, computerized solar evaluation studies, energy management and

implementation, mechanical/electrical design. Solar design: Madeira School, Greenway, Va., and the Princeton Education Center, Blairstown, N.J. **Geographical range:** worldwide. **Contact:** George Rainer, P.E., A.I.P., Director, Energy Resource & Planning Div.: phone: (212) 354-8240.

ENERGY PLANNING INC., *P.O. Box 387, 100 Bluff Drive, East Rochester, N.Y. 14445.* Designers and construction managers for industrial energy management systems, energy conservation programming, air conditioning, industrial ventilation. "Many years experience in field." **Geographical range:** greater Rochester area. **Contact:** Fred J. Belluscio. P.E., vice president; phone: (716)385-4280.

NORTH COUNTRY ENGINEERS, *Sandy Creek, N.Y. 12145.* Consulting engineers and design service for heat recovery systems and alternate energy systems: solar, wind, methane. Three years studying solar—designing twenty tract houses for solar. **Geographical range:** Northeast. **Contact:** Daniel E. French, Robert F. French; phone: (315) 387-3411.

OTTAVIANO TECHNICAL SERVICES INC., *150 Broad Hollow Rd., Melville, N.Y. 11746.* Two-day solar energy seminars held in major cities throughout the U.S. Costs, payback, case histories, new products, applications, and procedures for the contractor, engineer, and owner in industrial, municipal, and residential domestic heating and air conditioning. **Geographical range:** nationwide. **Contact:** Catherine Doty; phone: (516) 271-1911.

PATTERSON ENERGY GROUP, *185 Magnolia Ave., Floral Park, N.Y. 11001.* Sales and installation for solar space and domestic hot water heating. **Geographical range:** concentrating on Long Island and Metropolitan area. **Contact:** James B. Patterson; phone: (516) 354-2160.

EDWARD PEDERSON, *109 Haffenden Rd., Syracuse, N.Y. 13210.* Architecture and planning firm engaged in alternate energy system integrated design. **Geographical range:** New York, Pennsylvania, New Jersey. **Contact:** Edward Pederson; phone: (315) 472-5016.

ROCKLAND SOLAR ENERGY CO., *6 Beaver Hollow Lane, Monsey, N.Y. 10952.* Solar consultant to heating/cooling contractors, builders, and government agencies. Several installations completed. **Geographical range:** Rockland and Orange Counties in N.Y., Bergen County in N.J. **Contact:** Ronald Cataldo; phone: (914) 352-5408.

M.P.S. ROE, INC., *644 Coffeen Street, Watertown, N.Y. 13601.* Heating contractors familiar with design and construction of passive and active solar installations, and with the design of energy-efficient housing and the integration of various energy forms (solar, wood, gas, oil, electric). **Geographical range:** northern New York. **Contact:** Larry Honeywell; phone: (315) 782-7180.

I. SHIFFMAN, P.E., *529 Central Ave., Scarsdale, N.Y. 10583.* Consulting mechanical/electrical engineers working with alternate energy systems and energy conservation for industrial, commercial, institutional, and residential facilities. **Contact:** I. Shiffman; phone: (914) 725-1750.

SOLAR CONNECTION, *37-12 Broadway, Astoria, N.Y. 11103.* Solar store providing products and services and books for and on solar energy. **Contact:** John J. Gill; phone: (212) 274-7000.

SOLAR STRUCTURES, INC., *7 Sundance Rd., LaGrangeville, N.Y. 12540.* Consulting, supplying, and installation of solar systems and buildings. Three years of experience. **Geographical range:** 100 mile radius of Poughkeepsie, N.Y. **Contact:** Harry Wenning, AIA; phone: (914) 223-3929.

PAUL M. STURGES, *P.O. Box 397, Stone Ridge, N.Y. 12484.* Energy consultant specializing in domestic and industrial heat recovery systems. Twenty years of innovation. **Contact:** Paul Sturges; phone: (9140 687-0281.

SYSKA & HENNESEY, *110 West 50th Street, New York, N.Y. 10020; 1720 Eye Street, N.W. Washington, D.C.; 1900 Ave. of the Stars, Century City, Los Angeles, Cal. 90067.* Consulting mechanical and electrical engineers with capability to provide solar energy studies and designs. In-house computer facility maintains design and energy analysis programs, including one that predicts solar collector performance based on hourly measured solar radiation and weather data. Solar energy systems designs include the Eastern Liberty Savings and Loan Assn., Washington, D.C. **Geographical range:** worldwide. **Contact:** in New York—Joseph Manfredi, phone: (212) 489-9200; in Washington—Robert Manfredi, phone: (202) 296-8940; in Los Angeles—Robert Logan, phone: (213) 553-5550.

T-K SOLAR DISTRIBUTORS, LTD., *103 Plandome Road, Manhasset, N.Y. 11030.* Solar consultants and marketing programs for solar equipment including sales, installation, and service. Marketing files on 324 companies in the solar industry. **Geographical range:** not given. **Phone:** (516) 482-1252.

MARK URBAETIS, *R.D.#2, Box 250, Rexford, N.Y. 12148.* Design and installation of pool heating systems, domestic hot water systems, and testing of collectors. Three years with large plastics firm as solar project coordinator; two years with own firm as designer and installer. **Geographical range:** northeast, especially eastern N.Y. and western Vermont and Massachusetts. **Contact:** Mark Urbaetis; phone: (518) 371-9596.

ALEX WADE, *Box 43, Barrytown, N.Y. 12507.* Architect will design energy saving and passive solar heating houses. Co-author of *Low-Cost, Energy-Efficient Shelter.* Designed many small, efficient, passive solar houses. **Geographical range:** northeastern U.S. **Contact:** Alex Wade; phone: (914) 758-5554.

WENNING ASSOCIATES, *623 Warburton Ave., Hastings on Hudson, N.Y. 10706.* Designers of systems and buildings. Have done both passive and active systems for past several years, and have completed solar house under HUD Demonstration Program. **Contact:** Harry Wenning, AIA; phone: (914) 478-4110.

North Carolina

BEECH MOUNTAIN SOLAR ENER—GY CO., *Box 262, Banner Elk, N.C. 28604.* Manufacture and distribute solar energy panels for hot water heaters and space heating; engineering and installation. **Geographical range:** North Carolina, Virginia and Tennessee. **Contact:** Edwin L. Lotz; phone: (704) 387-2375.

CAROLINA SOLAR EQUIPMENT, *P.O. Box 2068, Salisbury,N.C. 28144.* Design engineers and solar consultants; installers specializing in Thomason systems. Three years in the field. **Geographical range:** not given. **Contact:** Dan Fisher; phone: (704) 857-3383.

COLLABORATIVE INTERFACE DESIGNS, INC. *129 W. Trade Street, Charlotte, N.C. 28202.* Complete architectural/engineering services and solar consulting; research and development. Commercial, residential and educational solar applications. **Geographical range:** southeast U.S. **Contact:** Thomas L. Ainscough; phone: (704) 372-3984.

FWA ENGINEERS, INC. *5672 International Drive, Charlotte, N.C. 28211.* Engineering consultants for building system design and analysis; residential, commercial, institutional and industrial computer design and analysis of existing and new structures; computer energy studies; cost estimating; solar system design.ERDA grant to solar heat 60,000 square foot hospital. **Geographical range:** primarily southeast Unites States. **Contact:** Louis L. Abernathy or Edgar C. Jones; phone: (704) 364-8220.

INTEGRATED ENERGY SYSTEMS, *108½ Henderson Street, Chapel Hill, N.C. 27514.* Energy consulting firm: energy studies, feasibility studies, site analyses, bid documents, field supervision of installations, design and construction of solar systems, lectures, and preparation of federal grant applications. Five commercial and fifteen residential solar projects completed. **Geographical range:** the southeast and the west coast. **Contact:** Daniel R. Koenigshofer; phone: (919) 942-2007.

LORIEN HOUSE *P.O. Box 1112, Black Mountain, N.C. 28711* Have published two books dealing with solar. *Practical Sun Power* and *The Solar Energy Notebook,* each 8 by 10 inches. 56 pages. $4.00.

NORTH CAROLINA SCIENCE AND TECHNOLOGY RESEARCH CENTER. *P.O.Box 12235, Research Triangle Park, N.C. 27709.* Technical literature and information on energy, conservation, solar, and other technical disciplines. Access to NASA. Specialized bibliographies; five years special emphasis on energy and solar energy. **Geographical range:** southeastern United States. **Contact:** C. Leon Neal; phone: (919) 549-0671 or for eastern U.S. outside of North Carolina: (800) 334-8561.

NORTH CAROLINA SOLAR DE—VICES,INC., *Route 1, Box 328, Edenton, N.C. 27932.* Sales, installation, and engineering service for domestic and commercial hot water, residential and commercial space heating, and swimming pool heat. **Contact:** Robert S. Harrell; phone: (919) 482-8833.

WILLIAM H. RANKINS III, *P.O. Box 146, Swannanoa, N.C. 28778.* Consultant for solar space heating, water heating, industrial applications, agricultural. Nine years experience as mechanical engineer; 4 years as solar engineer. Geographical range: "anywhere the sun shines." **Contact:** Wiliam H. Rankins III; phone: (704) 274-0814.

SOLAR DEVELOPMENT CO., INC., *204 Chatterson Drive,Raleigh,N.C.27609.* Solar thermal design, manufacturing, and consulting by professional engineers; twenty-five years in field of power generation. heating, and building. **Geographical range:**no limitation.**Contact:**Gordon R. Winders; (919) 782-8275.

SOLAR, P.I.E., *P.O. Box 506, Columbus, N.C. 28722.* Solar products, information, and engineering; energy conservation. **Geographical range:** western North Carolina. **Contact:** Richard Pratt; phone: (704) 894-3411.

SOLAR PRODUCTS AND DEALER SERVICES, INC., *4505 Franklin Ave., Wilmington, N.C. 28401.* Wholesale solar energy products and energy conservation devices; retain engineer and architect for individual projects. **Geographical range:** North and South Carolina. **Contact:** Doyle E. Murray; phone: (919) 799-8397.

SOLAR TECHNOLOGY, *119 North Center St., Statesville, N.C. 28677.* Design, testing, evaluation of various air and liquid solar collectors. Designed and built solar house in April, 1976. **Geographical range:** 1,000 miles. **Contact:** Nelson E. Brown; phone: (704) 873-7959.

SUNSPACE, INC., *Box 71A, Rt. 5, Burnsville, N.C. 28714.* Passive energy/climatically responsible construction firm; design, construction, and educational services. **Geographical range:** Southeast U.S.—educational: NOrth Carolina—construction. **Contact:**(704) 675-5286.

SUNSPOT SOLAR PRODUCTS, INC., *146 East Main Street, Carrboro, N.C. 27510.* Design, sales, and installation of solar equipment; wholesalers and retailers of components for solar water, space, and swimming pool heating systems. Have installed six domestic water heating systems in first six months of operation. **Geographical range:** Research Triangle Area in the eastern Piedmont of North Carolina. **Contact:**John R.Meeker; phone: (919) 929-6896..

SUNSHELTER DESIGN, *13 Maiden Lane, Raleigh, N.C. 27607.* Design/consultation for environmentally sound architecture; emphasize use of passive systems in regional design applications. Have designed passive "Stony Creek I" house in Rocky Mount, N.C. **Geographical range:** eastern U.S. 32°N Lat to 40°N Lat. **Contact:** John M. Meachem; phone: (919) 755-0700.

Ohio

THE A. BENTLEY & SONS CO., *P.O. Box 956, 201 Belmont Ave., Toledo, Ohio 43695.* Engineers and contractors interested in contracting for the installation of solar energy systems.**Geographical range:** 150 mile radius of Toledo. **Contact:** John Hilton; phone: (419) 244-5561.

JERRY BERGMAN, PH.D., *Dept.EDFI, Bowling Green State University, Bowling Green, Ohio 43403.* Writer-researcher in the solar power area; interested in researching public attitudes on solar, atomic, and other forms of energy. **Geographical range:** unlimited. **Contact:**Jerry Bergman; phone: (419) 372-0151.

ENVIRONMENTAL DESIGN ALTERNATIVES, *2011 Rose Wood Dr., Kent, Ohio 44240.* Full architectural services for shelters that act as energy effective spaces. Energy conservative and solar structures have been completed. **Geographical range:** Ohio and western Penn. **Contact:** Douglas G. Fuller.

KFC SYSTEMS ANALYSIS, *1533 Eastgate, Toledo, Ohio 43614.* Consultants in engineering development, marketing, and energy conservation. **Contact:** Ken Cherry.

SHELDON LAZAN, *8260 Kingsmere Court, Cincinnati, Ohio 45231.* Consulting engineer to solar energy projects. Has designed two solar-heated residences, a solar-heated manufacturing plant, and a solar flat-plate collector for Solar Sun, Inc. **Geographical range:** southwest Ohio. **Contact:** Sheldon Lazan; phones: (513) 733-4300, home: 522-5709.

FULLER MOORE, ARCHITECT, *7348 Buck Paxton Road, College Corner, Ohio 45003.* Architectural design, specialization in solar; consultant to architects on both passive and active solar building design. **Geographical range:** full architectural services—Ohio/Indiana region; consulting—anywhere. **Contact:** Fuller Moore; phone: (513) 796-3683.

SOLARGON ASSOCIATES, *R.R.1, 12-359 Bailey Rd., Grand Rapids, Ohio 43522.* Consulting, engineering, design, and instruction in solar pool heating, space and hot water heating, and total energy systems for domestic and commercial applications. **Contact:** Richard A. Kujawski.

SOLAR HOME SYSTEMS, INC., *12931 West Geauga Trail, Chesterland, Ohio 44026.* Architectural engineering with specialization in solar energy applications for residential, commercial, and institutional buildings. Collector design, material selection, and system simulation. Five years in solar research plus work on thirteen houses. **Geographical range:** Ohio—will consult over phone or travel if expenses are paid. **Contact:** Joseph Barbish; phone: (216) 729-9350 or 289-7020.

Pennsylvania

BALLINGER, *841 Chestnut Street, Philadelphia, Pa. 19107.* Architecture, engineering, feasibility analysis and design, integrated and retrofitted solar heating systems, commercial, industrial and institutional buildings. Have participated in three major NSF/ERDA/NASA projects. **Geographical range:** Eastern United States. **Contact:** Robert H. Rand, AIA; phone: (215) 629-0900.

ROBERT BENNETT, *6 Snowden Rd., Bala Cynwyd, Pa. 19004.* Solar engineering. **Contact:** Robert Bennett; phone: (215) MI6-1776.

BENNET SUN ANGLE CHARTS
6 Snowden Rd., Bala Cynwyd, Pa. 19004 Each chart shows position of sun at any time of day during year at a designated latitude, number of hours of available sunlight for any day. The charts simulate real motion of sun across the sky. Fifteen are available—one for each even-numbered latitude from 24°to 52°north. The 17 x 22 charts are rolled and shipped in a mailing tube. The complete set of fifteen is $30.00 ppd.; individual charts are $2.50 ppd. **Contact:** Robert Thomas Bennett, P.E.; phone: (215) MO7-7365.

ECO-ENERGY ASSOCIATES, *342 Franklin Street, Alburtis, Lehigh County, Pa.18011.* Design and installation service for solar energy and windpower equipment; includes design and specification of solar heating systems, heat loss studies of buildings, and wndpower site analysis. Have conducted one-day intensive solar workshops. **Geographical range:** eastern Pennsylvania. **Contact:** Carl Zipper; phone: (215) 435-1899.

STANLEY F. GILMAN, *505 Cricklewood Drive, State College, Pa. 16801.* Engineering consultant on the design of solar energy systems for the heating and cooling of buildings. Principal investigator of NSF/ERDA sponsored research project on solar energy assisted heat pump system since April 1974. **Geographical range:** no limitations. **Contact:** Stan Gilman; phone: (814) 234-0510.

PETER HOLLANDER, *301 N. 3rd. St., Philadelphia, Penn. 19106.*Solar photographer offers slide shows and individual photos; introductory set covers methods of collection and storage and sample structures across the country. Photos are late 1976 and have appeared in 13th edition of **Solar Heated Buildings, A Brief Survey,** by William Shurcliff. **Geographical range:** slides cover all of U.S.; sets mailed anywhere. **Contact:** Peter Hollander;phone (215) 925-5086.

ORGANICS UNLIMITED, *R.D.3,Boyertown, Penn. 19512.* Solar design consultation with contractors, architects, and

home-owners; fabrication of components for air systems. New in solar field: two systems for space heating and hot water under construction. **Geographical range:** design consultation—eastern Pennsylvania; fabrication service—components can be shipped anywhere. **Contact:** Robert Moser.

SHELTER DESIGN-BUILD, *Box 161A R.D.1, New Tripoli, Pa. 18066.* Design and construction services for arcitects, owners, and builders; primarily residential work using energy conservation techniques, passive and active solar heat, wood heat, and Clivus Multrum waste recovery systems. **Geographical range:** eastern Pennsylvania. **Contact:** Mike Ondra; phone: (215) 756-6112.

SOLAR, *129 Biddle Rd., Paoli, Pa. 19301.* Solar building and retrofitting. **Contact:** Donald K. Hull; phone: (215) NI4-8132.

SOLAR ENERGY ASSOCIATES, *1063 New Jersey Ave., Hellertown, Pa. 18055.* Engineering and installation of energy-efficient systems for residential, commercial, and agricultural applications for new and existing buildings. Solar home, hot water and pool heating systems; solar assisted heat pumps and solar absorption air conditioning; other energy alternatives. **Contact:** Albert Baker; phone: (215) 838-7460.

LAWRENCE G. SPIELVOGEL, INC. *Wyncote House, Wyncote, Pa. 19095.* Consultant on energy management in buildings: studies, analysis, audits, evaluations. Eighteen years experience in energy studies for buildings, designer of solar buildings for both heating and cooling, consultant on NSF, ERDA, NASA and HUD solar studies. Geographical range: registered engineer in 48 states. **Contact:** Lawrence G. Spielvogel; phone: (215) 887-5600.

SUN-WIND CO. OF PENNSYLVANIA, INC., *P.O. Box 16, Fairview Village, Pa. 19409.* Consulting architects and engineers to homeowners, builders, and developers, of solar, wind and wood energy systems. **Geographical range:** Delaware Valley Area of Penn. and N.J. **Contact:** Herman DeJong; phone: (215) 272-7392.

WHITE'S HORTICULTURAL CONSULTING SERVICE, *1311 Curtin Street, State College, Pa. 16801.* Professional horticulturist involved in solving heat-loss problems for greenhouses and in designing solar heated greenhouses. Will retrofit energy conservation systems to existing greenhouses. **Geographical range:** United States. **Contact:** John W. White.

CARL J. WILCOX ASSOCIATES, *2300 Marlborough Dr., York, Pa. 17403.* Consulting on wind turbines. Worked with the Smith-Putnam Wind Turbine Project. **Geographical range:** no limit. **Contact:** Carl J. Wilcox; phone: (717) 741-1866.

WOLFE AND GREENBERG, DESIGNERS, *3316 Arch St., Philadelphia, Pa. 19104.* Designing and consulting for passive and active solar energy systems, energy conservation, and wood burning back-up systems. Have worked on several solar projects. **Geographical range:** Delaware Valley, Tri-State Region. **Contact:** Martin B. Greenburg; phone: (215) 382-9474.

A.A. ZACHOW, *112 Locust Grove Road, Rosemont, Pa. 19010.* Consultant will assist home owner by teaching home energy conservation and do-it-yourself maintenance through classes and by surveys; will also analyze and design conservation methods for heating and cooling private homes and commercial applications. Consulting rates: $20 per hour; $150 per day; $600 per week. **Contact:** A.A. Zachow; phone: (215) LA5-7038.

Rhode Island

CRANSTON COMMUNITY ACTION PROGRAM COMMITTEE, INC., *30 Rolfe St., Cranston, R.I. 02910.* Have been funded to train unemployed persons in solar technology. The agency is willing to be a contractor for other agencies interested in undertaking similar programs. Have constructed a system and installed a number of manufactured systems. **Geographical range:** primarily Rhode Island. **Contact:** Daniel S. Waintroob; phone: (401) 467-9610.

South Carolina

ENERGY DESIGNS/ARCHITECTS, *201 Woodrow Street, Columbia, S.C. 29205.* Architectural design with emphasis on energy conservation and passive and active solar designs. Have designed a solar office building in Chester, South Carolina. **Geographical range:** S.C., N.C., Georgia. **Contact:** Dick Lamar; phone: (803) 799-7495.

SOLAR SOURCES CORP., *Bankers Trust Office Park, 6 Pope Ave., Hilton Head Island, S.C. 29928.* Information and marketing services for solar energy industry in Southeast. Slide shows, filmstrips, films, workshops, public meetings, technical writing and editing, brochures, editorial consulting, marketing stategies, public relations and working with media. **Geographical range:** southeastern U.S. **Contact:** Ren Frutkin; (803) 785-8819.

Tennessee

ENERGY CONVERTERS, INC., *2501 N. Orchard Knob, Chattanooga, Tenn. 37406.* Engineering and technical aid for residential and commercial installations; design service for special installations not employing modular components; computer analysis of solar energy collection for particular installation orientations. **Geographical range:** nationwide. **Contact:** James Lents or David Burrows; phone: (615) 624-2608.

Texas

THE ENERGY MIZERS, *2210 Hancock Drive, Austin, Tex. 78756.* Recommendation, sales, installation, and service of solar and energy saving devices for houses and business; physical survey of property—site analysis. Three years of R & D on program to help conserve energy, reduce costs, and improve living. **Geographical range:** Texas, south, southwest. **Contact:** Charles M. Burke; phone: (512) 458-5333.

ENERSOL COMPANY, *Suite 1800, First International Building, Dallas, Tex. 75270.* Wholesale equipment distributors for solar space and hot water heating, with offices in Flagstaff, Arizona; Albuquerque, New Mexico, and Dallas. **Contact:** J.O. Carnes; phone: (214) 748-8511.

MCDERMOTT HUDSON ENGINEERING, DIV. OF HUDSON ENGINEERING CORP., *5900 Hillcroft Avenue, P.O. Box 36100, Houston, Tex. 77036.* Architectural, engineering, and construction firm: engineering design, feasibility studies, architectural design, and project management services. Have done one of the southwest's first solar-heated commercial office buildings. **Geographical range:** worldwide. **Contact:** DR. D.G. Elliot; phone: (713) 782-4400.

SOLAR RESEARCH, INC., *5011 Duval St., Austin, Tex. 78751.* A consulting design firm for solar heating, cooling, hot water, and solar-assisted heat pumps; also prepare ERDA and HUD grant proposals and offer computer design aids to consulting engineers and architects. **Geographical range:** U.S. **Contact:** George Smith; phone: (512) 459-7686.

GEORGE E. WAY, *1448 Heights Blvd., Houston, Tex. 77008.* Energy conscious design and consulting services for active and passive solar applications. Three years design experience. **Geographical range:** Texas, Gulf Coast. **Contact:** George E. Way; phone: (713) 862-3393.

Vermont

MICHAEL BEATTIE, *13 Center St., Rutland, Vt. 05701.* Integral solar heated building design; low-cost, simple solutions. Several years' work with construction firms; five years' design experience. **Geographical range:** Vermont. **Contact:** Michael Beattie.

GENTLE ENERGY SYSTEMS, INC., *Starksboro, Vt. 05487.* Consultation, design, and construction of solar and/or wood heating systems for residential and small commercial applications; wind analyses for power potential. Plans to market a water heating unit adaptable to most wood heaters. **Geographical range:** Vermont. **Contact:** Frederic Lowen; phone: (802) 453-3546.

JOHNSON & DIX FUEL COMPANY, *25 Depot Ave., Windsor, Vt. 05089.* Solar contractors and engineers, full service work from design stage through completion of installation and start-up, including annual service contracts. Installed 500 square feet of collectors to date; long-established heating firm. **Geographical range:** Vermont and New Hampshire. **Contact:** Roger Maynard; phone: (802) 674-5505.

SOLAR ALTERNATIVES INC., *30 Clark St., Brattleboro, Vt. 05301.* Complete engineering and installation services for solar heating systems. Nine installations to date. **Contact:** Alain Ratheau; phone: (802) 254-8221.

THE SOLAR COLLABORATIVE, INC
Box 292
Warren, Vermont 05674
Solar system design, residential design, energy consulting. Registered architects who have done five solar houses. HUD grant designers/installers. Geographical range: northeastern United States. **Contact:** Glenn Gazley or Tom Cabot; phone: (802) 496-2983.

Virginia

DESIGN ALTERNATIVES, INC., *1448 N. Lancaster Street, Arlington, Va. 22205.* Technical assistance in architectural, engineering and sociological design, with particular emphasis on solutions to energy and resource problems in community settings. Six solar houses designed in last six years. **Geographical range:** not given. **Contact:** Eugene Eccli; phone: (202) 223-6336.

WALTER L. GOTTSCHALK, M.E., M.S.E.E., *The Rosemary, 198 West Main St., Orange, Va. 22960.* Alternate energy sources design including solar, wind, wood, and water. Many such systems designed, built, and analysed. **Geographical range:** central Virginia, eastern Atlantic States. **Contact:** Walter Gottschalk; phone: (703) 672-2731.

THE METREK DIVISION, THE MITRE CORPORATION, *1820 Dolley Madison Boulevard, McLean, Va. 22101.* Conducts research in energy, environment, and natural resources fields for the federal government and other agencies. Does systems R & D engineering, analysis, planning, and project design with emphasis on unconventional sources of energy such as solar, geothermal, wind, and fluidized-bed conversion of coal. Have been active in the field since 1970, and have held national and international conferences on energy problems. **Geographical range:** national (1st priority); international (2nd priority). **Contact:** Dr. Richard S. Greeley, Technical Director, Division of Energy, Resources and the Environment; phone (703) 790-6771.

SOLAR CORPORATION OF AMERICA, *100 Main Street, Warrenton, Va. 22186.* SCA and its parent company, The InterTechnology Cop. (ITC), offer system engineering design of solar space heating, cooling, and domestic hot water heating plants. SCA can also be contracted to manufacture its standard line of flat plate collector and install "turn-key" solar systems. In the field since 1968. **Geographical range:** continental United States. **Contact:** Walter Sutton; phone: (703) 347-7900.

TORRENCE DREELIN FARTHING & BUFORD INC., *Suite 600 Seaboard Building, Richmond, Va. 23230.* Consulting engineers and architects; feasibility studies and design of solar energy systems for heating and cooling of buildings and for heating domestic and process hot water. Have completed major solar projects. **Geographical range:** East coast of U.S. **Contact:** Richard P. Hanking, Jr.; phone: (804) 358-9111.

Wisconsin

WILLIAM HURRLE, *908 Elmore, Green Bay, Wis. 54303.* Builder, design engineer, and journalist working on solar projects. **Geographical range:** Minnesota, Wisconsin. **Contact:** William Hurrle; phone: (414) 435-9978.

Foreign

APPLIED RESEARCH OF AUSTRALIA PTY. LTD., *13 Durant Rd., Croydon Park, South Australia 5008.* Solar design and consultation; research and development of prototype solar equipment. **Contact:** P.J. Hastwell; phone: 08-46 2757.

WIEBE FOREST GROUP LTD., *604 24A Street N.W., Calgary, Alberta, Canada T2N 2S4.* Mechanical and electrical consultants specializing in HVAC. Involved in solar building. **Geographical range:** British Columbia, Alberta, and Saskatchewan. **Contact:** John Wiebe, P.E.; phone (403) 283-6658.

CORE INTERNATIONAL, SOLAR-GETICS DIV., *P.O. Box 86, Station Snowden, Montreal, Quebec, Canada H3X 3T3.* Represent Canadian and American manufacturers as well as offering design consultation to industry, government, institutions, or private individuals for feasibility studies of wind and direct solar energy systems. Have done project work with Brace Research Institute and the United Nations. **Geographical range:** Quebec, Ontario, and the maritime provinces of Canada. **Contact:** Roger Voisinet or Thomas Lawand; phone: (514) 739-2077.

PATRY, DE CHOCHOR, ALTMAN, *11 bis, Rue Toepffer, 1206 Geneve, Switzerland.* An investment company interested in financing existing business in the solar energy area. A private partnership under Swiss law. **Geographical range:** the European continent. **Contact:** Cecil Altmann; phone: (022) 47 67 22.

How do you describe your Mother?

It's just as difficult if she's a magazine.

MOTHER JONES isn't just a magazine about politics, literature, trivia, music (from R&B to Beethoven), home-cooking, feminism, poetry, art, the environment, movies, psychology, backpacking, pottery, and joy (and anger) — it's all those things and more.

MOTHER JONES is a magazine for the rest of us. For people who are surviving the age, but who want — and **expect** — more. For people who grew up in the Sixties and Seventies. Who've broken with the old society, but are still looking for the new. It's a kind of road map, compendium, home companion and provocation to thought — a catalog of possibilities for yourself and the society you won't find anywhere else. Or at least all in one place.

MOTHER JONES has attracted writers like Max Apple, Studs Terkel, Barbara Garson, Roger Rapoport, Vivian Gornick, Eugene Genovese, Margaret Atwood, Denise Levertov and Kirkpatrick Sale. And dozens of new young writers with the promise of a radically different magazine that is complex in times that have grown more simple-minded.

MOTHER JONES is a bit of the **Whole Earth Catalog,** but not quite; a bit of the old **Ramparts,** but not quite that either; in fact, it's a new blend of a whole fistful of magazines, newspapers, journals, catalogs, and books that we've grown to admire over the years, but that taken alone, never quite reflected the complexity, the richness or the range of our lives.

And you'll find stories in MOTHER JONES you won't see anywhere else: "A Case of Corporate Malpractice," which exposed the huge pharmaceutical company that made millions of dollars from a hazardous birth control device (nominated for one of the National Magazine Awards)...the Doonesbury cartoons most newspapers wouldn't print...Li-li Ch'en's moving memoir, "Peking! Peking!" — about coming of age in the China of the '40s, famine, first love, a missionary school, the onset of revolution (also nominated for a National Magazine Award)...the full story of the FBI's meanest, most effective agent provocateur...a frightening report on the dangers of liquefied natural gas...not to mention the random sampling of tidbits like the recipe for Oreo cookie filling... how much money the IRS could collect if dope were legalized ...China's hit parade...the 1800 poisons in your daily diet. And more...all in MOTHER JONES.

It's a unique magazine, designed for a unique reader. **And you can try it out with no obligation or commitment.** We'll send you the current issue of MOTHER JONES FREE. If you like it, we'll bill you for a year's subscription (9 additional issues) at just $8 — **a $4.00 saving.** But if you don't like it — for any reason — just write "cancel" across the bill when it comes. That's it. Hassle free. And the issue of MOTHER JONES is yours to keep, either way.

Why not take the chance? There's no obligation, and besides, isn't it about time you got back in touch with the rest of us? After all, there's no sense in living the Seventies alone.

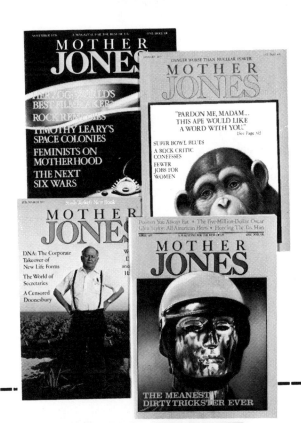

What Kind of Magazine Is SOLAR AGE?

If you are not a subscriber, examine in our Index for Volume I, 1976, reprinted on the opposite page and on a portion of the next the kind of lively, exciting, and informative reading that could lie ahead for you each month in SOLAR AGE. The "Petroleum Age" is coming to an end; the "Solar Age" has begun and in the magazine of that same name you will see not only what this means to you but how you can best shape your own life or lifestyle to this new energy era.

SUBSCRIBE TODAY
(Use Order Form in This Issue)

BACK ISSUES AVAILABLE
(First come, first served)

$2.50 per copy

REPRINTS
(In multiples of 100)

Inquire about rates

Write TODAY to

SOLAR AGE
Church Hill,
Harrisville, N.H. 03450

Solar Age:
Index for Volume I, 1976

WHO ARE YOU?

This catalog is a kind of present to the readers of *Solar Age* magazine, and to a lot of other solar-interested people. We saw a need, and we have tried to fill it. To continue to fill people's solar information needs as completely as possible, we'd like to get to know you better. Will you help?

Fill out the questionnaire and mail it back to us. Include your name only if you want to. Include your comments. Include suggestions. And in a future issue of *Solar Age* we will tell you what you've all told us: **SOLAR AGE, Box A, Hurley, N.Y. 12443**

I.   I am interested in solar energy as:
Homeowner □; apartment dweller □; interested layperson □; engineer □; researcher □; educator □; student □; architect □; builder □; planner □; banker/economist □; businessman □; politician/administrator □; contractor □; construction person (plumber, electrician, sheet metal worker, etc.) □; other_____

_____
_____
_____

II. I am □, am not □ now involved with solar energy professionally. I plan □, do not plan □ to become involved professionally.

III. I am □, am not □ now involved with solar energy personally. I plan □, do not plan □ to become involved personally. My involvement consists or will consist of using solar energy as: hot water heat □; space heat □; swimming pool heat □; cooling □; electricity □ (photovoltaics □, wind □); process heat □; other_____

_____
_____
_____

This use if for: my own house □; my place of business □ (office □; factory □); other people's houses □; commercial project (specify)_____

_____
_____

local or regional organization or project (specify)_____

_____
_____

other (specify)_____

_____
_____

I see this use as: immediate □; within 6 months □; within a year □; within 2 to 5 years □; within 5 to 10 years □; don't know □.

IV. I am most interested in (if more than one, number your greatest interest "1," your second interest "2," and so on): energy conservation □; solar heating □; solar cooling □; solar photoconversion □; solar thermal electricity □; wind □; bioconversion □; ocean wave or thermal conversion □; other_____

_____
_____

I am most interested in: houses □; small-scale individual projects □; local, community, or regional projects □; industrial or commercial projects □; governmental projects □; large-scale projects □.

I am most interested in reading about:
research results □; the results of personal experience □, business experience □; technical information □; products □; news (specify areas)_____

_____
_____

in-depth thinking and background □.

I would prefer articles:
by people about their own project □; by others □.

I would like to see interviews with:_____

_____
_____

I would like to see articles on: _____

_____
_____

I am □, am not □ at present a subscriber to *Solar Age*.

V.   I live in the
Northeast □
Southeast □
South □
Midwest □
Southwest □
Northwest □
West □

I live in a:
Single-family house □
Multi-family building □

I own □; rent □

My house has □, has not □ got energy conservation □, or solar use potential □. Don't know □.

I live in a:
City of more than 1 million people □
500,000 to 1 million □
100,000 to 500,000 □
25,000 to 100,000 □
5,000 to 25,000 □
Less than 5,000 □
Farm or rural area □

My income is:
Less than $10,000 □
$10,000 to $15,000 □
$15,000 to $20,000 □
$20,000 to $30,000 □
More than $30,000 □

I am □, am not □ married.
I have _____ children.
Number of people living in my house is_____

VI. Comments _____
_____
_____
_____
_____
_____

VII. (If you care to)   Name _____
Address_____
_____ Zip _____

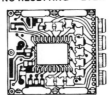

WIND
directory

Use of the wind is an old American custom that has suffered a lapse of some thirty to forty years, during that odd time between the Rural Electrification Act and the Arab Boycott when we thought that fossil fuel was cheap. Manufacturers of equipment for harvesting the wind almost vanished from the continent. Now they are back—but so far, only a few. This list was assembled with the help of Rick Katzenburg and the American Wind Energy Association (New Boston, N.H. 03070).

AERO—POWER
432 Natoma St.
San Francisco 94103
Contact: Tom Conlon
phone 415-777-4131

AMERICAN ENERGY ALTERNATIVES
P.O. Box 905
Boulder, Colo. 80302
Contact: John Sayler;
phone: 303-447-0820

EDMUND SCIENTIFIC CO.
380 Edscorp Bldg.
Barrington, N.J. 08007
Contact: Kevin Moran;
phone: 609-547-3488

GRUMMAN AEROSPACE
Bethpage, N.Y. 11174
Contact: Kenneth Speiser;
phone: 516-575-6205

HELION
P.O. Box 445
Brownsville, Cal. 95919
Contact: Jack Park;
phone: 916-675-2478

KEDCO, INC.
9016 Aviation Blvd.
Inglewood, Cal. 90301
Contact: Jack Park;
phone: 213-776-6636

PINSON ENERGY CORP.
P.O. Box 7
Marstons Mill, Mass. 02648
Contact: Herman Drees;
phone: 617-477-2913

REDE CORP.
P.O. Box 212
Providence, R.I. 02901
Contact: Ron Beckman;
phone: 401-751-7333

SENCENBAUGH WIND ELECTRIC
P.O. Box 17323
Palo Alto, Cal. 94306
Contact: Jim Sencenbaugh;
phone: 714-560-9452
phone: 415-964-1593

WIND POWER SYSTEMS, INC.
P.O. Box 17323
San Diego, Cal. 92117
Contact: Edmund L. Salter;
phone: 714-560-9452

WINDWORKS
P.O. Box 329
Rt. 3
Mukwonago, Wis. 53149
Contact: Hans Meyer;
phone: 414-363-4408

WINDPOWER CORP.
1207 First Ave. E.
Newton, Iowa 50208
Contact: Roy Brewer;
phone: 515-792-1301

WTG ENERGY SYSTEMS
Box 87
1 La Salle St.
Angola, N.Y. 14006
Contact: Alan Wellikoff;
phone: 716-549- 5544

ZEPHYR WIND DYNAMO CO.
P.O. Box 241
Brunswick, Maine 04011
Contact: Willard Gillette;
phone: 207-725-6534

Other companies of interest

AUTOMATIC POWER
P.O. Box 18738
Houston, Texas 77023
Contact: Robert Dodge;
phone: 713-228-5208

ENERTECH CORP.
P.O. Box 420
Norwich, Vt. 05055
Contact: William Drake;
phone: 802-649-1145

ENVIRONMENTAL ENERGIES INC.
P.O. Box 73
Front St.
Copemish, Mich. 49625
Contact: Timothy J. Horning;
phone: 616-373-2000

NATURAL POWER INC.
New Boston, N.H. 03070
Contact: Richard Katzenberg;
phone: 603-487-2426

NORTH WIND POWER CO.
P.O. Box 315
Warren, Vt. 05674
Contact: Donald Mayer;
phone: 802-496-2955

SUNFLOWER POWER
Rt. 1, Box 93-A
Oskaloosa, Kansas 66066
Contact: Steven Blake;
phone: 913-597-5603

If You've Been Reading

THE NATION

Every Week

For The

Past 112 Years

Then, starting back in 1865, right to the present era of Corporation-dominated America, you have not only kept fully abreast of political and social events . . . but you have also had a refreshing, independent, and venturesome perspective of the undercurrents—the dynamics—of events. Also, you've been getting a straightforward analysis and review of literature, music, poetry, theatre, TV, films, art and dance.

You have been reading comment by such of our contributors as William Butler Yeats, Ralph Nader, Henry James, Elizabeth Holtzman, Leon Trotsky, Carey McWilliams, John Dos Passos, Corliss Lamont, Andre Malraux, George McGovern, Thomas Mann, Emile Capouya, Robert Frost, Robert Sherrill, Emily Dickinson, and a sparkling lot of other writers and thinkers.

The Nation, America's oldest weekly, has been ahead of the news ever since its first issue in 1865

WOOD Directory

Use of wood for heat is an even older American custom, one that never really lapsed, because it wasn't long after those turn-of-the-twentieth-century Americans boarded up their Colonial fireplaces to install new, more efficient wood stoves that even more modern Amercians, centrally-heated, unboarded them again, for their charm. This list is a mixed bag including manufacturers, some distributors, and some retail people. It was assembled with the help of Jay Shelton, (The Woodburner's Encyclopedia, Vermont Crossroads Press, Waitsfield, Vt. 05673; $6.95), Andrew Shapiro of the Wood Energy Institute (Box 1, Waitsfield, Vt.), Bob Ross (Modern and Classic Woodburning Stoves and the Grass Roots Energy Revival, Overlook Press, Woodstock, N.Y. 12498; $10.), and Elizabeth Shaw (New England Yellow Pages of Solar Energy Development, New England Solar Energy Assn., Townshend, Vt. 05353; $5). Where the information was available, the type of wood-using equipment has been included.

ABUNDANT LIFE FARM
P.O. Box 63
Lochmere, N.H. 03252
Heating stoves

ASHLEY AUTOMATIC HEATER CO.
1604 17th Ave. S.W.
P.O. Box 730
Sheffield, Ala. 35660
Heating stoves, circulating heaters

ASHLEY SPARK DISTRIBUTORS, INC.
710 N.W. 14th Ave.
Portland, Ore. 97209

ATLANTA STOVE WORKS, INC.
P.O. Box 5254
Atlants, Ga. 30307
Cookstoves, heating stoves, circulating heaters, free-standing fireplace, fireplace accessories

ATLANTIC CLARION STOVE CO.
Brewer, Me. 04412

AUTOCRAT CORPORATION
New Athens, Ill. 62264
(618)475-2121
Cookstoves, heating stoves, circulating heaters

BELLWAY MANUFACTURING
Grafton, Vt. 05146
(802)843-2432
Furnaces and boilers

BIRMINGHAM STOVE & RANGE
P.O. Box 2647
Birmingham, Ala. 35202
(205)322-0371

BLACKS'S
58 Maine St.
Brunswick, Me. 04011

BLAZING SHOWERS
P.O. Box 327
Point Arena, Cal. 95468
(707)882-9956
Hot water heaters

BOW & ARROW STOVE CO.
14 Arrow St.
Cambridge, Mass. 02138
(617)492-1441
Cookstoves, heating stoves,

CANAQUA CO.
Box 6
High Falls, N.Y. 12440
or
1908 Baker St.
San Francisco, Cal. 94115
(415)346-0752

CARLSON MECHANICAL CONTR'S, INC.
Box 242
Prentice, Wisc. 54556
(715)564-2481 or (715)428-3481

CHIMNEY HEAT-RECLAIMER CORP.
Dept. Y. 53 Railroad Ave.
Southington, Conn. 06489
(203)628-4738 (Ext. Y) ·
Heat reclaimers

WALDO G. CUMINGS
Fall Road
East Lebanon, Me. 04027
(207)457-1219
or
20 Schuler St.
Sanford, Me. 04073
Furnaces and boilers

DAMPNEY COMPANY
85 Paris St.
Everett, Mass. 02149
(617)389-2805

DAMSITE DYNAMITE STOVE
RD 3
Montpelier, Vt. 05602
(802)223-7139
Heating stoves, furnaces and boilers

SAM DANIELS CO.
Box 868
Montpelier, Vt. 05602
(802)223-2801
Furnaces and boilers

DAWSON MFG. CO.
Box 2024
Enfield, Conn. 06082

DIDIER MFG. CO.
1652 Phillips Ave.
Racine, Wisc. 53403
(414)634-6633

DOUBLE STAR
c/o Whole Earth Access Co.
2466 Shattuck Ave.
Berkley, Calif. 94704
(415)848-0510
Heating stoves, free-standing fireplaces

DOVER CORP.
Peerless Div.
P.O. Box 2015
Louisville, Ky. 40201

DOVER STOVE COMPANY
Main St.
Sangerville, Me. 04479

DYNA CORP.
2540 Industry Way
Lynwood, Cal. 90262

EDISON STOVE WORKS
P.O. Box 493
469 Raritan Center
Edison, N.J. 08817
(201)225-3848
Heating stoves

ENERGY ALTERNATIVES, INC.
69 Amherst Rd.
Leverett, Mass. 01054
(413)549-3644

ENERGY ASSOCIATES
P.O. Box 524
Old Saybrook, Conn. 06475
(203)388-0081

ENWELL CORP.
750 Careswell St.
Marshfield, Mass. 02050
(617)837-0638
Furnaces and boilers

TODD EVANS, INC.
100 Cooper St. Box X
Babylon, N.Y. 11702

FABSONS ENGINEERING
P.O. Box F-11
Leominster, Mass. 01453

FIREPLACES (N.S. LIMITED)
Suite 215 Duke St. Tower
Halifax, Nova Scotia, B3JIN9

FIRE-VIEW DISTRIBUTORS
P.O. Box 370
Rogue River, Ore. 97537
(503)582-3351

FISHER STOVES, INC.
504 So. Main St.
Concord, N.H. 03301
(603)224-5091
Heating stoves

FISK STOVES
Tobey Farm
Box 935
Dennis, Mass. 02638
(617)385-2171
Barrel stoves

FOREST FUELS, INC.
7 Main St.
Keene, N.H. 03431
(603)357-3311

FRANKLIN FIREPLACES
1100 Waterbury Blvd.
Indianapolis, Ind. 46202

FUEGO HEATING SYSTEMS
P.O. Box 666
Brewer, Me. 04412
(207)989-5757

GARDEN WAY RESEARCH
P.O. Box 26 W
Charlotte, Vt. 05445
(802)425-2137
Heating stoves, fireplace accessories, wood splitters

L.W. GAY STOVE WORKS, INC.
Marlboro, Vt. 05344
(802)257-0180
Heating stoves

GEMCO
404 Main St.
Marlborough, N.H. 03455

GLO-FIRE
Spring & Summer Streets
Lake Elsinore, Cal. 92330
(714)674-3144

GREENBRIAR PRODUCTS, INC.
Box 473G
Spring Green, Wisc. 53588
Free-standing fireplaces

HDI IMPORTERS
Schoolhouse Farm
Etna, N.H. 03750
(603)643-3371

HEARTWOOD HEAT CO.
Stoney Hollow
Box 158
Glenford, N.Y. 12433
(914)339-4040

HEATILATOR
Box 409
Mount Pleasant, Iowa 52641
(319)385-9211
Pre-fabricated fireplaces

HEAT RECLAMATION DIVISION
939 Chicopee St.
G.P.O. Box 366
Chicopee, Mass. 01021
(413)536-1311 Telex 95-5342
Heat reclaimers

M.B. HILLS, INC.
Belfast, Me. 04915
(207)338-4120
Furnaces

HINCKLEY FOUNDRY
13 Water St.
Newmarket, N.H. 03875
(603)659-5804

HOME FIREPLACES
(MORSE CANADIAN IMPORTER)
Markham, Ontario L3R1GE
(416)495-1650
or:
971 Powell Ave.
Winnipeg, Manitoba R3H 0H4
(204)774-3834
Cookstoves, heating stoves, circulating heaters, free-standing fireplaces, pre-fabricated fireplaces

HUNTER ENTERPRISES ORILLIA LIMITED
P.O. Box 400
Orillia, Ontario, Canada
(705)325-6111
Circulating heaters, furnaces and boilers

HYDRAFORM PRODUCTS COMPANY
P.O. Box 2409
Rochester, N.H. 03867
(603)332-6128
Heating stoves

INGLEWOOD STOVE COMPANY
Rte. 4
Woodstock, Vt. 05091
(802)457-3238
Heating stoves

ISOTHERMICS, INC.
P.O. Box 86
Augusta, N.J. 07822
(201)383-3500
Heat reclaimers

JERNLUND PRODUCTS, INC.
1620 Terrace Dr.
St. Paul, Minn. 55113

KENENATICS
1140 No. Parker Dr.
Janesville, Wisc. 53545

KICKAPOO STOVE WORKS, LTD.
Rte. 1-A
LaFarge, Wisc. 54639
(608)625-4431
Heating stoves

KNT, INC.
P.O. Box 25
Hayesville, Ohio 44838
(419)368-3241 or (419)368-8791
Free-standing fireplaces, pre-fabricated fireplace

**KRISTIA ASSOCIATE
(JOTUL IMPORTERS)**
P.O. Box 1118
Portland, Me. 04104
(207)772-2112
*Heating stoves, free-standing fireplaces,
pre-fabricated fireplaces*

LANCE INTERNATIONAL
P.O. Box 562
1391 Blue Hills Ave.
Bloomfield, Conn. 06002
(203)243-9700
Fireplace accessories, heat reclaimers

W.F. LANDERS CO.
P.O. Box 211
Springfield, Mass. 01101
(413)786-5722
**LEYDEN ENERGY
CONSERVATION CORP.**
Brattleboro Rd.
Leyden, Mass. 01337

LOCKE STOVE CO.
114 West 11th St.
Kansas City, Mo. 64105
(816)421-1650
Heating stoves, circulating heaters

LONGWOOD FURNACE CO.
Gallatin, Mo. 64640
Furnaces

LOUISVILLE TIN & STOVE CO.
P.O. Box 1079
Louisville, Ky. 40201
(502)589-5380
Heating stoves

**LYNNDALE MANUFACTURING
COMPANY, INC.**
1309 North Hills Blvd.
Suite 207
North Little Rock, Ark. 72116
(501)758-9602

P.O. Box 1154
Harrison, Ark. 72601
(501)365-2378
Furnaces and Boilers

MAINE WOOD HEAT CO.
R.D. #1 Box 38
Norridgewock, Me. 04957
(207)696-5442

MALLEABLE IRON RANGE CO.
715 N. Spring St.
Beaver Dam, Wisc. 53916
(414)887-8131

MARATHON HEATER CO.
Box 165 R.D. 2
Marathon, N.Y. 13803
(607)849-6736

MARCO INDUSTRIES, INC.
P.O. Box 6
Harrisonburg, Va. 22801

MARKADE-WINNWOOD
4200 Birmingham Rd., N.E.
Kansas City, Mo. 64117
(816)454-5260
*Heating stoves, furnaces and boilers, free-
standing fireplaces, fireplace accessories,
barrel stoves*

MARTIN INDUSTRIES
P.O. Box 730
Sheffield, Ala. 35660
(205)383-2421
*Cooking stoves, heating stoves, circulating
heaters, free-standing fireplaces*

NEWMAC MFG., INC.
236 Norwich Ave.
Box 545
Woodstock, Ontario N4S 7W5, Canada
(519)539-6147
Furnaces and boilers

NICHOLS ENVIRONMENTAL
5 Apple Rd.
Beverly, Mass. 01915

**OLD COUNTRY APPLIANCES
(TIROLA IMPORTER)**
P.O. Box 330
Vacaville, Cal. 95688
(707)448-8460
Cooking stoves

PIONEER LAMPS & STOVES
71A Yesler Way
Pioneer Sq. Station
Seattle, Wash. 98104
Cooking stoves

PORTLAND STOVE FOUNDRY
57 Kennebee St.
Portland, Me. 04104
(207)773-0256
*Cook stoves, heating stoves, circulating
heaters, free-standing fireplaces, fireplace
accessories, barrel stoves*

**MECHANICAL PRODUCT
DEVELOPMENT CORP.**
Box 115
Swarthmore, Pa. 19081

**MERRY MUSIC BOX
(STYRIA IMPORTERS)**
20 McKown
Boothbay Harbor, Me. 04538
(207)633-2210
Cook stoves, heating stoves

**METAL BUILDING
PRODUCTS, INC.**
35 Progress Ave.
Nashua, N.H. 03060
(603)882-4271
Circulating heaters, pre-fabricated fireplaces

MODERN-AIRE
Modern Machine and Welding
Highway 2 West
Grand Rapids, Mich. 55744

MOHAWK INDUSTRIES, INC.
173 Howland Ave.
Adams, Mass. 01220
(413)743-3548
Heating stoves

**PRESTON DISTRIBUTING CO.
(POELE IMPORTER)**
10 Whidden St.
Lowell, Mass. 01852
(617)485-6303
Circulating heaters

RAM & FORGE
Brooks, Me. 04921
(207)772-3379
Heating stoves, furnaces and boilers

REM INDUSTRIES
408 C Simms Bldg.
Dayton, Ohio 45402

**RIDGEWAY STEEL
FABRICATORS, INC.**
Box 382
Bark St.
Ridgeway, Pa. 15853
(814)776-1323 or (814)776-6156
*Pre-fabricated fireplace, fireplace
accessories*

RITEWAY MANUFACTURING CO.
P.O. Box 6
Harrisonburg, Va. 22801
(703)434-7090
*Heating stoves, circulating heaters,
furnaces and boilers*

RO KNICH PRODUCTS, INC.
P.O. Box 311-E
No. Chicago, Ill. 60064

S/A DISTRIBUTORS
730 Midtown Plaza
Syracuse, N.Y. 13210

SCANDINAVIAN STOVES, INC.
(L. LANGE & CO. INPORTER)
Box 72
Alstead, N.H. 03602
(603)835-6029
Cook stoves, heating stoves

SCOT'S STOVE CO.
11 Ells St.
Norwalk, Conn. 06850
Heating stoves

SELF SUFFICIENCY PRODUCTS
1 Appletree Square
Minneapolis, Minn. 55420

**SHENANDOAH
MANUFACTURING CO., INC.**
P.O. Box 839
Harrisonburg, Va. 22801
(703)434-3838
*Heating stoves, circulating heaters,
fireplace accessories*

SOLAR SAUNA
Box 466
Hollis, N.H. 03049

SOTZ CORP.
23797 Sprague Rd.
Columbia Station, Ohip 44028

**SOUTHEASTERN
VERMONT COMMUNITY
ACTION, INC.**
7-9 Westminster St.
Bellows Falls, Vt. 05101
(802)463-4447
Heating stoves

**SOUTHPORT STOVES
(MORSO IMPORTER)**
(Division of Howell Corporation)
248 Tolland St.
East Hartford, Conn. 06108
(203)289-6079
Heating stoves, free-standing fireplaces

STURGES HEAT RECOVERY, INC.
P.O. Box 397
Stone Ridge, N.Y. 12484
(914)687-0281
Heat reclaimers

**SUBURBAN MANUFACTURING
COMPANY**
4700 Forest Dr.
P.O. Box 6472
Columbia, S.C. 29206
(803)782-2649
circulating heaters

SUNSHINE STOVE WORKS
R.D. 1 Box 38
Norridgewock, Me. 04957
(207)887-4580
Heating stoves

**TEKTON DESIGN CORP.
(TASO & KEDEL FABRIC-TARM
IMPORTER)**
Conway, Mass. 01341
(413)369-4685
Furnaces and boilers

THERMALITE CORP.
Dept. PS Box 69
Hanover, Mass. 02339

THERMO CONTROL WOOD STOVES
Central Bridge, N.Y. 12035
Heating stoves

THERMO-RITE
The Fireplace House
1950 Wadsworth
Denver, Colo. 80215

TORRID MANUFACTURING CO., INC.
1248 Poplar Place So.
Seattle, Wash. 98144
(206)324-2754
Free-standing fireplaces, heat reclaimers

TRIWAY MFG. INC.
7819 Old Highway 99
Box 37
Marysville, Wash. 98270

**UNITED STATES STOVE
COMPANY**
P.O. Box 151
South Pittsburg, Tenn. 37380
(615)837-8631
*Heating stoves, circulating heaters, free-
standing fireplaces*

VAPORPACK, INC.
Box 428
Exeter, N.H. 03833
(603)778-0509
Heat reclaimers

VERMONT CASTINGS, INC.
Box 126
Prince Street
Randolph, Vt. 05060
(802)728-355
Heating stoves

**VERMONT COUNTERFLOW
WOOD FURNACE**
Plainfield, Vt. 05667
Furnaces

VERMONT ENERGY PRODUCTS
100 Broad St.
Lydonville, Vt. 05851

**VERMONT IRON STOVE
WORKS, INC.**
The Bobbin Mill
Warren, Vt. 05674
(802)496-2821
Heating stoves

VERMONT SOAPSTONE CO.
Pekinsville, Vt. 05151
Heating stoves

**VERMONT WOODSTOVE
COMPANY**
307 Elm St.
Bennington, Vt. 05201
(802)442-3985
Heating stoves

E.G. WASBURNE & CO.
83 Andover St.
Danvers, Mass. 01923

WASHINGTON STOVE WORKS
P.O. Box 687
3402 Smith St.
Everett, Wash. 98201
(206)252-2148
*Cook stoves, heating stoves, circulating
heaters, free-standing fireplaces, fireplace
accessories, barrel stoves.*

WHITE MESA, INC.
110 Laguna N.W.
Albuquerque, N.M. 87104
(505)247-1066

WHITTEN ENTERPRISES
Arlington, Vt. 05250

WHITTIER STEEL & MFG., INC.
10725 S. Painter Ave.
Santa Fe Springs, Cal. 90670

**WHOLE EARTH ACCESS
COMPANY**
2466 Shattuck Ave.
Berkeley, Cal. 94704
(415)848-0510
Barrel stoves

WILSON INDUSTRIES
2296 Wycliff
St. Paul, Minn. 55114
(612)646-7214

**WOODBURNING SPECIALITIES
(HUNTER IMPORTER)**
P.O. Box 5
No. Marshfield, Mass. 02059
Circulating heaters, furnaces

YANKEE WOODSTOVES
Cross St.
Bennington, N.H. 03442
(603)588-6358
Heating stoves

Further reading

Being a selected list of books important to people interested in knowing about—or knowing more about—solar energy

Specifically solar

SOLAR DWELLING DESIGN CONCEPTS, by the AIA Research Corporation, for sale by the Superintendant of Documents, U.S. Government Printing Office, Washington, D.C. 20402, 1976, 146 pp., $2.30. An excellent book full of architects drawings and plans. By all means get hold of a copy.

THE SOLAR HOME BOOK, by Bruce Anderson, Cheshire Books, Church Hill, Harrisville, N.H. 03450, 1976, 297 pp., $7.50. The tale of our own Mr. Anderson's scorching affair with the sun, a thriller. (But, seriously, one of, if not *the* best.)

SUNSPOTS, COLLECTED FACTS AND SOLAR FICTION, by Steve Baer, Zomeworks Corporation, Box 712, Albuquerque, N.M., 1975, 115 pp., $3. The layperson should read this as an initiation to solar; the solar technologist should read it as an introduction to reality. Excellent.

SOLAR ENERGY FOR MAN, by B.J. Brinkworth, Halstead Press, 1972, 251 pp. Introduction to the principles underlying solar science and technology.

ENERGY FOR SURVIVAL: THE ALTERNATIVE TO EXTINCTION, by Wilson Clark, Anchor Press/Doubleday, 1975, 652 pp., $4.95. An important, if not essential, book.

NEW, LOW-COST SOURCES OF ENERGY FOR THE HOME, by Peter Clegg, Garden Way Publishing, Charlotte, Vt. 95445, 1975, 250 pp., $4.95 paper; $8.95 hardback. Covers solar heating and cooling, wind and water power, wood heat, and methane digestion.

SUN/EARTH, by Richard L. Crowther, Crowther/Solar Group, 310 Steele St., Denver, Colo. 80206, 1976, 232 pp. A clear study of basic solar design and energy conservation principles, easy to follow.

DIRECT USE OF THE SUN'S ENERGY, by Farrington Daniels, Balantine Books, 1964, 271 pp., $1.95. The Classic.

SOLAR HOMES AND SUN HEATING, by George Daniels, Harper and Row, 1976, 178 pp., $8.95. Attempts to explain solar energy to the layman and to provide enough information for the I'd-Rather-Do-It-Myself handyman to make valid decisions.

THE SURVIVAL GREENHOUSE, by James B. DeKorne, The Walden Foundation, P.O. Box 5, El Rito, N.M. 87539, 1976, 150 pp., $7.50. A serious hard-core book on starting literally from the ground up, providing construction and operation details of a pit greenhouse.

SOLAR ENERGY THERMAL PROCESSES, by John A. Duffie and William Beckman, John Wiley and Sons, 1974, 386 pp., $18. The best engineering treatment of solar available. Highly technical.

LOW-COST, ENERGY EFFICIENT SHELTER: FOR THE OWNER AND BUILDER, by Eugene Eccli, Rodale Press, 1975, 432 pp., $10.95 hardcover; $5.95 paper. The book's extensive coverage of building techniques, plans, materials, new technologies and retrofitting in older buildings is excellent.

SOLAR GREENHOUSES, THE FOOD AND HEAT PRODUCING, DESIGN, CONSTRUCTION, OPERATION, by Rick Fisher and Bill Yanda, John Muir Publications, P.O. Box 613, Santa Fe, N.M. 87501, 1976, $6. Thorough book by leading people in continuing evolution of the solar greenhouse. Highly recommended.

THE COMING AGE OF SOLAR ENERGY, by D.S. Halacy, Avon, 1963, 248 pp., $1.95. A non-technical overview of the worldwide potential for solar power.

SOLAR CELLS, by Harold Hovel, Academic Press, 111 Fifth Ave., New York, N.Y. 10003, 262 pp., $14.50. A technical work on the research and development of solar cells, problems and advances.

SOLAR HEATING AND COOLING: ENGINEERING, PRACTICAL DESIGN, AND ECONOMICS, by Jan Kreider and Frank Kreith, McGraw-Hill Book Company, 1976, 342 pp., $22.50. Provides an introductory look at the subject of solar heating and cooling.

SUNLIGHT TO ELECTRICITY, by Joseph A. Merrigna, The MIT Press, 28 Carelton St., Cambridge, Mass. 02142, 1976, 163 pp., $12.95. The author thinks the solar cell business will be a multi-

billion dollar industry by 2000. Useful as a concise introduction to the technically-oriented business person.

HANDBOOK OF HOMEMADE POWER, by The Mother Earth News, Bantam Books, 1974, 374 pp., $1.95. A smorgasbord of good ideas on all aspects of homemade power.

THE FUEL SAVERS, by Dan Scully, Don Prowler, Bruce Anderson, with Doug Mahone, TEA, Inc., Church Hill, Harrisville, N.H. 03450, 1976, 60 pp. A kit of solar ideas for existing houses.

SOLAR HEATED BUILDINGS: A BRIEF SURVEY, by W.A. Shurcliff, 19 Appleton St., Cambridge, Mass. 02138, 1977, 306 pp., $12. The 13th and final edition of Mr. Shurcliff's invaluable report on houses, schools and commercial buildings that are heated and/or cooled by the sun. No solar afficionado would be caught without a copy.

ENERGY, ENVIRONMENT, AND BUILDING, by Phillip Steadman, Cambridge University Press, 1975, 287 pp., $5.95. Nearly always precise and lucid, without burdensome jargon or irritating oversimplification.

ENERGY WE CAN LIVE WITH: APPROACHES TO ENERGY THAT ARE EASY ON THE EARTH AND ITS PEOPLE, edited by Daniel Wallace, Rodale Press, Inc. Emmaus, Penn. 18049, 1976, 150 pp., $3.95. An adequate, readable introduction to methods of appropriate technology.

DESIGNING & BUILDING A SOLAR HOUSE: YOUR PLACE IN THE SUN, by Donald Watson, Garden Way Publishing, Charlotte, Vt. 05445, 1977, 281 pp., $8.95. An excellent, worthwhile book.

General background

THE CLOSING CIRCLE: NATURE, MAN & TECHNOLOGY, by Barry Commoner, Alfred A. Knopf, Inc., 1971, 326 pp. Excellent ecological diagnosis of the planet, insisting that we are both malefactors and medical personnel. Suggests that we should concentrate on being the latter.

THE POVERTY OF POWER, by Barry Commoner, Knopf, 1976, 382 pp., $10. Mr. Commoner examines the energy crisis and lays it on the line as to how our lives will have to change.

ENERGY FUTURES: INDUSTRY AND THE NEW TECHNOLOGIES, by S.W. Herman and J.S. Cannon, of Inform, Inc., 25 Broad St., New York, N.Y. 10004, 1976, 760 pp. $265 to industry and government, $160 to educational institutions. This book covers almost every know alternative energy source, and for each it explains the potential rewards, the problems, and the technical approaches available. It tells which major industrial corporations are involved, what solutions they are trying to develop, where their work stands today, and what the expectations are for the near future.

THE OWNER-BUILDER AND THE CODE: POLITICS OF BUILDING YOUR HOME, by Ken Kern, Ted Kogan and Rob Thallon, Owner-Builder Publications, Box 550, Oakhurst, Cal. 93664, 1976, 168 pp., $5 postpaid. An informative but frightening history of the development of building codes, and the abuse of these codes. Chronicles the experience of homeowners who have built beatiful, unconventional houses, only to find them declared illegal.

DESIGN WITH NATURE, by Ian L. McHarg, Doubleday, 1969, 198 pp., $6.95 paperback. "A vision of organic exuberance and human delight, which ecology and ecological design promise to open up for us."--Lewis Mumford.

NEIGHBORHOOD POWER: THE NEW LOCALISM, by David Morris and Karl Hess, Beacon Press, Boston, 1975, 180 pp., $3.45. A primer on how to organize communities to be self-reliant. Mr. Morris writes a regular column—Local Report—for *Solar Age.*

DESIGN WITH CLIMATE, by Victor Olgyay, Princeton University Press, 1963, 190 pp., $28.50. A classic treatment of comprehensive arcitectural design concepts in terms of climatic conditions.

ENERGY PRIMER, by Portola Institute, 1974, 200 pp., $5.50. An excellent source book from the Whole Earth Catalog people.

SMALL IS BEAUTIFUL, by E.F. Schumacher, Harper and Row, 1973, 305 pp., $2.45. Economics as if people mattered, or doing more with less. One of the philosophical bases of alternate energy.

A LANDSCAPE FOR HUMANS: A CASE STUDY OF THE POTENTIALS FOR ECOLOGICALLY GUIDED DEVELOPMENT IN AN UPLANDS REGION, by Peter van Dresser, The Lightning Tree, P.O. Box 1837, Santa Fe, New Mexico 87501, 1972, 128 pp., $3.95. A discussion of the alternatives to continued urban-industrialization as the means of growth. Mr. van Dresser is one of the early solar pioneers whose approach to problems of energy has had great influence on the whole alternative technology movement.

Using wood

THE COMPLETE BOOK OF HEATING WITH WOOD, by Larry Gay, Garden Way Publishing, Charlotte, Vt. 05445, 1974, 128 pp., $3. Choosing fuelwoods, improving woodlands, modern stoves, wood furnaces, old stoves, hows and whys.

MODERN AND CLASSIC WOOD BURNING STOVES AND THE GRASS ROOTS ENERGY REVIVAL: A COMPLETE GUIDE, by Bob and Carol Ross, Overlook Press, Woodstock, New York, 1976, 143 pp., $10. "This well-illustrated book is written from a wood stove retailer's perspective, with emphasis on the selection and installation of wood stoves." Paul Sullivan, *Solar Age.*

THE WOODBURNER'S ENCYLOPEDIA, by Jay Shelton, Vermont Crossroads Press, Inc., Waitsfield, Vt., 1976, 156 pp., 140 illustrations, $6.95. "The best, most complete book available for a technical understanding of how to use wood most efficiently for heating purposes...for those with an engineering bent or with specific questions on wood heating." Paul Sullivan, *Solar Age.*

Periodicals

Solar Energy, a technical bimonthly from Pergamon Press for the International Solar Energy Society. Write: Subscription Fulfillment Manager, Headington Hill Hall, Oxford OX3 OBW, England. $71 yearly.

Solar Engineering Magazine, a trade magazine for the solar industry. Write: Solar Engineering Publishers, Inc., 8435 N. Stemmons Freeway, Ste. 880, Dallas, Texas 75247. $15 yearly (foreign: $17).

Solar Heating and Cooling Magazine, bimonthly from Gordon Publications, P.O. Box 2126-R, Morristown, N.J. 07960.

Solar Energy Intelligence Report, is a weekly newsletter covering the latest solar developments in Washington as well as within the solar industry. Write: Business Publishers, Inc., P.O. Box 1067, Silver Spring, Md. 20910. $90 yearly; $50 for six months.

Solar Energy Digest, a monthly report available from P.O. Box 17776, San Diego, Cal. 92117. $27.50 yearly.

Index

R

S

T

U

V

W

Y

Z

If you're building a solar collector

we've already built your collector frame

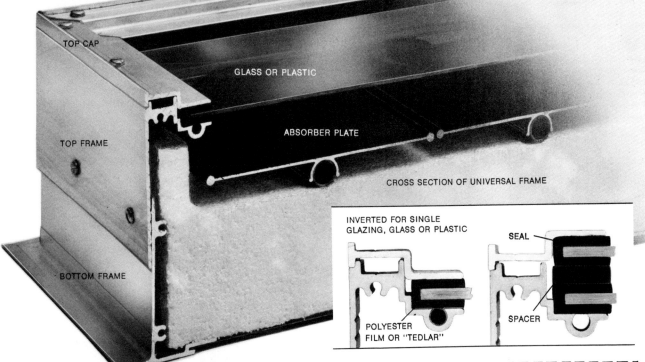

A UNIVERSAL ALUMINUM SOLAR FRAME. ONE SIZE FITS ALL UP TO 4' x 8'
Combining 25 years of experience in aluminum extrusion and fabrication with an in depth experience in supplying the solar industry*, we've perfected the universal aluminum solar collector frame. It is already being used by builders throughout the country, and is the practical solution to costly custom fabrication. Remember, aluminum is durable, lightweight, strong and its good looks will add value to your home.

Here's What You Get
Adjustable between 4½" and 6½" in height, the aluminum frame comes to you in straight lengths including cap extrusions (absorber plate if desired) ready to be cut to size and assembled. It will accommodate single or double glazing, glass or plastic (1/8" or 3/16"), film or a combination of glass, plastic or film.

*Member Solar Energy Association

NEW JERSEY ALUMINUM ... WE KNOW OUR WAY AROUND A SOLAR COLLECTOR.

NOTES

NOTES

NOTES

NOTES

NOTES

NOTES

SPECIAL CATALOG SUBSCRIPTION OFFER
Receive a $5.95 Discount

Yes — I understand that as a buyer of this $5.95 book I am entitled to a subscription to Solar Age Magazine at $5.95 off the regular subscription price.

Please enter my subscription for

☐ 1 YEAR (12 issues) at $14.05
    (Normally $20.00)

☐ 2 YEARS (24 issues) at $26.05
    (Normally $32.00)

☐ 3 YEARS (36 issues) at $34.05
    (Normally $40.00)

CANADA AND OVERSEAS ADD
$6.00 per year

TOTAL ORDER

$ _____

☐ CHECK ENCLOSED

☐ PLEASE BILL ME

G01

Name _____

Company _____

Address _____

City _____ State _____ Zip _____

☐ Credit Card No.

☐ Master Charge   Expiration Date ☐☐☐☐
☐ American Express
☐ BankAmericard (VISA)   Master Charge
                         Interbank No. ☐☐☐☐

INTRODUCTORY SUBSCRIPTION ORDER

Please enter my subscription to **SOLAR AGE** beginning with the next issue.

☐ 1 YEAR (12 issues)   at $20.00

☐ 2 YEARS (24 issues)   at $32.00

☐ 3 YEARS (36 issues)   at $40.00

CANADA AND OVERSEAS ADD
$6.00 per year

TOTAL ORDER

$ _____

☐ CHECK ENCLOSED

☐ PLEASE BILL ME

G01

Name _____

Company _____

Address _____

City _____ State _____ Zip _____

☐ Credit Card No.

☐ Master Charge   Expiration Date ☐☐☐☐
☐ American Express
☐ BankAmericard (VISA)   Master Charge
                         Interbank No. ☐☐☐☐

Back Issues!

Since January, 1976 we've been bringing in-depth reporting of the solar scene to businesses, home-owners and professionals.

If you've recently joined us as a subscriber, there may be past issues of SOLAR AGE you wish to order.

May 1976 Bioconversion. Swimming Pools.
July 1976 Passive Heat: Adobe. Trombe Walls.
August 1976 Farming the Sea. Photosynthesizing Fuel.
September 1976 Conservative Housing: Underground. Simple Approaches.
October 1976 Greenhouses: A Study in Four Climates. Simple Designs for Solar Hot Water Heating.
November 1976 All About Wood. Stove Comparisons.
December 1976 Standards, Part I. Mass Production. Index Vol. I.
February 1977 Process Heat.
March 1977 Standards, Part II. Retrofit.
April 1977 Cities: Energy & Urban Architecture.
May 1977 Absorbers: Copper, Steel, Aluminum.
August 1977 Trombe Walls.
September 1977 Energy & Jobs. Teaching Solar.
October 1977 New Mexico: "A solar state of mind"
November 1977 Greenhouses.
December 1977 HUD & ERDA. Index Vol. II.
Note: December issues contain an index of articles.

Name _____

Company _____

Address _____

City _____ State _____ Zip _____

☐ Credit Card No.

☐ Master Charge   Expiration Date ☐☐☐☐
☐ American Express
☐ BankAmericard (VISA)   Master Charge
                         Interbank No. ☐☐☐☐

Please send the issues I've checked.

May 76 ____	Oct 76 ____	Mar 77 ____	Sep 77 ____
July 76 ____	Nov 76 ____	Apr 77 ____	Oct 77 ____
Aug 76 ____	Dec 76 ____	May 77 ____	Nov 77 ____
Sep 76 ____	Feb 77 ____	Aug 77 ____	Dec 77 ____

Back issues $2.50 ea. ppd. 12 or more issues $2.00 ea.

ATTACH
STAMP
HERE

SOLAR AGE

P. O. Box 4934

Manchester, NH 03103

ATTACH
STAMP
HERE

SOLAR AGE

P. O. Box 4934

Manchester, NH 03103

ATTACH
STAMP
HERE

SOLAR AGE

P. O. Box 4934

Manchester, NH 03103